Understanding Biostatistics

Statistics in Practice

Series Advisors

Human and Biological Sciences
Stephen Senn
University of Glasgow, UK

Earth and Environmental Sciences
Marian Scott
University of Glasgow, UK

Industry, Commerce and Finance
Wolfgang Jank
University of Maryland, USA

Statistics in Practice is an important international series of texts which provide detailed coverage of statistical concepts, methods and worked case studies in specific fields of investigation and study.

With sound motivation and many worked practical examples, the books show in down-to-earth terms how to select and use an appropriate range of statistical techniques in a particular practical field within each title's special topic area.

The books provide statistical support for professionals and research workers across a range of employment fields and research environments. Subject areas covered include medicine and pharmaceutics; industry, finance and commerce; public services; the earth and environmental sciences, and so on.

The books also provide support to students studying statistical courses applied to the above areas. The demand for graduates to be equipped for the work environment has led to such courses becoming increasingly prevalent at universities and colleges.

It is our aim to present judiciously chosen and well-written workbooks to meet everyday practical needs. Feedback of views from readers will be most valuable to monitor the success of this aim.

A complete list of titles in this series appears at the end of the volume.

Understanding Biostatistics

Anders Källén

Department of Statistics, AstraZeneca R&D Lund, Sweden

A John Wiley & Sons, Ltd., Publication

Library of Congress Cataloging-in-Publication Data

Källén, Anders, author.
   Understanding biostatistics / Anders Källén, Department of Biostatistics,
AstraZeneca, Sweden.
       p. ; cm.
   Includes bibliographical references and index.
       ISBN 978-0-470-66636-4 (print) – ISBN 978-1-119-99268-4 (epdf) – ISBN 978-
1-119-99267-7 (obook) – ISBN 978-1-119-99350-6 (epub) – ISBN 978-1-119-99351-3
(mobi)
   1. Biometry. I. Title.
       [DNLM: 1. Biostatistics. 2. Models, Statistical. WA 950]
       QH323.5.K35 2011
       570.1′5195–dc22
                                                                            2010051070
A catalogue record for this book is available from the British Library.

Print ISBN: 978-0-470-66636-4
ePDF ISBN: 978-1-119-99268-4
oBook ISBN: 978-1-119-99267-7
ePub ISBN: 978-1-119-99350-6
Mobi ISBN: 978-1-119-99351-3

Set in 10/12pt Times by Thomson Digital, Noida, India

Contents

Preface

The fact that you use biostatistics in your work does not say much about who you are. You may be a physician who has collected some data and is trying to write up a publication, or you may be a theoretical statistician who has been consulted by a physician, who has in turn collected some data and is trying to write up a publication. Whichever you are, or if you are something in between, such as a biostatistician working in a pharmaceutical company, the chances are that your perception of statistics to a large extent is driven by what a particular statistical software package can do. In fact, many books on biostatistics today seem to be more or less extended manuals for some particular statistical software. Often there is only one software package available to you, and the analysis you do on your data is governed by your understanding of that software. This is particularly apparent in the pharmaceutical industry.

However, doing biostatistics is not a technical task in which the ability to run software defines excellence. In fact, using a piece of software without the proper understanding of why you want to employ statistical methods at all, and what these methods actually provide, is bad statistics, however well versed you are in your software manual and code writing. The fundamental ingredient of biostatistics is not a software package, but an understanding of (1) whatever biological/medical aspect the data describe and (2) what it is statistics actually contribute. Statistics as a science is a subdiscipline of mathematics, and a proper description of it requires mathematical formulas. To hide this mathematical content within the inner workings of a particular software package must lead to an insufficient understanding of the true nature of the results, and is not beneficial to anyone.

Despite its title, this book is not an introduction to biostatistics aimed at laymen. This book is about the concepts, including the mathematical ones, of the more elementary aspects of biostatistics, as applied to medical problems. There are many excellent texts on medical statistics but one cannot cover everything and many of them emphasize the technical aspects of producing an analysis at the expense of the mathematical understanding of how the result is obtained. In this book the emphasis is reversed. These other books have a more systematic treatment of different types of problems and how you obtain the statistical results on different types of data. The present volume differs from others in that it is more concerned with ideas, both the particular aspects concerned with the role of statistics in the scientific process of obtaining evidence, and the mathematical ideas that constitute the basis of the subject. It is not a textbook, but should be seen as complementary to more traditional textbooks; it looks at the subject from a different angle, without being in conflict with them. It uses non-conventional and alternative approaches to some statistical concepts, without changing their meaning in any way. One such difference is that key computational aspects are often replaced by graphs, to illustrate *what* you are doing instead of *how*.

The ambition to discuss a wide range of concepts in one book is a challenge. Some concepts are philosophical in nature, others are mathematical, and we try to cover both. Broadly speaking, the book is divided into three major parts. The first part, Chapters 1–5, is concerned with what statistics contribute to medical research, and discusses not only the underlying philosophy but also various issues that are related to the art of drawing conclusions from statistical output. For this we introduce the concept of the confidence function, which helps us obtain both p-values and confidence intervals from graphics alone. In this part of the book we mostly discuss only the simplest of statistical data, in the form of proportions. We need a background to have the discussion on, and this simple case contains almost all of the conceptual problems in statistics.

The second part consists of Chapters 6–8, and is about generalizing frequency data to more general data. We emphasize the difference between the observed and the infinite truth, how population distributions are estimated by empirical (observed) distributions. We also introduce bivariate distributions, correlation and the important law of nature called 'regression to the mean'. These chapters show how we can extend the way we compare proportions for two groups to more general data, and in the process emphasize that in order to analyze data, you need to understand what kind of group difference you want to describe. Is it a horizontal shift (like the t-test) or a vertical difference (non-parametric tests)? A general theme here, and elsewhere, is that model parameters are mostly estimated from a natural condition, expressed as an estimating equation, and not really from a probability model. There are intimate connections between these, but this view represents a change to how estimation is discussed in most textbooks on statistics.

The third part, the next four chapters, is more mathematical and consists of two subparts: the first discusses how and why we adjust for explanatory variables in regression models and the other is about what it is that is particular about survival data. There are a few common themes in these chapters, some of which are build-ups from the previous chapters. One such theme is heterogeneity and its impact on what we are doing in our statistical analysis. In biology, patients differ. With some of the most important models, based on Gaussian data, this does not matter much, whereas it may be very important for non-linear models (including the much used logistic model), because there may be a difference between what we think we are doing and what we actually are doing; we may think we are estimating individual risks, when in fact we are estimating population risks, which is something different. In the particular case of survival data we show how understanding the relationship between the population risk and the individual risks leads to the famous Cox proportional hazards model.

The final chapter, Chapter 13, is devoted to a general tie-up of a collection of mathematical ideas, spread out in the previous chapter. The theme is estimation, which is discussed from the perspective of estimating equations instead of the more traditional likelihood methods. You can have an estimating equation for a parameter that makes sense, even though it cannot be derived from any appropriate statistical model, and we will discuss how we still can make some meaningful inference.

As the book develops, the type of data discussed grows more and more complicated, and with it the mathematics that is involved. We start with simple data for proportions, progress to general complete univariate data (one data point per individual), move on to consider censored data and end up with repeated measurements. The methods described are developed by analogy and we see, for example, the Wilcoxon test appear in different disguises.

The mathematical complexity increases, more or less monotonically, with chapter number, but also within chapters. On most occasions, if the math becomes too complicated for you to understand the idea, you should move to the next chapter, which in most cases start out simpler. The mathematical theory is not described in a coherent and logical way, but as it applies locally to what is primarily a statistical discussion, and it is described in a variety of different ways: to some extent in running text, with more complex matters isolated in stand-alone text boxes, while even more complex aspects are summarized in appendices. These appendices are more like isolated overviews of some piece of mathematics relevant to the chapter in question. All mathematical notation is explained, albeit sometimes rather intuitively, and for some readers it may be wise to 'hum' their way through some more complicated formulas. In that way it should be possible to read at least half the book with only minor mathematical skills, as long as one is not put off by the existence of such equations as one comes across. (If you are put off by formulas, you need to get another book.) As already mentioned, at least some of the repetitive (and boring) calculations in statistics have been replaced by an extensive use of graphs. In this way the book attempts to do something that is probably considered almost impossible by most: to simultaneously speak to peasants in peasant language and the learned in Latin (this is a free translation of an old Swedish saying). But there is a price to pay for targeting a wide audience: we cannot give each individual reader the explanation that he or she would find the most helpful. No one will find every single page useful. Some parts will be only too trivial to some, whereas some parts will be incomprehensible to others. There are therefore different levels at which this book can be read.

If you are medically trained and have worked with statistics, in particular p-values, to some extent, your main hurdle will probably be the mathematics. Your priority should be to understand what things intuitively mean, not only the statistical philosophy but also different statistical tests. There is no specific predefined level of mathematics, above basic high-school math, that you need to master for most parts of the book. You only need some basic understanding of what it is a formula tries to say, in order to grasp the story, and you do not need to understand the details of different formulas. To understand a mathematical formula can mean different things, and all formulas definitely do not need to be understood by everyone. The non-trivial mathematics is essentially only that of differentiation and integration, in particular the latter, which most people in the target readership are expected to have encountered at least to some degree. An integral is essentially only a sum, albeit made up of a vast number of very, very small pieces. If you see an integral, it may well suffice to look upon it as a simple sum and, instead of getting agitated, leave such higher-calculus formulas to be read by those with more mathematical interest and skill.

On a second level, you may be a reader who has had basic training in statistics and is working with biostatistics. Being a professional statistician nowadays does not necessarily mean that you have much mathematical training. Hopefully you can make sense of most of the equations, but you may need to consult a standard textbook or other references for further details.

The third level is when you are well versed in reading mathematical textbooks, deriving formulas and proving theorems. For you, the main reason for reading this book may be to get an introduction to biostatistics in order to see whether you want to learn more about the subject. For you, the lack of mathematical details should not be a problem; most left-out steps are probably easily filled in. At this point I beg the indulgence of any mathematician who has ventured into this book and who sees that mathematical derivations are not completely rigorous but sacrificed for the sake of more intuitive 'explanation'. It must also be noted that this book is not an introduction to what to consider when you work as a biostatistician. It may be helpful in some respects, but there is most often an initial hurdle to such work, not addressed in this book, which is about being able to translate biological or medical insight and assumptions to the proper statistical question.

These three levels represent a continuum of mathematical skills. But remember that this book is not a textbook. We use mathematics as a tool for description, an essential tool, but we do not transform biostatistics into a mathematical subdiscipline. One aspect of mathematics is notation. Proper and consistent use of mathematical notation is fundamental to mathematics. In this book we do not have such aspirations, and are therefore occasionally slack in our use of notation. Our notation is not consistent between chapters, and sometimes not even within chapters. Notation is always local, optimized for the present discussion, sacrificing consistency throughout. On most occasions we use capital letters to denote stochastic variables and lower case letters to denote observations, but occasionally we let lower case letters denote stochastic variables. Sometimes we are not even explicit about the change from the observation to the corresponding stochastic variable. Another example is that is not always well defined whether a vector is a column vector or row vector, it may change state almost within a sentence. If you know the importance of this distinction, you can probably identify which it is from the context. This sacrifice is made because I believe it increases readability.

All chapters end with some suggestions on further reading. These are unpretentious and incomplete listings, and are there to acknowledge some material from which I have derived some inspiration when writing this book.

I am deeply grateful to Professor Stephen Senn for the strong support he has given to the project of finalizing this book and for the invaluable advice he has given in the course of so doing. It has been a long-standing wish of mine to write this book, but without his support it is very doubtful that it would ever have happened. I also want to give credit to all those (to me unknown) providers of information on the internet from which I have borrowed, or stolen, a phrase now and then, because it sounded much better than the Swenglish way I would have written it myself. In addition, I want to thank a number of present or past colleagues at the AstraZeneca site where I have worked for the past 25 years, but which the company has decided to close down at more or less the same time as this book is published, in particular Tore Persson and Tobias Rydén, who, despite conflicting priorities, provided helpful comments. Finally, I also want to thank my father and Olivier Guilbaud for input at earlier stages of this project.

This book was written in LATEX, and the software used for computations and graphics was the high level matrix programming language GAUSS, distributed by Aptech Systems of Maple Valley, Washington. Graphs were produced using the free software *Asymptote*.

The Cochrane Collaboration logo in Chapter 3 is reproduced by permission of Cochrane Library.

Anders Källén
Lund, October 2010

1

Statistics and medical science

1.1 Introduction

Many medical researchers have an ambiguous relationship with statistics. They know they need it to be able to publish their results in prestigious academic journals, as opposed to general public tabloids, but they also think that it unnecessarily complicates what should otherwise be straightforward interpretations. The most frustrated medical researchers can probably be found among those who actually do consult biostatisticians; they only too often experience criticism of the design of the experiment they want to do or, worse, have done – as if the design was the business of the statistician at all.

On the other hand, if you ask biostatisticians, they often consider medical science a contradiction in terms. Tradition, subjectivity and intuitive thinking seem to be such an integral part of the medical way of thinking, they say, that it cannot be called science. And biostatisticians feel fully vindicated by the hype that surrounded the term 'evidence-based medicine' during the 1990s. Evidence? Isn't that what research should be all about? Isn't it a bit late to realize that now?

This chapter attempts to explain what statistics actually contributes in clinical research. We will describe, from a bird's-eye perspective, the structure within which statistics operates, and the nature of its results. We will use most of the space to describe the true nature of one particular summary statistic, the p-value. Not because it necessarily is the right thing to compute, but because all workers in biostatistics have encountered it. How it is computed will be discussed in later chapters (though more emphasis will be put on its relative, the confidence interval).

Medicine is not a science *per se*. It is an engineering application of biology to human disease. Medicine is about diagnosing and treating individual patients in accordance with tradition and established knowledge. It is a highly subjective activity in which the physician uses his own and others' experiences to find a diagnostic fit to the signs and symptoms of a particular patient, in order to identify the appropriate treatment. For most of its history, medicine has been about individual patients, and about inductive reasoning. Inductive reasoning is when you go from the particular to the general, as in 'all crows I have seen have

Understanding Biostatistics, First Edition. Anders Källén.
© 2011 John Wiley & Sons, Ltd. Published 2011 by John Wiley & Sons, Ltd.

Box 1.1 The philosophy of science

What is knowledge about reality and how is it acquired? The first great scholar of nature, Aristotle, divided knowledge into two categories, the original facts (axioms) and the deduced facts. Deduction is done by (deductive) logic in which propositions are derived from one or more premises, following certain rules. It often takes the shape of mathematics. When applied to natural phenomena, the problem are the premises. In a deductive science like mathematics there is a process to identify them, but in empirical sciences their nature is less obvious. So how do we identify them?

Early thinkers promoted the idea of induction. When repeated observations of nature fall into some pattern in the mind of the observer, they are said to induce a suggestion of a more general fact. This idea of induction was raised to an alternative form of logic, inductive logic, which forced a fact from multiple observations, a view which was vigorously criticized by David Hume in the mid-eighteenth century.

Hume's argument started with an analysis of causal relations, which he claimed were found exclusively by induction, never deduction, and contains an implicit assumption that unobserved objects resemble observed ones. The causal connection is by induction, not deduction, and the justification of the inductive process becomes a circular argument, Hume argues. This was referred to as 'Hume's dilemma', something that upset Immanuel Kant so much that he referred to the problem of induction as the 'scandal of philosophy'. This does not mean that if we have always observed something in a particular situation, we should not expect the same to happen next time. It means that it cannot be an absolute fact, and instead we are making a prediction, with some degree of confidence.

Two centuries later Karl Popper introduced refutationism. According to this there are no empirical, absolute facts and science does not rely on induction, but exclusively on deduction. We state working hypotheses about nature, the validity of which we test in experiments. Once refuted, a modified hypothesis is formulated and put to the test. And so on. This infinite cycle of conjecture and refutation is the true nature of science, according to Popper.

As an example, used by Hume, 'No amount of observations of white swans can allow the inference that all swans are white, but the observation of a single black swan is sufficient to refute that conclusion'. It was a long-held belief in Europe that all swans were white, until Australia was discovered, and with it *Cygnus atratus*, the black swan.

Inductionism and refutationism both have their counterparts in the philosophy of statistics. In the Bayesian approach to statistics, which is inductive, we start with a summary of what we believe and update that according to experimental results. The frequentist approach, on the other hand, is one of refuting hypothesis. Each case is unique and the data of the particular experiment settle that case alone.

---

been black, therefore all crows are black'. It is the way we, as individuals, learn about reality when we grow up. However, as a foundation of science, induction has in most cases been replaced by the method of falsification, as discussed in Box 1.1. (It is of course not the case that medicine is exclusively about inductive reasoning: a diagnostic fit may well be put to the test in a process of falsification.)

Another peculiarity of medicine is ethics. Medical researchers are very careful not to put any patients at risk in obtaining the information they seek. This is often a complicating factor in clinical research when it interferes with the research objective of a clinical trial. For example, in drug development, at one important stage we need to show that a particular drug is effective. The scientific way to do this is by carrying out a clinical trial in which the response to the drug is compared to the response when no treatment is given. Everything else should be the same. However, in the presence of other effective drugs, it may not at all be ethical to withhold a useful drug for the sole reason that you want to demonstrate that a new drug is also effective.

Finally, there is the general problem of why it appears to be so hard for many physicians to understand basic statistical reasoning: what conclusions one may draw and why. To be honest, part of the reason why statistics is so hard to understand for non-statisticians is probably that statisticians have not figured it out for themselves. There is not one statistical philosophy that forms the basis for statistical reasoning, there are a number of them: frequentists versus Bayesians, Fisher's approach versus the Neyman–Pearson view. If statisticians cannot figure it out, how can they expect their customers to be able to do so?

These are some properties of medical researchers that statisticians should be aware of. Of course, they are not true statements about individual medics. They are statements about the group of medics, and statements about groups are what statistics is all about. This will be our starting point in Chapter 2 when we initiate a more serious discussion about the design of clinical trials. But before we do that we need to get a basic understanding of what it is statistics is trying to do. This journey will start with an attempt to describe the role of statistics within science.

1.2   On the nature of science

For almost all of the history of mankind the approach to health has been governed by faith, superstition and magic, often expressed as witchcraft. This has gradually changed since the period of the Enlightenment in the eighteenth century, so that doctors can no longer make empty assertions and quacks can no longer sell useless cures with impunity. The factor that has changed this is what we call science.

But what *is* science? We know what it does: it helps us understand and make sense of the world around us. But that does not define science; religion has served much the same purpose for most of mankind's history. Science is often divided into three subsets: natural sciences (the study of natural phenomena), social sciences (the study of human behavior and society), and mathematics (including statistics). The first two of these are empirical sciences, in which knowledge is based on observable phenomena, whereas mathematics is a deductive science in which new knowledge is deduced from previous knowledge. There is also applied science, engineering, which is the application of scientific research to specific human needs. The use of statistics in medical research is an example, as is medicine itself.

The science of mathematics has a specific structure. Starting from a basic set of definitions and assumptions (usually called axioms), theorems are formulated and proved. A theorem constitutes a mathematical statement, and its proof is a logical chain of applications of previously proved theorems. A collection of interlinked, proved, mathematical theorems makes up a mathematical theory of something. The empirical sciences are similar to this in many respects, but differ fundamentally in others. Corresponding to an unproved mathematical theorem is a hypothesis about nature. The mathematical proof corresponds to an experiment that tests the hypothesis. A theory, in the context of empirical science, consists of a number of not yet refuted hypotheses which are bound together by some common theme.

What we think we know about the world is very much the result of an inductive process, derived from experiences and learning. The difference between science and religion is not about content, but about the way knowledge is obtained. A statement can only be a scientific statement if it can be tested, and science is qualified by the extent to which its predictions are borne out; when a model fails a test it has to be modified. Science is therefore not static, it is dynamic. Old 'truths' are replaced by new 'truths'. It is like an enormous jigsaw puzzle in which pieces are constantly replaced and added. Sometimes replacement is with a set of new pieces that give a clearer picture of the overall puzzle, sometimes a piece turns out to be wrong and needs to be replaced by a new, fundamentally different, one. Sometimes we need to tear up an entire part of the jigsaw puzzle and rebuild it. The basic requirement of the individual pieces in this jigsaw puzzle is that each one addresses a question that can be tested for validity. Science is a humble practice; it tells us that we know nothing unless we have evidence and that our state of knowledge must always be open to scrutiny and challenge.

The fundamental difference between empirical sciences and mathematics is that a mathematical proof proves the hypothesis (i.e., theorem), whereas in empirical sciences experiments are designed to disprove the hypothesis. A particular hypothesis can be refuted by an observation that is inconsistent with the hypothesis. But the hypothesis cannot be proved by experiment – all we can say is that the outcome of the experiment is consistent with it.

Example 1.1   Like most people before modern times, the Greeks thought that the earth was the center of everything. They identified seven moving objects in heaven – five planets, the sun and the moon – and Ptolemy worked out a very elaborate model for how they move, using only circles and circles moving on circles (epicycles). The result was an explanation of the heavens (planets, at least) that fulfilled all the criteria of science. They made predictions that could be tested, and these never failed. When the idea of putting the sun at the center of this system emerged, it was not found to work better in any way; it did not produce better predictions than the Greek model. It was not until Johannes Kepler managed to identify his famous three laws that astronomers actually got a sun-centered description of the heavens that even matched the Greek version. This meant that there were two competing models with no one really ahead.

However, this changed with Isaac Newton. With his law of gravitation the science of the heavens took a gigantic leap forward. In one go, he reduced the complex behavior of the planets to a few fundamental and universal laws. When these laws were applied to the planets they not only predicted their movements to any precision measurable, they also allowed a new planet to be discovered (Neptune, in 1846). So many experiments were conducted over hundreds of years with outcomes consistent with Newton's theory, that it was very tempting to consider it a true fact. However, during the twentieth century some astronomical observations were made that were inconsistent with the mathematical predictions of the theory, and it is today superseded by Albert Einstein's theory of general relativity in cosmology. As a theory though, Newton's theory of gravitation is still good enough to be used for all everyday activities involving gravitation, such as sending people to the moon.

This example illustrates an important point about science which must be kept in mind, namely that 'all models are wrong, but some are useful', a quotation often attributed to the

English statistician George Box. Much of the success of Newton's physics was due to the fact that it was expressed in mathematical terms. As a general rule scientific theory seems to be least controversial when it can be expressed in the form of mathematical relationships. This is partly because this requires a rather well-defined logical foundation to build on, and partly because mathematics provides the logical tool to derive the correct predictions.

That one theory replaces another, sometimes with fundamental effects, is common in biology, not least in medicine. (On my bookshelf there are three books on immunology, published in 1976, 1994 and 2006, respectively. It is hard to see that they are about the same science. On the other hand, there is also a course in basic physics from 1950, which could serve well as present-day teaching material – in terms of content, if not style.) We must always consider a theory to be no more than a set of hypotheses that have not yet been falsified. In fact, mathematics also has an element of this, since a theorem that has been proved has been so only to the extent that no one has yet found a fault in the proof. There are quite a few examples of mathematical theorems that have been held to be true for a period of time until someone found a mistake in their proofs.

1.3   How the scientific method uses statistics

To produce objective knowledge is difficult, since our intuition has a tendency to see patterns where there is only random noise and to see causal relationships where there are none. When looking for evidence we also have a tendency, as a species, to overvalue information that confirms our hypothesis, and we seek out such confirmatory information. When we encounter new evidence, the quality of it is often assessed against the background of our working assumption, or prior belief, leading to bias in interpretation (and scientific disputes).

To overcome these human shortcomings the so-called scientific method evolved. This is a method which helps us obtain and assess knowledge from data in an objective way. The scientific method seeks to explain nature in a reproducible way, and to use these explanations to make useful predictions. It can be crudely described in the following steps:

1. Formulate a hypothesis.

2. Design and execute an experiment which tests the hypothesis.

3. Based on the outcome of the experiment, determine if we should reject the hypothesis.

To gain acceptance for one's conclusion it is critical that all the details of the research are made available for others to judge their validity, so-called *peer review*. Not only the results, but also the experimental setup and the data that drive the experimenter to his conclusions. If such details are not provided, others cannot judge to what extent they would agree with the conclusions, and it is not possible to independently repeat the experiment. As the physicist Richard Feynman wrote in a famous essay, condemning what he called 'cargo cult science',

> if you are doing an experiment, you should report everything that you think might make it invalid – not only what you think is right about it: other causes that could possibly explain your results; and things you thought of that you've eliminated by some other experiment, and how they worked – to make sure that the other fellow can tell if they have been eliminated.

A key part of the scientific method is the design, execution and analysis of an experiment that tests the hypothesis. This may employ mathematical modeling in some way, as when one uses statistical methods. The first step in making a mathematical model related to the hypothesis is to quantify some entities that make it possible to do calculations on numbers. These quantities must reflect the hypothesis under investigation, because it is the analysis of them that will provide us with a conclusion. We call a quantity that is to be analyzed in an experiment an *outcome measure*, because it is a quantitative measure of the outcome of the experiment. After having decided on the outcome measure, we design our experiment so that we obtain appropriate data. The statistical analysis subsequently performed provides us with what is essentially only a summary presentation of the data, in a form that is appropriate to draw conclusions from.

So, for a hypothesis that is going to be tested by invoking statistics, the scientific method can be expanded into the following steps:

1. Formulate a hypothesis.

2. Define an outcome measure and reformulate the hypothesis in terms of it. This involves defining a statistical model for the data. This version of the hypothesis is called the *null hypothesis* and is formulated so that it describes what we want to reject.

3. Design and perform an experiment which collects data on this outcome measure.

4. Compute statistical summaries of the data.

5. Draw the appropriate conclusion from the statistical summaries.

When the results are written up as a publication, this should contain an appropriate description of the statistical methods used. Otherwise it may be impossible for peers to judge the validity of the conclusions reached.

The statistical part of the experiment starts with the data and a model for what those data represent. From there onwards it is like a machine that produces a set of summaries of the data that should be helpful in interpreting the outcome of the experiment. For confirmatory purposes, rightly or wrongly, the summary statistic most used is the p-value. It is one particular transformation of the data, with a particular interpretation under the model assumption and the null hypothesis. It measures the probability of the result we observed, or a more extreme one, given that the null hypothesis is true. Thus a p-value is an indirect measure of evidence against the null hypothesis, such that the smaller the value, the greater the evidence. (Often more than one model can be applied to any given set of data so we can derive different p-values for a given hypothesis and set of data – as in the case of parametric versus non-parametric tests.)

Note that, as a consequence of the discussion above, the conclusion from the experiment is either that we consider ourself as having proved the null hypothesis wrong, or we have failed to prove it wrong. Never is the null hypothesis proved to be true. To understand why, look at the hypothesis 'there are no fish in this lake' which we may want to test by going fishing. There are two possible outcomes of this test: either you get a fish or you do not. If you catch a fish you know there is (or was) fish in the lake and have disproved the hypothesis. If you do not get any fish, this does not prove anything: it may be because there were no fish in the lake, or it may be because you were unlucky. If you had fished for longer, you may have had a catch and therefore rejected the null hypothesis. There is a saying that captures this and is worth keeping in mind: 'Absence of proof is not proof of absence.' Failure to reject a hypothesis does not prove anything, but it may, depending on the nature and quality of the experiment, increase one's confidence in the validity of the null hypothesis – that it to some degree reflects the truth. As such it may be part of a theory of nature, which is held true until data emerge that disprove it.

Failure to understand the difference between not being able to provide enough evidence to reject the null hypothesis and providing evidence for the null hypothesis is at the root of the most important misuse of statistics in medical research.

Example 1.2   In the report of a study on depression with three treatments – no treatment (placebo), a standard treatment, B, and a new treatment, A – the authors made the following claim: 'A is efficacious in depression and the effect occurs earlier than for B.' The data underlying the second part of this claim refer to comparisons of A and B individually versus placebo, using data obtained after one week. For A, the corresponding p-value was 0.023, whereas for B it was 0.16. Thus, the argument went, A was 'statistically significant', whereas B was not, so A must be better than B.

This is, however, a flawed argument. To make claims about the relative merits of A and B, these must be directly compared. In this case a crude analysis of the data tells us what the result should be. In fact, the first p-value was a result of a mean difference (versus placebo) of 1.27 with a standard error of 0.56, whereas the second p-value comes from a mean difference of 0.79 with the same standard error. The mean difference between A and B is therefore 0.48, and since we should probably have about the same standard error as above, this gives a p-value of about 0.40, which is far from evidence for a difference.

The mistake made in this example is a recurrent one in medical research. It occurs when a statistical test, accompanied by its declaration of 'significant' or 'not significant', is used to force a decision on the truth or not of the null hypothesis.

1.4   Finding an outcome variable to assess your hypothesis

The first step in the expanded version of the scientific method, to reformulate the hypothesis in terms of a specific outcome variable, may be simple, but need not to be. It is simple if your hypothesis is already formulated in terms of it, as when we want to claim that women on the average are shorter than men. The outcome variable then is individual height. It is more difficult if we want to prove that a certain drug improves asthma in patients with that disease. What do we mean by improvement in asthma? Improvement in the lung function? Fewer asthma symptoms? There are many ways we can assess improvement in asthma, and we need to be more specific so that we know what data to collect for the analysis. Assume that we want to focus on lung function. There are also many ways in which we can measure lung function: the simplest would be to ask the patients for a subjective assessment of their lung function, though usually more objective measures are used.

Suppose that we settle for one particular objective lung function measurement, the forced expiratory volume in one second, FEV_1. We may want to prove that a new drug improves the patient's asthma by formulating the null hypothesis to read that the drug does not affect

FEV_1. If we subsequently carry out an experiment and from the analysis of it conclude that there is an improvement in lung function as measured by FEV_1, we have disproved the null hypothesis.

The question is what we have proved. The statistical result relates to FEV_1. How much can we generalize from this and actually claim that the asthma has been improved? This is a non-trivial issue and one which must be addressed when we decide on which outcome measure to use to reflect our original hypothesis.

Quality of life is measured by having patients fill in a particular questionnaire with a list of questions. The end result we want from the analysis of such a questionnaire is a simple statement: 'The quality of life of the patients is improved'. In order to achieve that, the scores on individual questions in the questionnaire are typically reduced to a summary number, which is the outcome variable for the statistical analysis. The result may be that there is an increase in this outcome variable when the treatment is given. However, the term 'quality of life' has a meaning to most people, and the question is whether an increase in the summary variable corresponds to an increase in the quality of life of the patients, as perceived by the patients. This question necessitates an independent process, in which it is shown that an increase in the derived outcome variable can in fact be interpreted as an improvement of quality of life – a validation of the questionnaire.

The IQ test constitutes a well-known example. IQ is measured as the result of specific IQ tests. If we show that two groups have different outcomes on IQ tests, can we then deduce that one group is more intelligent than the other group? It depends on what we mean by intelligence. If we mean precisely what the IQ test measures, the answer is yes. If we have an independent opinion of what intelligence should mean, we first have to validate that this is captured correctly by the IQ test.

Returning to the measurement of FEV_1, for a claim of improvement in asthma, lung function is such an important aspect of asthma that it is reasonable to say that improved lung function means that the asthma has improved (though many would require additional support from data that measure asthma symptoms). However, if we fail to show an effect of FEV_1 it does not follow by logical necessity that no other aspect of the asthma has improved. So we deliberately choose one aspect of the disease to gamble on, and if we win we have succeeded. If we fail, we may not be any wiser.

1.5 How we draw medical conclusions from statistical results

Before we actually come to the subject of this section we need to consider the ultimate purpose of science, which is to make predictions about the future. What we see in a particular study is an observation. What we want from the study is more than that: we want statements that are helpful when we need to make decisions in the future. We want to use the study to predict what will be seen in a new, similar study. It is an observation that in a particular study 60% of males, but only 40% of females, responded to a treatment. Unless your sample is very large it is not reasonable to generalize this to a claim that 60% of males and 40% of females will respond to the drug in the target population. It may be the best predictor we have at this point in time, but that is not the same thing. What we actually can claim depends on the statistical summary of the data. A more cautious claim may be that in general males respond better to the treatment than females. To substantiate this claim we analyze the data under the null hypothesis that there is no difference in the response rates for males and females.

Suppose next that we want to show that some intervention prolongs life after a cancer diagnosis. Our null hypothesis is that it does not. We assume that we have conducted an appropriate experiment (clinical trial) and that the statistical analysis provides us with $p = 0.015$. This means that, if there is no effect at all of the intervention, a result as extreme as that found in the experiment is so unlikely that it should occur in only 1.5% of all such clinical trials. This is our confidence in the null hypothesis (not to be confused with the probability of the null hypothesis) after we have performed the experiment.

That does not prove that the intervention is effective. No statistical analysis proves that something is effective. The proper question is: does this p-value provide sufficient support to justify our starting to act as if it is effective? The answer to that question depends on what confidence is required from this particular experiment for a particular action. What are the consequences if I decide that it is effective? A few possibilities are:

- I get a license for a new drug, and can earn a lot of money;

- I get a paper published;

- I want to take this drug myself, since I have been diagnosed with the cancer in question.

In the first case it is really not for me to decide what confidence level is required. It is the licensing authority that needs to be assured. Their problem is on the one hand that they want new, effective drugs on the market, but on the other hand that they do not want useless drugs there. Since all statistics come with an uncertainty, their problem is one of error control. They must make a decision that safeguards the general public from useless drugs, but at the same time they must not make it impossible to get new drugs licensed. This is a balancing act, and they do it by setting a significance level α such that if your p-value is smaller than α, they agree that the drug is proved to be effective. The significance level defines the proportion of truly useless drugs that will accidentally be approved and therefore the level of risk the licensing agency is prepared to take (if we include almost useless drugs as well, the proportion is higher). Presently one may infer that the US licensing authority, the Food and Drug Administration (FDA), has set the significance level at $0.025^2 = 0.000625$ when it comes to proving efficacy for their market, for reasons we will come back to.

The picture is similar if you want to publish a paper. In general there is an agreed significance level of 5% (two-sided) for that process. If your p-value is less than 5% you can publish a paper and claim that the intervention works. But that does not prove that the intervention works, only that you can get a paper published that claims so. The significance level used by a particular journal is typically not explicitly spelt out, since a remark by the eminent statistician R.A. Fisher led to the introduction of the golden threshold at 5% a long time ago (see Box 1.2), making it unnecessary to argue about it. That is really its only virtue – there is no scientific reason why it should not be 6% or 0.1%. In relation to this particular threshold we now also have some jargon, the term 'statistical significance' , which is discussed in some detail in Box 1.3.

In the last situation in the bullet list above, the case where you had that particular cancer yourself, you really decide your own significance level. It may be very high, depending on how desperate you are. A significance level of 20% may be good enough for you. It may depend on side-effects and alternative options.

A situation where the interpretation of the p-value as a measure of confidence and its relation to what to do next becomes apparent, is in drug development. Clinical drug development

---

Box 1.2 The origin of the 5% rule

The 5% significance rule seems to be a consequence of the following passage in the book *Statistical Methods for Research Workers* by the inventor of the *p*-value, Ronald Aylmer Fisher:

> in practice we do not always want to know the exact value of P for any observed χ^2, but, in the first place, whether or not the observed value is open to suspicion. If P is between .1 and .9 there is certainly no reason to suspect the hypothesis tested. If it is below .02 it is strongly indicated that the hypothesis fails to account for the whole of the facts. ... A value of χ^2 exceeding the 5 per cent. point is seldom to be disregarded.

It is important that in Fisher's view a *p*-value below 0.05 does not force a decision, it only warrants a further investigation. Larger *p*-values are not worth investigating (note that he does not actually say anything about values between 0.05 and 0.1). On another occasion he wrote:

> This is an arbitrary, but convenient, level of significance for the practical investigator, but it does not mean that he allows himself to be deceived once in every twenty experiments. The test of significance only tells him what to ignore, namely all experiments in which significant results are not obtained.

Nowadays we use the 5% rule in a different way. We use it to force decisions in single studies, referring to an error-rate control mechanism on the ensemble of studies, following a philosophy introduced by Jerzy Neumann and Egon Pearson (see Box 1.3).

---

is a staged process in which we sequentially try to answer more and more complex questions such as:

- Is the drug effective at all?

- What is the appropriate dose for this drug?

- Is the appropriate dose effective enough to get the drug licensed?

The monetary investment that needs to be made in order to answer these questions is usually very different. Moreover, the more confidence we want to have in the answer to a particular question, the more money it costs to get that confidence, because larger studies need to be performed. The decision on what confidence we need that a drug is effective at all before conducting a dose-finding study, could then depend on the cost of the latter. Or, rather, a balance between that cost and the loss in time to market, which in itself is a cost. The bottom line is that it may be strategically right for a pharmaceutical company to do a small study which only can produce limited confidence in efficacy, say a one-sided *p*-value at 10%, before gambling with a larger dose-range study, in order to save time.

In view of the present avalanche of statistical *p*-values pouring over us – by one estimate some 15 million medical articles have been published to date, with 5000 journals around the

---

Box 1.3 The meaning of the term 'statistical significance'

There are two alternative ways of looking at p-values and significance levels which are related to the philosophy of science. Here is a brief outline of these positions.

The p-value builds confidence. R.A. Fisher originally used p-values purely as a measure of inductive evidence against the null hypothesis. Once the experiment is done there is only one hypothesis, the null, and the p-value measures our confidence in it. There is no need for the significance level; all we need to do is to use the p-value as a measure of our confidence that it is correct to reject the null hypothesis. By presenting the p-value we allow any readers of our results to judge for themselves whether the test has provided enough confidence in the conclusion.

The significance level defines a decision rule. The Neyman–Pearson school instead emphasizes statistical hypothesis testing as a mechanism for making decisions and guiding behavior. To work properly this setup requires two hypotheses to choose between, so the Neyman–Pearson school introduces an alternative hypothesis, in addition to the null hypothesis. A decision between these is then forced, using the test and a predefined significance level α. The alternative is accepted if $p < \alpha$, otherwise the null hypothesis is accepted. Neyman–Pearson statistical testing is aimed at error minimization, and is not concerned with gathering evidence. Furthermore, this error minimization is of the long-run variety, which means that, unlike Fisher's approach, Neyman–Pearson theory does not apply to an individual study.

In a pure Neyman–Pearson decision approach the exact p-value is irrelevant, and should not be reported at all. When formulated as 'reject the null hypothesis when $p < \alpha$, accept it otherwise', only the Neyman–Pearson claim of $100\alpha\%$ false rejections of the null hypothesis with ongoing sampling is valid. This is because α is the probability of a set of potential outcomes that may fall anywhere in the tail area of the distribution of the null hypothesis, and we cannot know ahead of time which of these particular outcomes will occur. That is not the same as the tail area that defines the p-value, which is known only after the outcome is observed.

This dualism between Fisher's inductive approach to p-values and the error control of Neyman and Pearson is really about what p-values imply, not what they are. For Fisher it is about inductive learning, for Neyman and Pearson it is about decision making. For Fisher, the Neyman–Pearson view is not relevant to science, since one does not repeat the same experiment over and over again. What researchers actually do is one experiment, from which they should communicate information, not force a yes–no decision.

---

world constantly adding to that number – a strict adherence to a rule such as 'if $p < 5\%$ I can say I have an effect, otherwise not', is a bit primitive, to say the least. Assume (probably incorrectly) that all statistical analyses done are done in a correct manner. Then 5% of all cases investigated where there is no true effect or association, are out there as false effect or relationship claims. We cannot, using statistics, guarantee that there are no false 'truths' in circulation, and this level may be appropriate. But most hypotheses tested are part of a bigger context, a theory. If the result we present is a trivial modification of, or an add-on to, what is already known, we may need less assurance than if the result may set an earthquake in motion and have a major impact on society. Ultimately the judgement about the correctness of the null hypothesis will depend on the existence of other data and the relative plausibility of the alternatives.

In fact, in a medical context it is probably a good idea to be a little relaxed about the first ground-breaking result. Let it be reproduced before you actually believe it. This only means that you work with a lower significance level when you draw your conclusion from such results, whereas for reports that more or less only confirm previous reports you may work on a higher significance level. In essence this means that you take a more inductive evidence approach in your use of p-values, as compared to a strict decision-theoretic one (see Box 1.3).

The very low significance level the FDA have set for proving efficacy, referred to earlier, is an example of this. In order to prove efficacy in the eyes of FDA you need to do so in two independent studies, each with a (two-sided) test at 5%. Since licensing efficacy only goes in one direction, this means that their significance level within a particular study is half of this, 2.5%, and they will only accept that the drug is an effective treatment if both studies succeed. That a treatment with no effect whatsoever should pass this hurdle then occurs with a probability as low as 0.000625. (Actually, this is a debatable point, because there is some lack of clarity about how many unsuccessful related studies are allowed. The presence of such studies obviously impacts on this probability calculation.)

The discussion in this section, about the separation between statistical results and the conclusion to be drawn, seems not to be clear to many statisticians in the pharmaceutical industry or health authorities. How else can we explain the rise in the late 1990s of the non-inferiority trial? This – to my mind peculiar – concept is discussed in Box 1.4. The mistake made with the non-inferiority trial concept is precisely a confusion about the relation between a statistical result, in this case the confidence interval for a particular parameter, and the conclusion we draw from that result. As discussed above, any conclusion should be drawn in a particular context. One such context can be that a health authority allows a particular result to mean that efficacy is demonstrated beyond any reasonable doubt, and grants you a license to sell the drug. In another situation the result may be part of a decision to switch standard treatment at a particular hospital. In a third example it may provide sufficient evidence to test the new drug on a particular patient. Each of these situations calls for a decision, and for each decision we need a standard of proof. Once that is decided, the action should be taken without reference to a statement like 'A is not inferior to B', only to the actual result, the confidence interval. The problem, in a nutshell, is that one tries to build the whole decision process into the study, so that the study result forces a definite decision, instead of viewing the result of the study as a step in this process.

It is somewhat ironic that the non-inferiority study was modeled on so-called bioequivalence studies. A bioequivalence study is a particular type of pharmacokinetic study which drug makers run when they want to change some aspect of how a tablet is manufactured. Such studies follow rather precise rules in terms of how they should be analyzed: the 90% confidence interval of a particular mean ratio should lie between 0.8 and 1.25. If that is the case, the new formulation can replace the old one. The key difference between this type of study and the non-inferiority study is that for the bioequivalency study the result has a very specific follow-up action: you can switch to the new formulation. The bioequivalency result in itself is of no independent interest.

Box 1.4 The non-inferiority trial

The non-inferiority trial originally addressed the following specific problem. In order to prove efficacy, we need to prove that the new drug is better than taking no treatment. However, in many disease areas giving no treatment may be unethical; cancer treatments for which there are available alternative and established treatments may serve as an example. One way to approach this would be to take the new drug, A, and compare it with a standard treatment, B, which we agree is effective. If the difference in response between A and B is not too large, the argument goes, then A must also be effective. Such a trial was called a non-inferiority trial and its logic went like this: prespecify how much inferior A can be to B without casting doubts on A being effective. If our study achieves this objective we can claim that A is effective.

Unfortunately, that is not exactly true. Instead of using the argument to claim that A is effective, one claims that A is not inferior (in efficacy) to B. So the result becomes a statement about the relative merits of A and B, instead of the original intent to use B as a tool to declare A effective. The criterion that is typically used is that a confidence interval of a mean difference must stay within certain bounds. The study designers construct those limits, and the study logic dictates that if they succeed in getting the confidence interval within those limits, they are allowed to draw the conclusion that A is not inferior to B.

The problem here is that everyone needs to agree that the prespecified limits imply that A is not inferior to B. If the limits are widely agreed, there is no need to prespecify them – they would be universally accepted anyway. If they are not, it may be that the conclusion differs depending on its consequences. For some purposes it may be good enough, for others it may not.

Apart from the logical problem, there is an executional problem that is as important: how do we know that the trial could have picked up a difference? This is referred to as *assay sensitivity* and is a distinguishing feature between this type of trial and the superiority trial. If there is no assay sensitivity in a superiority trial, the trial will be unsuccessful, whereas for a non-inferiority trial it may be successful. This means that with a non-inferiority trial we also need to provide evidence that this particular trial was sufficiently sensitive; that the control behaved also in this trial as it had done in previous trials where it had shown efficacy. This is very much the same as referring to historical controls.

1.6 A few words about probabilities

Before we proceed we need to say a few words about probabilities. To set the scene, consider the following example.

Example 1.3 You meet a woman in the street who you know has two children, one of whom is a boy playing in your son's soccer team. What is the probability that her other child is a girl? The chances are that you will say 50%. The argument is deductive: there are two choices, a boy or a girl, and there are the same number of boys and girls in the community. Is this a correct way of arguing?

The answer is no. The probability required should refer to an empirical statement: out of all two-children families with *at least* one boy, in what percentage is the other one a girl? With the appropriate model assumption, such as that a child in any family has the same probability of being a boy as being a girl, we can design an experiment to test the claim. Take two unbiased coins (with each coin representing a child so that heads (H) corresponds to a girl and tails (T) to a boy) and toss them, say, 100 times. Each time there is at least one H, note on paper if the other is a T. Out of your 100 experiments there will be some, say N, with at least one H, and out of these in a certain number of cases, say n, the other is a T. The number n/N is then an estimate of the probability that the other child is a girl. If you do this experiment, you will probably end up with a number closer to 2/3 than to 1/2. In fact if you do it on a computer instead, using a random number generator, with a very large number of experiments, you will get rather close to 2/3.

So you are advised to reject your hypothesis that the probability is 1/2. We will discuss why in a short while.

The type of probabilities we discuss here are relative frequencies, not observed relative frequencies but theoretical ones – entities that in principle can be estimated by observed frequencies. The concept of probability is actually non-trivial, and we will return to it at the end of this chapter. For now we assume that it is simple to define.

Probabilities are computed for events. If we denote an event (like that the other child is a girl) by A, we denote the probability that it occurs in a particular experiment by $P(A)$. In the previous example this is 2/3, which is the frequency if we do the experiment an infinite number of times. If A denotes an event, we denote by A^c the complement of that event (i.e., that it does not occur), and $P(A^c)$ is then the probability that A does not occur. It is computed as $P(A^c) = 1 - P(A)$, since we are dealing with relative frequencies.

Example 1.4   You are participating in a game show, in which the host has placed a car behind one of three doors and a goat behind each of the other two doors. The game host instructs you to choose one door by pointing at it. When you have done so, he opens one of the other two doors to reveal a goat. After you have seen that goat, you are given the opportunity to switch doors. You win whatever is behind the door you select.

The problem is simple: should you switch doors, or does it matter at all? The chances are that you think it does not matter. You have two doors to choose between, so there should be a 50% chance to find the car behind whichever door you selected first. Actually the probability is only 1/3 that it is behind the door you selected first, so the correct strategy is to switch doors. This particular problem is called the Monty Hall problem, and some of its history can be found in Box 1.5.

We now have two examples of what may well be counterintuitive probabilities. Intuition is perhaps nothing but a reflection of personal experience, and the reason why these examples appear counterintuitive may be a lack of the appropriate experience. In the first example we have that a family with precisely two children has one of the following structures: $(B, B), (B, G), (G, B), (G, G)$, where B denotes boy, G denotes girl and the pair is written as (oldest, youngest). Moreover, if boys and girls are equally likely, we have the same number of

Box 1.5 The Monty Hall problem

The game discussed in Example 1.4 appeared in the 1990s in TV shows all over the world, and was loosely based on an American game show called *Let's Make a Deal*, hosted by Monty Hall. This game show epidemic had its origin in a letter to the column *Ask Marilyn* in the American journal *Parade* in February 1990. The columnist, Marilyn vos Savant, received the following question:

> Suppose you're on a game show, and you're given the choice of three doors: Behind one door is a car; behind the others, goats. You pick a door, say No. 1, and the host, who knows what's behind the doors, opens another door, say No. 3, which has a goat. He then asks you 'Do you want to pick door No. 2?' Is it to your advantage to switch your choice?

Marilyn offered the correct solution, thereby provoking a debate involving some 10 000 readers, 92% of whom, including (legend has it) several hundred mathematics professors, said she was wrong. In fact, many harsh statements about the level of education in the country were made.

The original game show was, however, fundamentally different: Monty Hall did not let the participant switch door. The door was opened only to build excitement.

these different family types, so each of them constitute 25% of all families. It was part of the conditions of the problem that there was one boy in the family, but no more information than that. That means that the structure of the family in question is one of $(B, B), (B, G), (G, B)$, and each of these have the same probability. In two of these we have a girl, so the probability that the other child is a girl is 2/3. There is an important subtle point here: if we instead know that *the oldest child* is a boy, there is 50% chance the other child is a girl. So the assumption must be spelt out in detail.

Before we leave this, let us repeat the discussion in a slightly different way. Let A be the event that a randomly chosen child from a two-child family is a girl, and let C be the event that the child chosen comes from a family with at least one boy. We then have that $P(A) = 1/2$ and $P(C) = 3/4$, and the probability we are interested in is the conditional probability that A occurs when we know that C has occurred, a probability we denote by $P(A|C)$. This is the frequency of A events among the C events, for which we have

$$P(A|C) = \frac{P(AC)}{P(C)} = \frac{1/2}{3/4} = \frac{2}{3}.$$

Here AC is the event that both A and C occur (i.e., the event that we have one boy and one girl in the family), an event which has probability 1/2. (The formula above, written as $P(AC) = P(A|C)P(C)$, implies a very basic probabilistic statement called Bayes' theorem, which relates the transposed conditional probabilities $P(A|C)$ and $P(C|A)$ to each other. To derive it we utilize the symmetry $P(AC) = P(CA)$; see Box 4.2.)

What about the Monty Hall problem? The situation at the start of the game can be described as one of the triplets $(C, G, G), (G, C, G)$ and (G, G, C). Here the position denotes a particular door; G denotes a goat and C denotes the car. Each of these are equally likely, and therefore each has probability 1/3. This means that the probability is 1/3 that you picked the correct door from the start, and therefore 2/3 that you did not. Since it is more likely you picked the wrong door, you should switch if you are given the opportunity. Of course this may not win you the car in an individual game. But if you play it many times, with this strategy you will win it in about 67% of the games, as opposed to only in 33% if your strategy is not to switch door.

The reason why this was initially considered counterintuitive is that we often assume that if we have n choices, each choice has probability $1/n$ of being the correct one. But we may have information that invalidates this, just as picking a horse to bet on at random at the trot is a worse strategy than getting some knowledge about the fitness and qualities of the different horses before you make your bet.

There is one fundamental difference between the two examples we discussed above, both of which gave the probability 2/3. In the game show, if the rules are adhered to, our argument provides a correct probability and therefore the correct game strategy. In the example with the children, however, there are assumptions that are made in the computations that may not hold true in real life. The assumption is that the three pairs (B, G), (G, B) and (B, B) all occur with the same frequency in the relevant population. This may not be true, not only because the ratio boys/girls may not be precisely one, but also because family planning strategies may lead to unequal probabilities for the different pairs. So what we have in this case is not necessarily a true description of the world, only a model of it.

The reason for bringing up these examples is to point out how important it is that you understand the context in which you compute probabilities. Statistics is about probabilities, and ignorance around the context can not only produce bias in the results, but also lead to misleading or erroneous p-values. It may well be that a conditional probability that is involved, but which one may be less apparent. When probabilities are computed by the uninformed, disaster may strike, as the sad case of Sally Clark, outlined in Box 4.3, shows. Another interesting, but not disastrous, example might be the discovery of the basic genetic laws (see Box 1.6).

1.7   The need for honesty: the multiplicity issue

The story about p-values may appear rather simple: we start with a hypothesis, collect data and compute the p-value. However, there are a few important assumptions in this process that need to be understood in order for the analysis to provide credible conclusions. The key assumption is that you compute one p-value and that you have clearly identified *a priori* when and how you do that. This is because it is important to make sure that your choice is not data-driven. The reason for this can be summarized in the following sentence: 'The value of the p-value is influenced by the history behind its computation.' This section and the next will illustrate the importance of bearing this in mind.

The particular issue to be discussed here is called the multiplicity problem. Recall that when the p-value is below a certain, prespecified, significance level, we reject the null hypothesis. The multiplicity problem refers to the simultaneous application of this rule to a set of null hypotheses. It is one of the problems that many medical workers consider an unnecessary complication invented by statisticians in order to make it more difficult for physicians to draw the 'appropriate' conclusions.

Box 1.6 Did Mendel cheat?

In one of his experiments, the monk Gregor Mendel, the father of genetics, crossed two species of pea which, when cultivated, had shown themselves to be constant in color. One species was red, the other was white. The locus for color had two alleles: A for red and a for white, of which A is dominant (so that both AA and Aa become red and only aa white). In one experiment Mendel had 600 red colored peas in what is called an F2-generation, which means that the proportion of homozygotes (genotype AA) should be 1/3 (Aa is twice as common as AA). Thus Mendel expected 200 homozygotes, and counted to 201. A very good result!

Or was it? How did Mendel determine that a particular red pea is a homozygote? His method was to investigate the color of 10 offspring, obtained by self-fertilization. If all were red, he declared the parent to be a homozygote, otherwise to be a heterozygote (genotype Aa). The problem with this decision rule is that by chance alone a heterozygote pea can produce 10 red offspring! In fact, the probability for this is $(3/4)^{10}$, so the total probability of declaring a particular pea a homozygote (call that event B) is

$$P(B) = P(B|AA)P(AA) + P(B|Aa)P(Aa) = 1 \cdot \frac{1}{3} + \left(\frac{3}{4}\right)^{10} \cdot \frac{2}{3} = 0.371.$$

This means that in 600 red colored peas we expect, using Mendel's method, to declare 222.6 to be homozygotes, including 22.6 misclassifications. To obtain 201 is therefore rather unlikely!

However, this is not really a statistical problem. The same problem occurs in medicine, for example with screening activities. Consider the situation where a physician is carrying out a routine health check-up on a patient. As a part of this he takes a 'lab status': he draws blood which he sends to a laboratory. In return he gets measurements of a number of chemicals in various blood compartments. In order to assess the clinical implications of these numbers, to understand their relation to health, the laboratory also provides reference ranges for each of the measurements. These reference ranges define (we assume) an interval within which 95% of measurements from healthy individuals will fall.

Suppose we have on the list N different values. What is the probability that a healthy individual will be considered healthy after the physician has read through the list of test results? In other words, what is the probability that all N measurements will lie within their reference limits? We simplify the discussion by making the unrealistic assumption that these N measurements are independent of each other. (Two events are independent if the probability of both occurring equals the product of the individual probabilities.) Because of this assumption, the probability is 0.95^N that all values lie within their respective reference limits for a healthy individual. This means that the probability of at least one value lying outside its normal reference limit is given by

$$P_N(\alpha) = 1 - (1 - \alpha)^N, \quad \text{where } \alpha = 0.05.$$

This function $P_N(\alpha)$ is plotted in Figure 1.1 for a few choices of N.

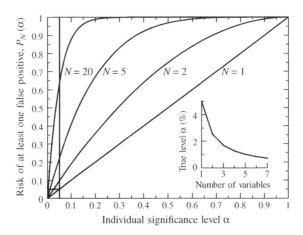

Figure 1.1   The graph of $P_N(\alpha)$ for a few values of N. In the small inset graph we see the significance level at which we need to do individual tests, in order to preserve the overall significance level at 5%.

If we declare a person healthy precisely when all values fall within their respective reference range, and we do this in a nationwide screening program, the number $P_N(0.05)$ will provide us with the percentage of healthy subjects who will wrongfully be found to be not healthy. This number is given by the intersection between the vertical line in Figure 1.1 and the curve describing the relevant function $P_N(\alpha)$. If the whole population is healthy, this is the fraction that will be declared sick. This number grows fast with N, which explains why we should not adhere strictly to a decision rule like this. When used for general health test purposes, the interpretation of data must be made with much more common sense. One looks for patterns, or uses the observation for follow-up testing to confirm or reject a hypothesis generated by the screening data.

Multiple p-values work in exactly the same way. In fact, it is more than an analogy, it is the same math. The number $P_N(0.05)$ gives us the probability that we have at least one statistically significant test when performing N independent tests at the significance level $\alpha = 0.05$. This is therefore the true significance level for the procedure, if you make individual statistical tests at the 5% level. It follows that if we still want to do this panel of N independent tests, but that our overall risk of being wrong must not exceed α, we must find the number x that solves the equation $\alpha = 1 - (1 - x)^N$, namely $x = 1 - (1 - \alpha)^{1/N}$, which is approximately α/N. Graphically this is the same as finding the α-value that corresponds to the intersection of the curve with the line $y = 0.05$. These values are illustrated in the small picture in Figure 1.1. In particular, if we make two experiments we need to compare the p-value to the 2.5% level in order to draw our conclusion at the overall significance level 5%.

This correction (which does not require independence to hold true) is called the Bonferroni correction and is based on the assumption that we should compare all our tests to a common significance level. It can be improved upon by distributing the risk α at our disposal unevenly among the tests.

A natural follow-up question to the multiplicity problem is to ask how many tests one can make. This is addressed in Box 1.7.

Box 1.7 How many tests can we do?

A natural follow-up question to the discussion on multiplicity is how many tests we can do. There are two extreme answers. Either you can say that each test controls its own error rate, and that multiple testing therefore is not a problem. Alternatively, you can argue that the multiplicity issue is there as soon as more than one test, world-wide, has been done. So only one test is allowed, and that was done ages ago.

Ultimately, this is another instance of using *p*-values to guide behavior. What matters is what action the result triggers. If we use individual *p*-values in the inductive way of mainly measuring how extreme the signal–noise ratio is, we do not need to adjust significance level, because we do not really use it. When used in this way, *p*-values are sometimes called exploratory *p*-values.

The need to adjust for multiplicity arises when we make a claim. It is here the whole history must be accounted for in the computation. It is the validity of the claim that must be addressed based on data. This means that we somehow need to restrict how many claims we make per study, but statistics does not define the rules that should govern this.

A standard way of settling the history problem is to prespecify what tests to do and how the significance level is kept under control. Though a tool that solves one aspect of the problem, it must be exercised with care. It is not legitimate to specify 200 hypotheses for a study and then report the successful ones as having been 'prespecified'. To mitigate this problem various multiplicity control procedures have been invented in order to control the overall error rate. Though statistically sound, they are sometimes hard to understand in a non-biostatistical world. Effectively they mean that if we walk through the different *p*-values in a prespecified way, we are allowed to make a claim from the final *p*-value, but if we walk through the same *p*-values in a different, not prespecified way, that claim is not valid. The information about the particular variable in question is fixed, it is how well we predicted the outcome of other variables that defines the validity of the claim. Personally I have full sympathy with those who find it hard to understand how the appropriate medical action can hinge on what some statistician happened to prespecify.

1.8 Prespecification and *p*-value history

The multiplicity problem that we discussed in the previous section is one example that illustrates the need for a *p*-value to capture the full story. In the multiplicity case we need to say that we did all *N* tests in order to find one that is statistically significant. If we report only the significant *p*-values and omit to mention the others, we are cheating. The following example is related to this.

Example 1.5 It was noted at a particular workplace that there was an unusually high frequency of children with severe malformations born to women who had worked there during their pregnancy. When this observation was made, workforce records over the previous 5 years were collected, which showed that during that period 50 pregnant women had been employed there and as many as 5 of their babies were born with a severe malformation.

How likely is this result? It is known that in the general population about 2% of children are born with a malformation of (at least) such severity. We can then compute the probability of observing 5 or more such malformations among the 50 pregnancies. We simply have to list all possibilities. If we denote a malformed child by M and a healthy one by H, we list all sequences of Ms and Hs of length 50. A particular sequence with precisely k Ms has probability $p^k(1 - p)^{n-k}$, where n = 50 and p = 0.02. If we therefore sum the probabilities for all such sequences, we find that the probability of getting precisely k malformed babies is given by

$$\binom{n}{k} p^k(1 - p)^{n-k}.$$

Here $\binom{n}{k}$, which is called a binomial coefficient, is the number of ways we can pick a set of k elements from a bigger set of n elements. The distribution given by these probabilities is called the binomial distribution and denoted by $Bin(n, p)$. From this we can compute the probability of getting an observation that is at least 5 for a $Bin(50, 0.02)$ distribution. It turns out to be very unlikely; it is 0.0032. Can we from this conclude that something in the environment is harmful to pregnancies?

The validity of such a conclusion depends on how the workplace that was investigated was chosen. Did we chose it at random, or did we first note the unusually high number and then start the investigation? In the former case, we are correct in drawing the conclusion above. However, as the situation is described it is most likely we have the second case. Also rare events occur by chance, and if we look for one and then use the fact that what we have found is rare to prove something, we are making a circular argument. In fact, for a $Bin(50, 0.02)$ distribution, we have the following probabilities for different outcomes that can occur:

Outcome	0	1	2	3	4	≥ 5
Probability	0.364	0.372	0.186	0.0607	0.0145	0.00321

From this it follows that if we, for example, have 300 workplaces at which 50 women employees worked during their pregnancies, we expect 109 of these to have no malformed babies, 111 to have exactly one, 56, 18, 4 to have respectively 2, 3 and 4 and 1 to have at least 5 malformed babies, respectively. So, even though it is a rare event, we do expect this rare event to occur in about one in 300 such workplaces.

The key point here is that we need to carefully plan the experiment in order to guarantee that the statistical model we use is appropriate, so that we can draw valid conclusions. If, as in the case above, we observe a rare event, we must put this observation into proper perspective. By definition rare events do occur, though rarely, and we must ask ourselves why we came to observe this rare event – what mechanism underlies its detection? Would we react in the same way if it was another rare event that occurred? Is there a large number of rare events such that if any one of them occurred, we would have hit the alarm bell? Such considerations contribute to the appropriate probability model, and since we in general do not have control over this, the simple solution is to avoid making claims from unplanned observations. Instead the appropriate action in the case above would be to choose one or more similar workplaces and assess the outcome at these. We would then use our observation only as a trigger for a carefully planned experiment in which statistics can be used appropriately. If this is not a possible way forward, one has to find other means than statistics to prove the point that there is some environmental hazard at that particular workplace.

How do we then make certain that a reported p-value really has followed the rules of the game? The standard answer is: prespecify. This means that we should write a protocol before we have collected and analyzed our data, outlining what we are going to do. If in this protocol we specify the hypothesis we want to test, and the data to use, we are in a good position to use statistics in a proper way.

We see the use of p-values in courts of law, albeit in a disguised form. A piece of evidence, which is an event, is presented by the prosecutor together with a more or less explicit calculation of the probability of this event occurring for an innocent person. If this probability is small, it is used as evidence against the accused. This is in complete analogy with how p-values are used in science, and therefore embeds the same problems. As an analogy to the discussion above we have the following example.

Example 1.6 Assume that a match in two DNA profiles occurs only once in 10 000 instances. Consider the following two situations.

1. A woman has been raped and foreign DNA has been obtained from her. Based on witness statements, a man has been identified and arrested. A DNA test shows a match to the sample from the woman and he is brought to trial.

2. A woman has been raped and foreign DNA has been obtained from her. The sample is compared against a database consisting of DNA from 20 000 men. A match is found and the man in question is brought to trial.

What is the value of the DNA test in these two situations? In the first case the probability of a match for an innocent subject is 1 in 10 000. In the second case we can compute the probability that at least one of the 20 000 in the database provides a match by chance alone. The result is $1 - (1 - \frac{1}{10000})^{20000} \approx 1 - e^{-2} = 0.86$, which is considerably larger than the first probability.

1.9   Adaptive designs: controlling the risks in an experiment

A classical way of destroying your significance level is to repeatedly look at your data, and to decide to stop when you are ahead. Consider the following experiment: we want to compare two treatments that are truly equal in all respects. We do not know that, and decide to pick 40 subjects and randomly allocate half to one treatment and the other half to the other treatment. However, we decide to randomize the study in such a way that for each pair of patients included in the study, we assign one of them to each of the two treatments. In order to be more efficient, we also decide to analyze our data after each pair, and to stop the experiment if the p-value drops below 5%. What is now the probability that we falsely end up claiming that there is a difference between the two drugs?

Figure 1.2 shows what the true significance level $P(\alpha)$ would be, if we make each individual test at the significance level α. This corresponds to the $N = 20$ curve in Figure 1.1, except that in this case there is a dependence between the 20 tests done; each new test adds the observation from one new pair of subjects to the old data. In the graph two special observations are illustrated:

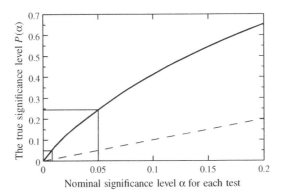

Figure 1.2   The true significance level $P(\alpha)$, as a function of the nominal significance level α of the individual tests, for the procedure discussed in the text, in which we analyze after each of the 40 pairs. The dashed line shows the corresponding function when we only look at data only after study completion.

- The value $P(0.05) = 0.25$, which gives the true significance level for an efficacy claim when individual tests are done at the conventional 5% level.

- The solution $\alpha = 0.007$ to the equation $P(\alpha) = 0.05$, which tells us at what level the individual tests should be done in order to protect the overall significance level at 5%.

The latter observation illustrates that we can in fact look repeatedly at data, if we so wish. But we need to be careful, so that we protect the overall significance level.

Adaptive designs for clinical trials are designs in which, on at least one occasion during the study execution, we take a look at the data and make a decision on the further conduct of the study. Depending on how this is done, it may or may not have consequences for how we distribute our α, the significance level. The example just given is an extreme type, called a *sequential design*. To be useful one must design decision rules in such a way that the total amount of α one can spend is properly distributed between the different looks, and in the process take into account the dependence between different tests.

A variation of this is called a *group sequential design* in which we only look a few times before the final readout. The problem is the same, we need to distribute our α, but more simply. However, we may not always like the end result, as the following real-life example illustrates.

Example 1.7   The TORCH study was a 3-year study in chronic obstructive pulmonary disease (COPD) patients with mortality as outcome. Even though there were four treatments in the study, one comparison was of primary interest: that of a certain combination product versus placebo. It had all the virtues of a good clinical design (double-blind, randomized, etc.) and it was planned to recruit 1510 patients per arm.

To carry out a 3-year study on mortality in which one treatment may be effective poses an ethical problem. If the effect is large, can we really wait for the study to come to completion? This ethical consideration forced the study designers to include two interim analyses, the first to be performed when 358 deaths had occurred in the study and the second when 680 had occurred. On each occasion a test for efficacy was performed, and the study was to be stopped if a predefined significance level was reached, which depended on which interim analysis it was. In order to preserve the overall significance level it was decided (these are not the exact numbers used, but serve our purpose) to do the first test at significance level 0.0006 and the second at level 0.012, so that the final test would be done at level 0.038. This preserves the overall significance level, because

$$0.0006 + (1 - 0.0006)0.012 + (1 - 0.0006)(1 - 0.012)0.038 = 0.05.$$

As it happened, the study proceeded to completion, since neither of the interim analysis passed the test. Taking the data at face value, the primary comparison produced a p-value of 4.0%. However, it needed to be below 3.8% in order to preserve the overall significance level at 5%, so the study missed its objective: there was not sufficient evidence in the study to conclude that the mortality rate will be decreased if patients are treated with the combination product. In passing, we may note that the 'true' p-value, adjusted for the two interim analysis, was

$$p = 0.0006 + (1 - 0.0006)0.012 + (1 - 0.0006)(1 - 0.012)0.040 = 0.052,$$

very close to the conventional cut-off of 5%.

This example is really instructive on many levels. On the one hand, how can it be that the 5% level is so set in stone that a FDA Advisory Board can decide that 5.2% is not sufficient evidence for an effect on mortality, when the logical expectation from effects previously demonstrated by the combination treatment is that some benefit should be expected (the present medical paradigm is that COPD worsens as patients get exacerbations, and if the drug decreased the rate of these, it should really also prolong life). But accepting that, it takes some deep thinking to understand why we should be punished for introducing interim analyses that assure that if the effect is extremely obvious, we should not drag out the study unnecessarily but instead make the drug available to, among others, the patients in the placebo group. It is not hard to understand why some non-statisticians sometimes consider statistics to be more mysticism than science.

1.10 The elusive concept of probability

The word 'probability', along with some of its synonyms such as 'chance' and 'risk', is part of our everyday language. We have already had a first discussion around this concept, indicating that it is not as trouble-free a concept as it may appear. But the nature of the problem is wider than previously indicated, and lies at the heart of the difference between the two dominant schools in statistics, the frequentist school and the Bayesian school.

Serious thinking about probabilities started in connection with games, in particular in France in the seventeenth century, where playing games for money was one of the major occupations of the nobility. In this situation many problems can find a solution by the type of combinatorial argument that was used for the Monty Hall problem earlier. This means that we define the probability of an event A to be $P(A) = g/m$, where m is the number of possible outcomes of the game, and g the number of successful outcomes (satisfying the specific criteria that define A). The underlying assumption is that all possible outcomes are equally likely, and what we do is make a list of all possible outcomes and count the proportion of successful ones.

However, this combinatorial definition is only useful at the gambling table, if what we compute is also what actually occurs. To determine if this is the case, we need to play a large number of games, say n, and count the number of times the event A occurs, call it n_A. We then expect that the proportion n_A/n, of occasions when the event A has occurred, approaches $P(A)$ as we increase the number of experiments. The frequentist school of statistics essentially derives the concept of probability from this property. The problem with this definition is how we determine $P(A)$ if we cannot do infinitely many experiments.

Probability theory comes to the rescue. One of its key statements is the law of large numbers which says that if you define how much error you can tolerate (defined by a small number $\epsilon > 0$) and you increase the number of experiments, your observed frequency will home in on the desired probability:

$$P\left(\left|\frac{n_A}{n} - P(A)\right| > \epsilon\right) \to 0 \quad \text{as } n \to \infty.$$

Probability theory is a topic in mathematics. It does not concern itself with reality, but instead needs a starting point to build the logic from – the premises or axioms. Many important results in probability theory were derived without such a foundation, and it was not until the early 1930s that mathematicians decided how to define what a probability is. These were the axioms of Kolmogorov which define a probability as something that measures events in such a way that the total measures one and the probability of the sum (union) of exclusive events is the sum of the individual probabilities.

But the mathematical definition of a probability is of no help when you want to assign probabilities to everyday events. Here the frequentist approach is useful, but its real usefulness is in the way it builds traditional (frequentist) statistics. Consider a coin, which may be biased. Denote the probability that it falls heads by π. How do I determine π when I cannot toss the coin infinitely many times? The frequentist statistician turns this question around and tells you to toss it a finite number of times, say n. That does not allow you to give a definite answer as to what π is, but you can make statements about what π almost certainly is not. Such statements are made with a certain degree of confidence which represents your knowledge about π. You may, for example, make a statement about whether you believe the coin is unbiased (which means that $\pi = 0.5$), or not. Your confidence in this statement can then be expressed in the p-value, which is a probability in the tradition of the frequentist definition above, namely the number of experiments with an unbiased coin that will produce the outcome you observed or 'worse'. This p-value is used to express your confidence in the statement that the coin is unbiased. Similarly, you can express confidence in the actual parameter value π using confidence intervals, to be described in more detail in later chapters.

But confidence is not a probability. It lies between 0 and 1, and it is intimately related to probabilities, but the relation is that my confidence in a particular statement about a parameter is at least 30% if more than 30% of all experiments are expected to provide a point estimate that fulfills that statement. The main problem is that any particular experiment we do is unique, and we cannot determine true probabilities with any accuracy from a single experiment. What do we do if an experiment cannot be repeated? The frequentist approach has problems with assigning probabilities to non-repeatable events, and may therefore say there is no such thing.

Bayesian statisticians take a different view. They differ from the frequentists already in their view of what a probability is and they acknowledge a wider use of the probability concept which includes what is usually called subjective probabilities. This leads to a very different view on what one should be doing in statistics and science. Whereas the frequentist is someone who is 100% focused on science as the process of falsifying hypotheses, the Bayesian views science more as an inductive process.

To see the difference, take a simple example. Brazil and Argentine are about to play a World Cup qualifying soccer game. What is the probability that Brazil will win? To estimate this, the frequentist will need to find a sample of matches already played which is such that the new game can be considered a random sample from this set. One way would be to look at all previous games between the nations. But is such an estimate relevant? The new match is not played with the same players as the previous matches, and the motivation for the two teams on this particular occasion may differ for the two teams depending on table position, whether it is a home game, etc. To account for such factors, the frequentist needs to make a more elaborate model of the previous games. When he has done that, he will arrive at an estimate of the probability that Brazil wins a game of this sort, accounting for factors such as which is the home team, table positions, etc. But it will refer to some historical average and some of the factors that matter cannot be taken care of, including the particular coaches and players participating on this occasion.

Actually you may, as a frequentist, deny there is a relevant probability concept at all for this situation because there is no relevant space to take a sample from. Alternatively you may say that since I can construct a thought experiment in which this game is played simultaneously in parallel universes, there should be a sensible probability, though its estimation still poses unsurmountable problems.

The Bayesian takes a different view. He says that there is a probability that Brazil will win, but it should not be calculated as a frequency. Instead each individual has a subjective estimate of this probability. To get an overall probability we may take the average of these subjective estimates, which is more or less what the betting companies do when they set the odds. So the Bayesian can actually talk about the probability of sunny weather tomorrow, even though that is not a repeatable experiment. Or the probability of a new Ice Age within the next millennium.

The different views on the nature of probability between a frequentist and a Bayesian really boil down to whether it is a real, physical tendency of the event to occur, or just a measure of how strongly you believe it will occur. They do agree that whatever it is, it should follow Kolmogorov's axiom system, which is the starting point of all the mathematical calculations in probability theory.

The mathematics of Bayesian statistics is the inductive (recursive) computation of new probabilities from old ones, accounting for new data. It therefore needs a starting point, the specification of an *a priori* probability, or prior for short. This is typically obtained from a consideration of available data on the matter. The issue is that for a given problem, there are in general multiple ways to assess such data, and choosing one is a matter of judgement; different people may assign different prior probabilities.

This difference in point of view has many implications. When comparing two hypotheses and using some data, frequency methods would typically result in the rejection or non-rejection of the null hypothesis at a particular significance level, and frequentists would all agree (in the best of all possible worlds) whether this hypothesis should be rejected or not at that level of significance. Bayesian methods would suggest that one hypothesis was more probable than another, but individual Bayesians might differ about which was the more probable and by how much, if they use different *a priori* probabilities. Bayesians would argue that this is right and proper – if the contemporary knowledge is such that reasonable people can put forward different, but plausible, priors and the data do not swamp the prior, then the issue is not resolved unambiguously on available knowledge. They would argue that any approach that aims at producing a single, definitive answer in these circumstances is flawed.

In the mainstream application of biostatistics the Bayesian view is seldom listened to. Most experimenters want a clear-cut answer, in black and white, and the uncertainty imposed by the use of p-values is at the limit of what they can take. To actually acknowledge that there are different interpretations of the results, depending on your prior view of the matter, is usually considered an unwelcome complication, best given a wide berth.

1.11   Comments and further reading

There are a number of great scientists mentioned in this book, who have contributed to the science of clinical trials. I have in general avoided discussing these scientists, their lives, deeds and the context they worked in. For such a treatise, see the book *Dicing with Death* (Senn, 2003), which also discusses much of what we address in our first three chapters.

Most books on philosophy probably have something to say about the philosophy of science. In his autobiography (Popper, 1976), Karl Popper includes a chapter on the philosophy around induction and falsification. The quote by Richard Feynman is taken from the last chapter of what may be called his autobiography (Feynman and Leighton, 1992). Wootton (2007) gives a historian's view of the impact of physicians on people's welfare in history. For more on the suspicion of fraud indicated in Box 1.6, see Fisher (1936). (R. A. Fisher was not only the inventor of modern statistics, but also a first rank geneticist.)

We have chosen to describe the statistical output in this chapter in terms of the p-value. This is not necessarily a choice made because it is a very good summary, but because of its role in the medical literature. There are a lot of ways in which p-values are misinterpreted (Gigerenzer, 2004), some (but not all) of which we have discussed in this chapter, and in the statistical community there is often a very negative attitude (Royall, 1997) toward its use. Hopefully we can explain what it means without entering into a debate on what the statistical summaries in the medical literature should or should not be. The citations by Fisher in Box 1.2 were from Fisher (1979, p. 80) and (Fisher, 1929), respectively.

The exact nature of the FDA rule discussed in the text is unclear, and our discussion may not be fully valid. In fact, it is not clear that anyone knows what the rule really is; it is probably somewhat flexible (Senn, 2007, Section 12.2.8). We know that the rule stems from the FDA interpretation of the 1962 amendment to the Federal Food, Drug, and Cosmetics Act, which required 'adequate and well controlled investigations'. However, a further amendment in 1997 permits the FDA to require only one such study, as long as there is other substantial evidence for the benefit of the drug. This seems to mean that, by law at least, approval is not only about a low significance level. To what extent this has had any impact on the FDA approval process is unclear (at least to me).

The comment about the non-inferiority study type may not go down well with every statistician, because finding the non-inferiority margin has provided food for numerous statistical publications, including regulatory guidelines from health authorities. The non-inferiority study type was designed to solve one problem: that absence of statistical significance was taken as proof of equality. However, the solution is almost as bad. As far as I know the concept was introduced in connection with a wider attempt to harmonize the regulatory requirements all over the world, in a document labeled ICH E10 (International Conference on Harmonisation, 2000).

There is much more to be said about the nature of probabilities and its implications for a proper treatment of p-values. Often the need for a probabilistic discussion stems from lack of information, as in the game show. If we only had complete information, we would often not need probabilities, like if we only knew all the initial conditions when we toss a coin, we can predict the outcome with certainty. In fact, you could argue that there are few cases when there are pure random events. The notable exception are some deep aspects of contemporary quantum physics. Deterministic processes may appear probabilistic to us, simply because we cannot obtain sufficient knowledge to explore the deterministic nature of the problem, a subject mathematicians discuss in chaos theory. Accepting that we need to compute probabilities, it becomes important to understand the conditions under which the computed probability is valid. A very rare event occurs, seen from a prospective vantage point, with a very small probability. Retrospectively the probability is one. To compute that probability when we know that it has occurred is basically meaningless. However, we should not confuse that with what we do when we compute p-values. These are probabilities for the outcome, computed under an assumption, and we use the p-value as indirect evidence for or against that assumption. Note that the p-value computes the probability of the outcome given that the null hypothesis is true, not the transposed conditional, the probability that the null hypothesis is true given the outcome we have observed.

When it comes to error control with multiple testing, the original suggestion by Bonferroni on how to allocate parts of the available α (significance level) to the different tests was improved upon considerably by Sture Holm in 1979. He showed that the testing could be done in a stepwise manner in the order of increasing individual p-values, where these p-values were compared with successively larger fractions of α. After that it took another 28 years for the next major step, made independently by Guilbaud and Strassburger-Bretz, which was the development of confidence intervals corresponding to Holm's and related testing procedures. A modern review of this subject, by Dmitrienko et al. (2010), covers not only basic/traditional approaches but also novel ones. In connection with the multiplicity problem we also touched upon one of the present hypes in medical statistics, the adaptive designs. We will say no more on this subject, and refer the reader who wants to learn more to the vast literature on the subject (there are also plenty of conferences he or she can go to). Both Whitehead (1997) and Chang (2008) offer useful starting points. The illustration in Example 1.7 is an adaptation of the main result in Calverley et al. (2007).

References

Calverley, P.M., Anderson, J.A., Celli, B., Ferguson, G.T., Jenkins, C., Jones, P.W., Yates, J.C. and Vestbo, J. (2007) Salmeterol and fluticasone propionate and survival in chronic obstructive pulmonary disease. *New England Journal of Medicine*, **356**(8), 775–789.

Chang, M. (2008) *Adaptive Design Theory and Implementation Using SAS and R,* CRC Biostatistics Series. Boca Raton, FL: Chapman & Hall/CRC.

Dmitrienko, A., Tamhane, A.C. and Bretz, F. (2010) *Multiple Testing Problems in Pharmaceutical Statistics,* CRC Biostatistics Series. Boca Raton, FL: Chapman & Hall/CRC.

Feynman, R.P. and Leighton, R. (1992) *Surely You're Joking Mr Feynman! Adventures of a Curious Character*. London: Vintage.

Fisher, R. (1929) The statistical method in psychical research. *Proceedings of the Society for Psychical Research*, **39**, 189–192.

Fisher, R. (1936) Has Mendel's work been rediscovered?. *Annals of Science*, **1**, 115–137.

Fisher, R. (1979) *Statistical Methods for Research Workers* 10th edn. Edinburgh: Oliver & Boyd.

Gigerenzer, G. (2004) Mindless statistics. *Journal of Socio-Economics*, **33**, 587–606.

International Conference on Harmonisation (2000) *ICH Harmonised Tripartite Guideline: Choice of Control Group and Related Issues in Clinical Trials*. Geneva: International Conference on Harmonisation.

Popper, K. (1976) *Unended Quest: An Intellectual Autobiography*. London: Fontana/Collins.

Royall, R. (1997) *Statistical Evidence: A Likelihood Paradigm,* vol. 71 of *Monographs on Statistics and Applied Probability*. London: Chapman & Hall.

Senn, S. (2003) *Dicing with Death: chance, risk and health*. Cambridge: Cambridge University Press.

Senn, S. (2007) *Statistical Issues in Drug Development*. Chichester: John Wiley & Sons, Ltd.

Whitehead, J. (1997) *The Design and Analysis of Sequential Trials*. Chichester: John Wiley & Sons, Ltd.

Wootton, D. (2007) *Bad Medicine: Doctors Doing Harm since Hippocrates*. Oxford: Oxford University Press.

2

Observational studies and the need for clinical trials

2.1 Introduction

Having discussed what statistics contributes to a scientific experiment, we will now address some basic aspects of some of the investigations to which biostatistics is applied, the clinical studies. Clinical studies generate new data about human diseases, except when they analyze data that already exist. In general the purpose of a clinical study is to clarify the relationship between a particular exposure, possibly a newly developed drug, and a particular outcome. We can broadly divide such studies into epidemiological studies, in which we need to accept the data as nature provides them, and experimental studies, in which we try to control the environment within which new data are generated. The degree of confidence in the conclusions from the statistical analysis differs between these two study types, because they differ in their ability to control for alternative explanations for the effects seen. Of key importance here is the notion of the confounder. In this chapter we will discuss different types of clinical studies and to what extent confounders may complicate the interpretation of the results. Related to this discussion is the issue of bias, to which we will turn in the next chapter. Here we will outline the main ways in which we can perform clinical trials and will try to point out the scope and limitations of the extremes, which are the case–control studies and the controlled randomized clinical trials.

2.2 Investigations of medical interventions and risk factors

The evaluation of the effect of a particular intervention on the progress of a disease is something man has probably been doing for ages. In most cases such assessments have been case stories: someone improved after a particular intervention, so someone else tried it as well. With many medical conditions there is a natural tendency to improve and what appeared to be an intervention-driven improvement may have been the natural course of the disease.

Understanding Biostatistics, First Edition. Anders Källén.

It is, however, in human nature to ascribe such effects to the intervention and some kind of group pressure then establishes certain types of interventions. This process was greatly facilitated by the emergence of a special trade to carry out these interventions, physicians. The fact that many diseases regress naturally means that even harmful effects of interventions may not have been discovered. Partly because no one looked for them, apart from the physicians, few could collect enough material to make an objective assessment about a particular intervention, and even if you had your doubts, where were the alternatives that could give the same hope?

The effects of some interventions are just so obvious that case stories are more than sufficient as evidence. When insulin was extracted from the pancreas of oxen in 1922 and injected into children dying from diabetic ketoacidosis lying in a large hospital ward, the effect was so stunning that before the discoverers Best and Banting had reached the last dying child, those treated first woke up from their coma. When the first antimicrobial drugs, the sulfonamides (sulfa), appeared in the 1930s, their effects on streptococcal infections were dramatic in a number of serious conditions, including the important woman-killer of the day, puerperal fever (childbirth fever), as well as bacterial meningitis. So great was the effect that there was a sulfa craze, with hundreds of manufacturers producing thousands of tons of myriad forms of sulfa. This, in combination with nonexistent testing requirements, led to a disaster in the fall of 1937, when at least 100 people in the US were poisoned with the additive diethylene glycol. This led to the passage of the Federal Food, Drug, and Cosmetic Act in 1938, which authorized the FDA to oversee the safety of drugs. That was the first of a series of crises that formed the FDA of today. The effect of penicillin on many infections was equally stunning when it arrived a decade later, as is the effect of opium on pain. More recently, the effects of organ transplants in patients with kidney, liver, or heart failure, or hip replacements in patients with arthritic pain are so striking that carefully controlled test are mostly superfluous.

Even though there are more charming little stories like these, most treatments do not have such immediate and dramatic effects in all individuals. Similarly, not all risk factors lead to a particular outcome in all exposed subjects. Sulfa could not repeat its stunning effects in, for example, pneumonia (lung inflammation), but is still worth administering it. It was not effective in scarlet fever (which is sensitive to penicillin). Another complication in the study of many diseases is that they have varying clinical pictures with spontaneous improvements. Also in such a serious disease as pulmonary tuberculosis, remarkable recoveries could occur with bed rest alone as treatment, and a few case stories cannot be accepted as proof of efficacy for a new treatment.

To address the question whether a particular exposure or intervention has a certain effect, we would like to study an individual subject and see if the effect occurs with the exposure but not without it. Everything else should be exactly the same on the two occasions. This experiment means that just before the exposure, the world should split into two identical, parallel worlds which differ only in that in one the subject is exposed and in the other he is not. No random events are allowed to occur which would mess up the interpretation of the experiment. Even though parallel universes are part of contemporary thinking in understanding quantum physics, this is not an experiment that can be done in real life.

The key problem is the elimination of random events. We can never completely eliminate these. In biology the looked-for effect can occur as a consequence of the exposure but it can also occur out of nowhere, as a purely random event. (The word 'random' usually only means that we do not understand the underlying cause, but that does not affect the argument, since such a cause is different from the exposure.) In the presence of random events we may well

---

Box 2.1 Epidemiology

Epidemiology is the science that describes the occurrence of human diseases. The most important of these in earlier history were infectious diseases, in particular the epidemics that have affected mankind in waves over the last few millennia, and epidemiology was originally about understanding them. Today epidemiology has been broadened to include both a description of the distribution of disease and various determinants of disease frequency. It follows that, on the one hand, it describes the occurrence of disease in terms of various population characteristics such as sex and age and, on the other hand, it tries to identify risk factors for different diseases by comparing subjects exposed to these factors with subjects not exposed. Studies of the latter kind are etiological in nature, which means that they try to understand what causes the disease. But like astronomy, epidemiology is largely an observational science with little room for planned and controlled intervention by the investigator. The standardization of environmental conditions of experimental studies is impossible to achieve for the epidemiologist, which means that actual causal evidence is hard to achieve by such means.

Epidemiology is a subdiscipline of medicine. However, it relies heavily on the use of statistical methods, and there is therefore a subdiscipline of biostatistics that concerns itself with the particular problems of epidemiological studies. These problems stem from the fact that in epidemiology there are substantially more issues with sampling, competing risks, confounders and bias than there are with purely experimental procedures, which can be set up to eliminate most of these problems.

---

find the effect without the exposure, but not with the exposure, and also when there is a true increase in the probability of the effect after the intervention or exposure.

In order to attribute the effect to the exposure we must understand how much the effect occurs at random. This means we need to come to grips with what may be called the background noise, which in turn means that we need to study more than one individual. Even if we study the same individual twice we cannot make sure that the random effects are exactly the same on both occasions. But with careful design we can make sure that what differs on the two occasions has a random nature. This means more background noise as compared to the split-world experiment, but in principle the same situation. More importantly, this experiment can be done. One noise-reducing aspect of an experiment might be that we use the same subject on both occasions, but that is in no way necessary. We can let some individuals have the exposure and others not have it. This means more possible alternative explanations for the effect, factors related to the particular individuals, but as long as were are careful in how we choose the subjects for the two groups, this can be considered as noise. This experimental design may make the signal we are looking for less obvious, but does not distort it.

We therefore need to give up trying to draw definite conclusions from the study of individual subjects and instead study groups of patients. Such groups can be defined by whether they have been exposed or not, but they cannot be equal in all other respects. We can allow for random events, as long as they are random, because we can handle them in the statistical analysis. The price is that we get statements about groups instead of individuals, but if the conditions are right, statistics can address the question we asked in the first place, that of the relation between the exposure and the effect, but on a group level. Note that since we draw our conclusions on a group level, not all individuals in the group need to have responded – an overall effect may also be demonstrated if only a subgroup responds.

Broadly speaking, there are two types of studies:

(a) experimental studies (e.g., clinical trials) in which we generate new data regarding the effect of an exposure;

(b) observational studies in which we only study data regarding the exposure which are already generated in 'their natural habitat'.

The difference is that in an experimental study we can control who is exposed and who is not, whereas in an observational study that is given. This implies that there are some very fundamental differences between the two study types when it comes to confounders.

In epidemiological studies we often want to investigate cause–effect relationships. We want to identify a particular factor (e.g., smoking) as a cause of a particular effect (e.g., lung cancer). However, causality statements are problematic in medicine. There are two main reasons for this: what does it actually mean, and have we got the correct cause?

A certain risk factor for a disease is identified by demonstrating that the disease incidence increases in the presence of the risk factor. This does not mean that each individual exposed to the risk factor will contract the disease. As an example, consider the condition COPD that we studied in Example 1.7, which smokers develop. Is smoking the cause of COPD? Since (almost) only smokers get the disease, it seems to be an (almost) necessary factor, but, on the other hand, only about 15–20% of smokers develop COPD in their lifetime, so it cannot be sufficient. Does that mean that smoking is only a catalyst, that the cause is a combination of smoking and, say, some genetic makeup? Alternatively, it might be that smoking always would lead to COPD, if people lived long enough, in which case it might be considered to be a true cause. Which of these is the case we may never know, and to avoid too complicated a discussion about what causality would mean for us, we adopt a more practical definition: a risk factor for a disease is a factor which, when introduced into a population, increases the incidence of that disease.

The previous discussion of cause–effect relationships was in the context of observational studies. In an experimental study where we examine a specific intervention in a randomized, double-blind study, the cause–effect interpretation may be simpler. But, as already noted, all claims are about groups. From a clinical study with a drug intervention we can never claim that a particular patient benefited from the drug, only that the group as a whole did so. The concept of responders, much discussed in many clinical trials, is basically meaningless. This very fact was the precise reason why there was much opposition to statistical methods in the early days of clinical research. As one prominent physician, Thomas Lewis, expressed it in a book written in 1934:

> it is to be recognized that the statistical method of testing treatments is never more than a temporary expedient, and that but little progress can come of it directly: for in investigating cases collectively, it does not discriminate between cases that benefit and those that do not, and so fails to determine criteria by which we may know beforehand in any given case that treatment will be successful.

But the fact that individual responses cannot be predicted does not mean that group responses are not important for the individual, as medical progress using statistical tools has shown.

Box 2.2 John Snow and the origin of cholera

Epidemiologists often refer to the British physician John Snow as the father of epidemiology, because of his investigation into an outbreak of cholera in Soho, London, in 1854. Most renowned is his identification of a specific public water pump on Broad (now Broadwick) Street as the source of a particular sub-epidemic. His argument regarding that pump was convincing enough to persuade the local council to remove the handle from it. However, as noted by Snow himself, at the time when this action was taken, the epidemic had more or less subsided.

More important for our discussion is a 'natural experiment' which he also conducted in order to prove his basic thesis that it was not bad air, miasma, that was the cause of cholera, but rather sewage-contaminated water. He was able to demonstrate that water that was delivered to homes by companies that took it from sewage-polluted sections of the River Thames had an increased incidence of cholera. As control group he used a company that took its water from a different section, free of sewage. Snow's key data are summarized in the following table:

	Sewage water	Clean water
Cholera deaths	4 093	461
Population	266 516	173 749

This shows that the relative risk for death in cholera is 5.8 with 95% confidence interval (5.3, 6.4).

The study in question is a cohort study. Technically it is not randomized, but which water supplier a household had was very much decided in the past and had a random signature to it. All social classes and relevant areas were supplied by all the companies involved, which left the water itself as the only remaining explanation for the outbreak.

To identify this opportunity to use these data in order to remove alternative explanations was part of Snow's genius.

However convincing to us, the explanation of an oral–fecal mode of transmission of the disease was repulsive to Snow's contemporaries and was not accepted. Actions taken were more driven by political need than scientific insight, but we should keep in mind that neither Snow nor his contemporaries knew anything about microorganisms and diseases. The casual relation was finally settled when the cholera bacterium were isolated by Robert Koch in 1883, providing the final piece of evidence for the waterborne transmission of the infection.

2.3 Observational studies and confounders

In order to discuss the basic types of observational studies we consider a large population. We can divide this population into subpopulations depending on whether the subjects have certain properties or not. The simplest case is when there is only one property and we divide the population into those who have it and those who do not have it. Notation-wise we let a capital letter, such as A, denote both the property and the subpopulation that has the property. Similarly, we let A^c denote both the absence of the property and the subpopulation without the property. The case with only one such factor essentially amounts to the estimation of a binomial parameter.

The next simplest case is when we have two properties, which we denote by E and C for reasons to be explained shortly. This gives us a partition of the whole population into four subpopulations:

	E	E^c
C	CE	CE^c
C^c	$C^c E$	$C^c E^c$

Here CE denotes those that have both properties, whereas, for example, CE^c denotes those that have property C but not property E.

The purpose of this partition into subpopulation is that we want to see if there is some relationship between the two properties. In such a case, E is some kind of exposure, whereas C is some kind of outcome/response. C stands for case and refers to individuals with the particular response. To make the discussion more concrete we may think of the exposure (E) as smoking and the response (C) as lung cancer. This particular relationship is of some historical importance in epidemiology (see Box 2.3).

A cross-sectional study is a study in which we sample from the whole population. This gives us estimates of all the probabilities for the four subgroups. Such studies are therefore sometimes called prevalence studies. In some cases cross-sectional studies can contain data on more or less complete populations, such as studies on birth registrations in Sweden or data collected by the Centers for Disease Control in the US. (Actually, we never study the whole population in biostatistics, because we always want to generalize to future events. We carry out studies so that we can take action and change the future.)

In order to show a relation between the exposure and the outcome, recall that the conditional probability $P(C|E)$ denotes the size of the CE group within the E group. If we have that $P(C \mid E) > P(C \mid E^c)$, the risk for the outcome is larger if a subject is exposed than if he is not exposed; the risk for lung cancer is larger for a smoker than for a non-smoker. The two events are independent if the risk of the outcome is the same in the exposed and in the non-exposed population.

However, an association does not prove that smoking *causes* lung cancer. The smokers may differ from the non-smokers in other, non-measured (or even non-measurable) respects that are the actual cause of, or at least contribute to, the difference. It may be that smokers are mainly people with a certain lifestyle, a lifestyle which also contains some other factor that is the true cause of lung cancer. Other factors than the one under study that affect the outcome are called *confounders*. In order to imply a cause–effect relationship we need to adjust for such confounders in some way.

Example 2.1   It is an accepted fact, as will be further discussed in Example 2.2 below, that the risk of bearing a child with Down's syndrome increases with the age of the mother at the time of pregnancy. It follows that if we have studied the occurrence of Down's syndrome and relate it to the parity (the number of live-born children the woman has delivered including this one), we expect to find an association. Not because parity in itself matters, but because mothers in general are older when they have their fourth child, as compared to when they have their first.

Box 2.3 Doll, Hill and the association between smoking and lung cancer

---

Smoking became the dominant mode of consuming tobacco world-wide in the early twentieth century. Before this death due to lung cancer was rare, but an increase prompted a number of case–control studies to look for potential explanations, including the role played by tobacco smoking. Early studies were performed in Germany during the Second World War, with results that were largely ignored, partly because of the close relationship between German doctors in general and the Nazi party. So when the British Medical Research Council organized a conference in 1947 to discuss the reason for the increased mortality in lung cancer, it was other data that led Richard Doll and Bradford Hill to design a case–control study to investigate its association with smoking.

Their results were published in 1950. They had sampled cases of lung cancer from four British hospitals, and used as controls patients who had been admitted to these hospitals for other reasons. Their data for men only were as follows:

	Smoker	Non-smoker
Case	647	2
Control	622	27

Thus amounts to an estimated odds ratio of 14.0 with 95% confidence interval (4.2, 87.3). The authors took care to exclude alternative explanations for the finding, and drew the conclusion that smoking is an important factor for lung cancer.

Skeptics, usually backed by the tobacco industry, argued (correctly) for many years that this type of study cannot prove causation, but to design a proper double-blind, randomized study to prove beyond all reasonable doubt the cause–effect relation is out of the question, for obvious practical and ethical reasons.

However, Doll and Hill constructed a prospective cohort study that decided the argument (more or less) – *the British Doctors Study*. In October 1951 the researchers wrote to all registered male physicians (actually to all physicians, 59 600 in total, but women were few and usually ignored) in the UK and asked them to fill in a simple questionnaire, which included smoking habits. Two thirds of doctors responded. This study was published a few years later, and showed that not only lung cancer but also, for example, myocardial infarction and respiratory disease occurred markedly more often in smokers than in non-smokers. (Similar results were obtained about the same time in the US in a large study sponsored by the American Cancer Society.) Since then follow-up reports, published every 10 years, have added more information on and details of this relationship. Though not formally proof for a smoking–lung cancer causality, to most people this study is the final decisive argument.

Similarly, we may obtain an association between the risk of Down's syndrome and the consumption of sedatives during pregnancy from which we may conclude that taking sedatives is a risk factor in itself, when the underlying reason is that more sedatives are used by older women, and therefore older mothers.

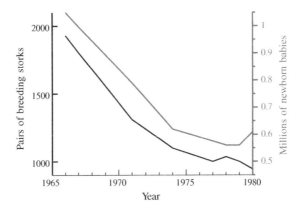

Figure 2.1   This illustration of the disconnect between covariation and causality was published in *Nature* in April 1988 (page 495). The article consisted more or less of only a graph similar to the one above, accompanied by the text 'There is concern in West Germany over the falling birth rate. The accompanying graph might suggest a solution that every child knows makes sense.'

In a cross-sectional study the only way we can handle confounders is to build them into a statistical model when we make the statistical analysis. But that may not capture the whole story, and it can never account for unknown confounders. It is therefore not necessarily valid to draw the conclusion that the exposure in question is a risk factor for the outcome if a relationship is found.

An alternative study design is to choose a sample of exposed individuals and another sample of non-exposed individuals and follow them for some time to see if they exhibit the response of interest. Such a study is called a prospective cohort study, but it can also be performed retrospectively. What matters is that two groups are defined, one exposed and one non-exposed. If we only collect the two groups from what is essentially a cross-sectional study, we have gained nothing. Here, however, we have the option of being careful in our construction of the groups. We can, for example, match individuals in the two groups for known confounders. This may be a step in the right direction, but adjustment for known confounders does not necessarily adjust for unknown confounders.

A quicker way to address the particular question if there is an increased risk of lung cancer in smokers, is to do a case–control study. This means that we first identify a sample of cases and then identify a sample of controls (with no lung cancer). Here we can match the controls with respect to key prognostic factors for the effect (i.e., confounders), such as age and sex, but otherwise select them at random from the appropriate population. Such a study is by necessity retrospective, since the effect must already have happened. By matching we hope to eliminate the effect of some known confounders from the analysis, getting us one step closer to a cause–effect relationship. Other known factors we can try to control for by building them into the statistical analysis appropriately. But again, unknown confounders cannot be eliminated in this way.

Sometimes effects are first identified in rather small case–control studies, as with the lung cancer example. In order to obtain more conclusive evidence, there may then be a need to perform larger, controlled studies. When this is done the effects seen in the case–control

---

Box 2.4 Hormones and heart disease in women: a confounder problem

---

Hormone replacement therapy (HRT) is given to women around the menopause in an attempt to lessen the discomfort, in particular hot flushes, that is commonly experienced and which is caused by a decline in circulating sex hormones, mainly estrogen and progesterone.

During the 1980s a significant number of observational studies seemed to indicate a link between HRT and a reduction in the incidence of coronary heart disease (CHD) in women. In one review of 31 studies, 25 found a numerical benefit of HRT, of which 12 were 'statistically significant', and a meta-analysis indicated a risk reduction of about a 50%. Credible mechanisms were advanced to explain why this might be, and a consensus arose that HRT was protective against CHD. This observation was picked up by the Women's Health Initiative (WHI), which focuses on defining the risks and benefits of strategies that could potentially reduce the incidence of diseases in postmenopausal women. In 1991–92 the WHI designed a double-blind, placebo-controlled clinical trial in order to get conclusive evidence.

In total 16 608 US women were randomized and allocated evenly between an estrogen-plus-progesterone and a matching placebo group, with subjects planned to be followed for 8.5 years. However, a safety committee stopped the study after an average follow-up period of 5.2 years, because there were indications of a negative risk–benefit ratio for the patient. The data were inspected, and although there was no evidence of an effect on all-cause mortality, the relative risk (hazard ratio) for CHD was estimated to 1.29 – a 29% increase, instead of a decrease.

Subsequent analysis has shown that the group of women opting for HRT were predominantly from higher socio-economic groups which had, on average, better diet and exercise habits. The studies had falsely attributed the benefits of these factors to the HRT itself. For other health outcomes, such as breast cancer, colon cancer, hip fracture, and stroke, results from observational studies were similar to those of the WHI study, suggesting that if confounding explains the apparent protective effect of HRT against CHD, the associations with these other outcomes were not similarly confounded.

---

studies are not always confirmed, possibly because there was some important confounder missed in the original studies. Box 2.4 outlines one such story, about hormone replacement therapy and heart disease in women. To demonstrate a cause–effect relationship by these types of studies in itself is impossible; one needs to build the story in conjunction with other data.

In summary, we have three main types of observational studies:

1. The cross-sectional study, where we sample from the whole population. This study design allows us to estimate all probabilities in the 2×2 table and therefore also the four conditional probabilities $P(C|E)$, $P(C|E^c)$, $P(E|C)$, $P(E|C^c)$.

2. The cohort study, which can estimate $P(C|E)$ and $P(C|E^c)$. This study type does not allow the estimation of the probabilities $P(E|C)$ or $P(E|C^c)$.

3. The retrospective case–control study, which can estimate the probabilities $P(E|C)$ and $P(E|C^c)$, but not $P(C|E)$ or $P(C|E^c)$.

Box 2.5 Odds ratios and sampling methods

Let $O(A) = P(A)/P(A^c)$ be the odds for a property A and let S denote a subpopulation from which we sample patients. Then we have (appealing to Bayes' theorem; see Box 4.2) that

$$O(A|S) = \frac{P(A|S)}{P(A^c|S)} = \frac{P(S|A)P(A)}{P(S|A^c)P(A^c)} = R\,O(A),$$

where R is the relative risk of being sampled for subjects with the event compared to those without it. If we apply this to the exposed subgroup we find that $O(A|E, S) = R(E)O(A|E)$, where $R(E)$ is the relative risk of being sampled in the exposed subpopulation. The odds ratio for sampled subjects is therefore proportional to the overall odds ratio, with the proportionality constant being $R(E)/R(E^c)$. If the relative probability of being sampled is independent of whether the subject is exposed or not, we therefore get the same odds ratio in the subpopulation as in the whole population. This explains why it does not matter if we have a cohort or case–control study when we compute the odds ratio.

If we have different degrees of exposure, which we denote by x, we can assume a relationship of the form

$$O(A|x) = e^{\alpha + x\beta},$$

where the parameter β is of primary interest. This defines the so-called logistic model, and we see that we can apply this not only to a cohort study, but also to a case–control study, as long as we are only interested in β. This is because α will be replaced by $\alpha + \ln(R)$ in the case–control study. In the case of the case–control study this means that the intercept is a pure nuisance parameter, and it can be eliminated by performing a conditional logistic regression instead (for more on this, see Section 9.4).

We now return to our original question: is the relative risk

$$RR = P(C\,|\,E)/P(C\,|\,E^c)$$

greater than one? This can be estimated both in the cross-sectional study and in the cohort study, but it cannot be estimated in the case–control study. So how can we use that study design to answer the question?

The answer is based on the observation that the odds $p/(1-p)$ is an increasing function of the probability p, which implies that RR is greater than one precisely when the odds ratio

$$OR = \frac{P(C\,|\,E)/P(C^c\,|\,E)}{P(C\,|\,E^c)/P(C^c\,|\,E^c)}$$

is greater than one. By a stroke of luck (or, rather, by Bayes' theorem) this number can also be written

$$OR = \frac{P(E\,|\,C)/P(E^c\,|\,C)}{P(E\,|\,C^c)/P(E^c\,|\,C^c)},$$

which is a set of probabilities that can be measured in a case–control study.

This observation highlights a symmetry between the cohort and the case–control study: in the case–control study we fix the outcome (C) and can consider many types of exposures, whereas in the cohort study we fix exposure (E) but can consider many types of outcomes. Loosely expressed, by switching C and E we go from one type of study to the other. This means that case–control studies are optimal for detecting potential risk factors for a particular outcome, such as a specific disease, because we can study many potential risk factors. In order to further understand other consequences of such an identified exposure, a cohort study is useful, in which we study multiple outcomes of the exposure. If we can even perform a randomized cohort study, this can be used to actually prove the cause–effect relationship in the sense we have discussed above.

Sometimes an odds ratio may be directly interpreted as a relative risk in an epidemiological context. In an observational study the proportions we discuss are often, but not always, the result of observing a population for a period of time in order to see if a particular event occurs. If individuals in the population turn into cases with a constant intensity, the prevalence odds in steady state equals the disease incidence multiplied by the disease duration. This will be further addressed in the final section of this chapter. If we have two groups, exposed and non-exposed, and we assume that the disease duration is unaffected by the exposure, this means that the prevalence odds ratio really is the ratio of the intensities of case generation in the two groups. Note the basic assumptions here: we are in steady state and the disease duration is unaffected by the exposure.

2.4   The experimental study

An experimental study in humans is usually referred to as a clinical trial and is a special case of the prospective cohort study, in which we can decide on who is exposed and who is not. It often refers to some kind of intervention, which may be a drug treatment, but may also be something else, including a prevention strategy, a diagnostic test or a screening program. According to most medical historians, one of the first studies of this type was performed when James Lind decided that the treatment for scurvy on long sea voyages should be oranges and lemons (see Box 2.6). The ability to decide who goes into which group makes a huge difference, since we can then devise methods that guarantee that potential confounders, both known and unknown, get, on average, evenly distributed between the groups. However, not all exposures can be analyzed in this way. It is, for example, usually not considered ethical to study the effects of harmful exposures in experimental studies.

The method that is used to guarantee group comparability is randomization. This means that we apply some kind of random mechanism when we allocate individual subjects to the groups. The main advantage with this is that even though imbalance in confounders may still prevail in the individual study, the statistical machinery allows us to describe what we actually learn. We will have more to say about randomization in the next chapter.

Parallel group trials are experimental counterparts of observational cohort studies. There is also a counterpart to the case–control study, with what may be called the case-crossover study in between. In such a study we select cases, but instead of selecting a matching set of controls we use as control material historical data on the selected cases, obtained prior to the time of exposure. The advantage with this design is that it eliminates some potential confounders, such as sex, but age and anything that goes with it will by necessity be a

---

Box 2.6 James Lind and the remedy for scurvy

The invention of the modern clinical trial is often attributed to the Scottish doctor James Lind and his investigation into different treatments for scurvy. Scurvy is a condition that we now know is due to vitamin C deficiency and was in those times one of the major killers in the high seas navies. Symptoms start with swollen gums and tooth loss, followed by an inability to work and finally, and slowly, death. When Lind joined the Navy as a surgeon's mate in the mid eighteenth century there was no accepted (and working) treatment available. Here is a summary of what happened next, according to most contemporary medical historians.

Lind first scanned the existing literature, noting that descriptions of the disease were either by lay seamen or by doctors who had never been to sea. Based on the observation that scurvy was rare on land and common among seamen (but not officers) he suspected something in the diet and decided to do a controlled experiment himself while serving as ship's surgeon on HMS *Salisbury* in May 1747. James Lind assembled 12 of his patients at similar stages of the illness, accommodated them in the same part of the ship, and ensured that they had the same basic diet (he controlled for some potential confounders). He then allocated two sailors to receive each of six treatments that were in use for scurvy – cider, sulfuric acid, vinegar, seawater, nutmeg, or two oranges and a lemon. Lind then observed that: 'The most sudden and visible good effects were perceived from the use of the oranges and lemons.' In fact, after six days one of the two given oranges and lemons was fit for duty, and the other fit enough to nurse the others. The test was continued for two weeks, but with no improvement in any of the others.

Of the treatments Lind compared, the Royal College of Physicians favored sulfuric acid while the Admiralty favored vinegar – Lind's fair test showed that both these authorities were wrong. In 1753 Lind published a monograph, *A Treatise of the Scurvy*, on his results, but it was not until 1795 that the British Navy supplied lemon juice to its ships. By the turn of the century this deadly disease had all but disappeared.

---

confounder; whether it matters is another question. The design may not be much used in epidemiological research, but in an experimental situation it may be possible to go one step further and repeat the experiment on the subject both with and without the exposure. This may be possible when the exposure is a drug treatment. By doing the experiment twice, with and without drug, we can use the subject as his or her own control. However, if we do the experiment without the drug first, followed by an experiment with the drug, we have a confounding factor – does the outcome depend on whether it is the first or the second assessment? To control for this, we also randomize this type of study, but now with respect to the sequence in which the treatments are taken (exposure followed by no exposure, or no exposure followed by exposure).

When investigating the efficacy of a drug, the clinical trial with all the characteristics of randomization, blinding, etc., is superior to the alternative of an observational study from a scientific point of view. However, that does not mean they are uncontroversial, at least not when one tries to give them a design that is optimal with respect to their objective. The problem here is of an ethical nature. Patients expect to get the best treatment from their physicians, but

---

Box 2.7 Ethics and The Declaration of Helsinki

The ethics of clinical research is outlined in the Helsinki Declaration, which is morally – though not legally – binding on physicians. They still have to follow local law, but will be held to the higher standard.

The events that triggered the Declaration of Helsinki were the inhumane Nazi experiments on humans carried out before and during the Second World War by physicians such as Josef Mengele. These experiments were the subject of some of the postwar Nuremberg trials in 1947, during which it was realized that there was no generally accepted code of conduct for the ethical aspects of research on humans. When the verdict on the so-called 'Doctors Trial' was given in August 1947, an opinion on acceptable medical experimentation on human beings was also formulated as the ten points of the Nuremberg Code.

Based on the Nuremberg Code, the World Medical Association adopted in Helsinki, Finland, in June 1964 a set of ethical principles for human research which came to be known as the Declaration of Helsinki. It has since undergone a number of revisions (six by 2009), as well as a few clarifications, and has grown from 11 to 35 paragraphs. It defines the ethical principles of human research and has led to the appearance of ethical review boards that need to approve clinical studies, in addition to the approval by regulatory health authorities.

Many of the principles in the declaration are uncontroversial, but there is one important discussion which is of fundamental importance to the design and conduct of clinical trials. It is a statement that effectively excludes the use of a pure placebo control in cases where proven therapeutic methods exist. This is highly controversial and has never been fully accepted by the FDA because it essentially eliminates the possibility of proving absolute efficacy, as opposed to relative efficacy. This will be further discussed in Box 3.6.

---

in a clinical trial you may need to study an inferior treatment (which may be no treatment at all) in order to show that a new treatment is superior. Instead we may be forced to study the new drug on top of existing effective treatments. The better the background treatment is in its efficacy, the harder it is for the new treatment to demonstrate that it is effective, because there is less room for improvement. We therefore need to study more patients to document an effect in such a study, compared to one where the patients in the control arm receives no treatment. To expose any patient at a stage when you do not know if the new drug is at all useful is an ethical concern in itself, so there is a compromise that needs to be reached here. Taken to its logical extreme, any new drug could only be studied on top of the best of the present treatments, which may make it hard to prove beyond reasonable doubt the efficacy of a new drug in (short-term) trials of reasonable size.

A clinical study is not necessarily of benefit to the patients participating in the study, but performed with the intention to obtain information that will benefit future patients. For some diseases, especially chronic disease, this may be the same patient. To what extent the patients who choose to participate in clinical trials are representative of patients in general is seldom well understood, which means that inference to the general population is not clear-cut. This is a discussion we will pick up in the next chapter.

2.5   Population risks and individual risks

Earlier in this chapter we mentioned that there is a difference between an individual response and the overall response in a group. In this section we will illustrate this with an example in which the risk depends on a measurable covariate. In so doing we will introduce the distribution function as a way to describe random data.

Suppose we want to see to what extent a particular event A occurs in a population. The event can be that a patient responds to a treatment in a clinical trial; in an epidemiological context it may be that he contracts a particular disease. The parameter $p = P(A)$ is then the proportion of subjects for which this particular event occurs. If the event is of a negative nature, p will be referred to as a *population risk* or *group risk*. However, this risk must be carefully distinguished from the *individual risk*, which is the risk of the event for a particular individual, given that person's genes, environment, age, etc. The group risk will be made up of the individual risks, weighted according to the distribution of the risk factors in the population. Obviously, this difference between group and individual probability also applies to the case where the event is positive, like a treatment effect.

We will assume that the strength of the risk factor can be expressed as a real number and the distribution of these values by a function, the cumulative distribution function (CDF), which is defined as

$$F(x) = \text{the fraction of values that are at most } x.$$

An entity X that is described by a CDF in this way is called a *stochastic variable*, a term that is sometimes helpful, less often necessary. The CDF is the function $F(x) = P(X \leq x)$ and by definition $F(x)$ is an increasing function which starts at the value zero and ends at the value one (i.e., $F(-\infty) = 0$ and $F(\infty) = 1$). The CDF plays a prominent role in statistics.

We will discuss in some detail an example in which the event A is that a newborn child has Down's syndrome and the risk factor X is the age of the mother. The individual risk (having a child with Down's syndrome) will then be a function of the mother's age, and the group risk, the proportion of babies with Down's syndrome, will be determined from the age distribution of pregnant mothers. Returning to generalities, we let

1. $p(x) = P(A|X = x)$ denote the probability of the outcome A for an individual for whom the risk factor takes the value x,

2. $F(x)$ denote the CDF for the risk factor in the population.

Note that $p(x)$ is also a group average – it is the average of all individuals in the population for whom the value of X is x. In that sense we never come down to pure individual risks, only an assessment of individual risks based on group data. In the total population, the probability of the outcome A is the weighted sum of individual risks:

$$P(A) = \int_{-\infty}^{\infty} p(x)dF(x).$$

This is the probability that a randomly sampled individual from the population will have the outcome A. (For a discussion on the integral notation, see Box 4.8.)

If we look at another population, for which the risk factor is distributed according to a different CDF, we will have the same individual risks, but a different group risk. The group risk may well differ between two studies, if the study populations differ in their distribution of risk factors. Similarly, the degree to which patients respond to a particular treatment may differ considerably between two studies if the patients in the studies have different propensity to respond, depending on disease severity, concomitant medication, etc. This also means that in order to generalize a treatment, or risk, estimate, from a particular study to a wider population, the population that was studied must be representative of that wider population.

Example 2.2   The risk of a mother aged 35 years or older having a child with Down's syndrome is many times greater than for a mother in her twenties. The reason may be that the risk that an egg contains the chromosomal abberation that causes Down's syndrome depends on how old the egg is (they are made before the mother's birth). Despite this, it is often said that most children with Down's syndrome are born to younger mothers. Let us see why (and whether this is true).

The risk is described by a function $p(x)$ which is shown as the dashed curve in Figure 2.2a. Also shown in this graph are the probability densities $p_Y(x)$ by mother's age in Sweden in

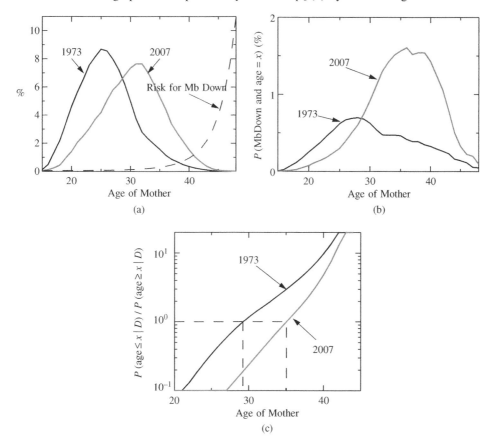

(a)

(b)

(c)

Figure 2.2   Do younger mothers give birth to most children with Down's syndrome? (a) shows the distribution of the age of mothers together with the age-dependency of the risk that the child is born with Down's syndrome, (b) shows the age distribution × risk and (c) shows the odds for a younger mother (with varying cut-offs defining what 'young' means), given that the child was born with Down's syndrome. Further details are given in Example 2.2.

two different years, $Y = 1973$ and 2007. Note that these densities are quite different; the one for 2007 lies further to the right than that of 1973; women tend to get pregnant later today than they did back then.

The graph in Figure 2.2b shows, for a range of ages, the probability that a randomly sampled individual is both age x and gives birth to a child with Down's syndrome, which is the function $q_Y(x) = p_Y(x)p(x)$. This is not a conditional probability, it is the probability that both criteria are fulfilled. These curves have a maximum, because older mothers are so few that, despite their increased risk, they do not provide many cases. By summing $q_Y(x)$ over ages $x \geq a$, we can compute the probabilities that a pregnancy provides a child with Down's syndrome with a mother of an age greater than a. Dividing each of these numbers with the corresponding proportion of pregnancies with Down's syndrome, we get the conditional probability that the mother is older than a, given that the child was born with Down's syndrome. The odds of these probabilities are given in Figure 2.2c. We see that in 1973 the ratio is one when the mothers age is 29.2 years, so if we make the cut-off at an age above this, we will find that more children with Down's syndrome are born by mothers younger than this age than by those that are older. For the year 2003 we find that the cut-off is substantially greater, at 35.1 years.

In summary, in 1973, when mothers were younger, there were more children born with Down's syndrome to mothers aged below 30 than above. This is typically what is meant by younger mothers. Their numbers compensated for the highly increased risk for the older mothers. In 2007 the age distribution of mothers has its mode (maximum) above 30 and the statement that younger mothers give birth to most children with Down's syndrome is no longer true.

As a final observation on this subject, we observe that the predicted number of children with Down's syndrome is larger than the number actually born. This is because older pregnant women often undergo amniocentesis to determine whether the child has the chromosome abnormality, based on which they may decide to have an abortion.

2.6   Confounders, Simpson's paradox and stratification

We noted earlier that when we investigate an exposure–effect relationship in an observational study, the identification of confounders is very important. However, when we adjust the analysis for such confounders, seemingly odd phenomena may occur. One extreme case, *Simpson's paradox*, relates to situations where the conditional results are the opposite of the overall result.

As an example we consider a medical study which compares the success rates of two treatments for kidney stones. It was a comparison of two surgical methods used during two different time periods: one was open surgery (denoted A below) for the period 1972–80, for which there was an overall success rate of 78% (273/350); the other was a less invasive method called percutaneous nephrolithotomy (B) used during the period 1980–85, which had a success rate of 83% (289/350). This corresponds to an odds ratio of 0.75 for A versus B, which seems to indicate that B should be the favored treatment.

However, when the data are split up according to stone size, dichotomized to small stones having a diameter less than 2 cm and large stones with a diameter larger than 2 cm, a different picture appears:

Box 2.8 Simpson's paradox described geometrically

To understand Simpson's paradox geometrically, consider the graph below, which describes the pairs (Success, Failure) as arrows from the origin: A and B represent the treatments and indices 1 and 2 the small and large stone groups, respectively. The slopes of the arrows are the corresponding odds. The slope for A_i is greater than that of B_i for each i. The combined table data are obtained by vector addition, and we see that $A_1 + A_2$ lies below $B_1 + B_2$.

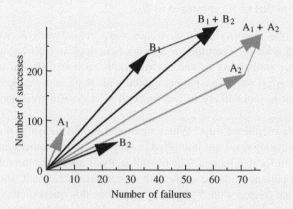

The key to understanding why this can occur is to note that the group slopes are unchanged if we make the arrows longer, so this is true of the odds ratio as well. The slope of the total, however, is highly dependent on how long the original arrows are.

| | Small stones | | Large stones | | Total | |
	A	B	A	B	A	B
Success	81	234	192	55	273	289
Failure	6	36	71	25	77	61
Odds	13.5	6.5	2.7	2.2	3.5	4.7

Dividing the data into sub-tables based on the values of other measured variables, in this case stone size, is called *stratification*. Individual sub-tables constitute the strata.

We see that the odds ratio is 2.1 for small stones and 1.2 for large stones, so the data favor A over B for each of the groups, which is the opposite of the result above. This phenomenon, that the odds ratios are on opposite sides of one in sub-tables compared to the combined table, constitutes *Simpson's paradox*. The paradox is that you come to different conclusions depending on how you look at data, and it becomes important to understand why this is, as well as the precise meaning of the different analyses.

One way to understand Simpson's paradox is to look at it from a mathematical perspective. A geometrical explanation is given in Box 2.8, which shows that it holds true for ratios in general. It will therefore apply equally well to odds and probabilities. It also explains that there is no paradox, as long as the math is explained correctly. Loosely speaking, we need two effects to act together:

1. The confounding factor has a large effect on the odds (there is a large angle between the pair A_1, B_1 and the pair A_2, B_2 in the graph in Box 2.8).

2. The relative sizes of the groups which are combined when the confounder is ignored differ substantially between strata (the length of A_1 in comparison to B_1 differs substantially from that of A_2 compared to B_2).

In the table above the majority of small stones were treated with B, whereas the larger stones were treated with A. Therefore B appears beneficial overall because small stones, which have a higher probability of success (the odds ratio is 2.9), are more likely to be treated with B. Indeed, if we have a patient treated with B, he is more likely to have a small stone, and therefore is more likely to end up with a successful treatment. As the overall test says.

But which result should we trust? Which treatment should we use? Of course we should use A, because the question we are interested in relates to a new patient and our problem is to choose a treatment for that patient. The overall analysis addresses the following question: if we are to pick a patient from the study with a successful treatment, should we pick one treated with A or one treated with B? In many situations this question has the same answer as the one we are interested in, but not in this case, because what is true for the patients in our study may not be true for a new patient. This insight also tells us how to compute the odds ratio for the new patient. In fact, A has a 93% success rate for small stones and a 73% success rate for large stones. Moreover, 51% of stones are small. So the probability of success when treated with A should be $0.51 \cdot 0.93 + 0.49 \cdot 0.73 = 0.83$. Similarly, the probability of success when treated with B should be 0.78, so the odds ratio is 1.4 for A relative to B.

So the right answer in this case is: go for the stratified analysis!

This last computation highlights the problem of which proportions to report from a stratified analysis. The crude rates may be somewhat misleading and we may need to standardize the proportions to some common standard to get a meaningful estimate. In the calculation above we used an internal standard, to get a fair comparison between the treatments. In other cases we may want to refer to some global, external, standard. As an example, assume we have a risk factor that has different effects in different age classes, and we want to see the overall effect in a particular population for which the age distribution is different from that of a particular study population. We then need to weight the age-specific risks of the study according to the age distribution in the target population.

However, the stratified analysis is not always the right thing to do. Suppose we take the same numbers as above, but now we assume that B works by reducing stone size, and that our data were obtained from an ultrasound examination some time after initiation of treatment. Under this (somewhat unrealistic) assumption the stratified analysis has no meaning, since it is only a result of the *modus operandi* of B. In this case we definitely should go for the overall analysis – and we should never have done the stratified analysis, because now the stone size is not a confounder in any meaningful sense of the word.

Box 2.9   Cornfield's inequalities

When the association between smoking and lung cancer was found in the 1950s, an immediate question was whether an unknown factor might significantly change the association in a Simpson's paradox type of reversal. For a suggested confounder, and data that allow us to compute risks, there are necessary conditions on the prevalence difference or ratio of the confounder in the exposed and unexposed groups. These conditions, called Cornfield's inequalities, are based on the simple observation that if the outcome C is actually independent of the exposure E conditionally on a confounder X, we have for exposed individuals that $P(C|E) = P(C|X)P(X|E) + P(C|X^c)$ $(1 - P(X|E))$, where $P(X|E)$ is the prevalence of the confounder in the exposed group, and similarly for the unexposed. There are two useful expressions that can be derived from this. The simplest is the observation that the observed relative difference $\Delta_E = P(C|E) - P(C|E^c)$ is related to the confounder relative difference $\Delta_X = P(C|X) - P(C|X^c)$ by

$$\Delta_E = (P(X|E) - P(X|E^c))\Delta_X.$$

If all factors are positive, this shows that a necessary condition for the confounder to explain all the association is that the prevalence difference $P(X|E) - P(X|E^c) \geq \Delta_E$.

Alternatively, considering observed and confounder relative risks RR_E and RR_X, we get

$$RR_E = \frac{P(X|E)RR_X + 1 - P(X|E)}{P(X|E^c)RR_X + 1 - P(X|E^c)}.$$

Under the assumption that the relative risks are at least one, this shows that the confounder relative risk must be larger than the observed relative risk and, more importantly, that the prevalence ratio $P(X|E)/P(X|E^c)$ must be larger than the observed relative risk. If this is not the case the proposed confounder cannot fully explain the association seen.

Simpson's paradox is more or less exclusively a problem for observational studies, because clinical trials often have balanced groups and imbalances are necessary to generate the paradox. In an epidemiological context there is always a fear that the results we see are not true, because there may exist some unknown confounder that will reverse the present conclusion. This is only a reflection of the general fact that new evidence could lead to a change in our opinion about what the truth is. That is what evidence is for, and only illustrates that science is not a stationary state, but a dynamic process. One should note, however, that there is an observation, called Cornfield's inequality (see Box 2.9), which helps to decide if a particular confounder can reverse the effect seen. Essentially it says that for a particular confounder to have the power to fully explain the association observed, not only does it need to have a larger effect than the association observed, but also the prevalence difference or ratio (depending on which way you want to express the result) must be larger than the observed effect.

At this point we should also note that the stratum-specific odds ratios and the overall odds ratio do not have a simple relationship; the latter is not a simple, weighted, average of the former. In fact, also when the odds are all the same in the different strata, the overall odds ratio may be different. In order to give a useful pooled estimate from a stratified analysis, epidemiologists therefore choose other measures, including the Mantel–Haenszel pooled odds ratio, which we will discuss later. The same is true of relative risks; geometrically it is a consequence of the ratio of ratios situation discussed in Box 2.8.

Another point of confusion about confounders and the subgroups they define is called *stage migration*. It originally referred to what appeared to be a paradox when one looked into cancer survival by disease stage. Cancer stage is an indicator of a tumor's anatomic dissemination and is a key predictor of cancer survival. In its simplest form, a tumor may be classified as 'localized' (Stage I) if there is no evidence of spread of the tumor beyond where it originated, 'regional' (Stage II) if there is evidence of tumor spread to adjacent tissues or lymph nodes but no further, and 'metastatic' (Stage III) if there is evidence that the tumor has spread to other organs. The lower the stage, the better the prognosis. The apparent paradox is that you may find, when you compare two studies, that they demonstrate the same overall survival, but that in one study all the individual stages have improved survival compared to the other study. This looks like a paradox, because you expect the sum to increase when all the terms in the sum is increased.

Again this is not a paradox if properly expressed. A good insight is often attributed to the US humorist Will Rogers in the following quotation: 'When the Okies left Oklahoma and moved to California, they raised the average intelligence level in both states.' Put in a medical context, assume that you use some test and divide a population into those that are healthy and those that are sick. The former group has on average a longer expected survival than the latter. Next assume that a new diagnostic tool is introduced which identifies some of those formerly believed to be healthy to be sick. With this new classification the group of healthy will have a longer expected survival than before, since that group has lost some of its more fragile individuals. Also the group of sick will have an increased expected survival time, since it has been augmented by a few almost healthy individuals. No individual has changed expected survival, so the total average stays the same. The important point is that if we compare the group of sick individuals in the two classifications, we would infer that now, after the introduction of the new diagnostic tool, they have a longer survival. It is not a faulty conclusion *per se*, but the survival has not changed for any individual.

A well-known conundrum related to stage migration is the so-called low-birthweight paradox, which is discussed in some detail in the next example. It refers to the fact that data tell us that a baby with low birthweight has a greater chance of being alive if the mother is a smoker, which seems to imply that maternal smoking benefits the baby.

Example 2.3 The following table shows the relation between infant mortality and maternal smoking in a region of Sweden in a particular year:

	Mother	
Infant	Non-smoker	Smoker
Stillborn	40	6
Liveborn	10 665	1 277
P(Dead) (%)	0.374	0.468

As we have learnt to expect, this shows that a smoking mother has an increased risk of having a stillborn baby. The number of stillborns is small, so the precision in this data is not that great, but it is consistent with data from many sources and parts of the world. However, if we look only at children weighing less than 2.5 kg at birth, we seem to get a different message.

| | Mother | |
Infant	Non-smoker	Smoker
Stillborn	19	1
Liveborn	371	73
P(Dead) (%)	4.87	1.35

Now the proportion of stillborn babies is larger for non-smoking mothers. Somehow it seems that a low-birthweight baby increases its survival chance by having a smoking mother. Does this mean that maternal smoking protects children with low birthweight?

There is no argument that the last table tells us that if we randomly choose a newborn with a low birthweight (less than 2.5 kg), then the probability that it is alive is larger if the mother is a smoker than if not. It is the conclusion from this observation that the mother's smoking habit has a beneficial effect on the babies survival chances that is wrong. This paradox is essentially only another example of stage migration. The underlying reason is that maternal smoking has a negative effect on the baby's birthweight. The staging is the dichotomy into low and high birthweight, where smoking means that we move some of the high-birthweight babies into the low-birthweight group, carrying with them their original risk of being stillborn. So we dilute the risk of death in the low-birthweight group by putting more 'healthy' babies there.

For the mathematically more interested reader we can build a model which captures a few, but key, aspects of the situation.

- Denote the distribution of the birthweight for babies of non-smoking women by $F(x)$. Also assume that smoking shifts this distribution to the left, so that the corresponding distribution for smoking women is $F(x + \Delta)$ for some Δ. (Our data estimate Δ to be 199 grams.)

- Assume that the weight distribution for stillborn babies is the same for smoking and non-smoking mothers.

To what extent the second point holds true is not important to us, because we want to describe the effect of the first point, which we know is true, on the subgroup analysis made above.

For illustration we will use Gaussian approximations to all distributions; these CDFs are shown in Figure 2.3(a). Using the model above, we can construct subgroup tables similar to the one above but for a general cut-off point x instead of the special case $x = 2.5$ kg only. Let W denote the birthweight and D denote the event that the baby is stillborn. Within one of the groups of mothers, the fraction of stillborn babies among those with a birthweight at most x is given by

$$P(D|W \leq x) = \frac{P(D \text{ and } W \leq x)}{P(W \leq x)} = \frac{P(W \leq x|D)P(D)}{P(W \leq x)},$$

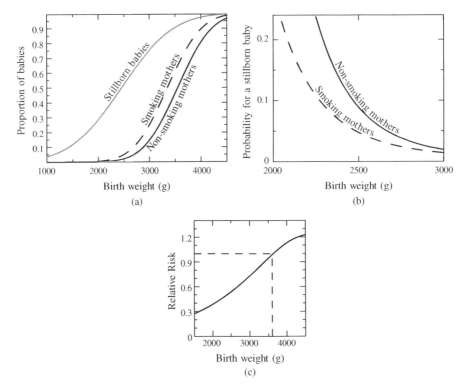

Figure 2.3   A model for the low-birthweight paradox. (a) The assumed birthweight distributions for smoking and non-smoking mothers as well as the birthweight distribution for stillborn babies. (b) The derived probability for a stillborn baby given an upper limit for the baby's weight, by the mother's smoking status. (c) The relative risk for a stillborn child given different cut-offs for baby weights. Further details in the text.

where $P(W \leq x|D)$ is assumed independent of group (the gray curve in Figure 2.3(a)), $P(D)$ is the overall probability of a stillborn baby and the denominator $P(W \leq x)$ is the distribution described in the first point above. This probability, as a function of the cut-off point x, is plotted for the two groups of mothers in Figure 2.3(b). (The numbers do not quite fit with the original subgroup table because of our simplifying assumption.) We see that this probability is smaller for smoking mothers than for non-smoking mothers, irrespective of which cut-off limit we choose. There is nothing special about the particular cut-off limit of 2.5 kg. The relative risk for a stillborn baby for smokers versus non-smokers is given by

$$RR(x) = \frac{F(x)}{F(x + \Delta)} RR(\infty),$$

where $RR(\infty) = 1.23 > 1$ is the overall relative risk. The graph of this function is plotted in Figure 2.3(c). We see that it decreases as we decrease the cut-off point, which means that the smaller we make it, the stronger the 'paradox'. It only becomes favorable for non-smokers if we take the cut-off at about 3.6 kg.

2.7 On incidence and prevalence in epidemiology

There is often confusion about what is meant by the rate at which an event occurs and the probability that it occurs. In fact, the two words are too often used interchangeably. This is not a good habit, since rate is like speed, something that happens per time unit. Loosely speaking, the rate at which an event occurs is an entity such that if we multiply it by some short time interval in which the rate is essentially constant, we (approximately) get the probability that the event occurs in that particular interval. If the event is adverse to the individual experiencing it, the rate is often called the hazard for the event. Another term that is often used is intensity.

Obviously there is a relationship between the rate at which events occur, and the probability that a randomly selected individual has experienced the event when observed. To model this situation we assume that there is a constant intensity λ with which the event occurs. Let $Q(t)$ denote the fraction of subjects that have *not* experienced the event at time t. This fraction decreases at a rate which is proportional to itself, which we mathematically write as a differential equation $Q'(t) = -\lambda Q(t)$. (This is actually the definition of the intensity λ.) Solving the equation, we find (since $Q(0) = 1$) that $Q(t) = e^{-\lambda t}$, so that the probability that the event has occurred in the interval $(0, T)$ is given by

$$p = 1 - Q(T) = 1 - e^{-\lambda T}.$$

This is the relation between the rate of occurrence λ and the probability of the event occurring during an observation time T.

There are two important concepts in epidemiology which are related to this discussion, the disease prevalence and the disease incidence. The typical situation is described in the diagram on the right, where H refers to the healthy individuals in the population, D denotes death and I is the population with 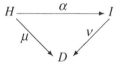 the disease. The rate α is the *disease incidence* (for a healthy individual) and μ and ν are the death rates for healthy and sick individuals, respectively. This diagram corresponds to two differential equations, each built from the observation that the rate of change of individuals in the population is the net effect of influx (the rate at which new people enter the population) and outflux (the rate at which people are removed from that population). A healthy individual can either get the disease or die without having contracted it, whereas a person with the disease must have been healthy once, and is assumed to leave this disease state only by dying. We assume that the disease is chronic, and the discussion is therefore not relevant for most infectious diseases. Mathematically this is formulated as the two differential equations

$$H'(t) = \beta - (\alpha + \mu)H(t), \quad I'(t) = \alpha H(t) - \nu I(t),$$

where β represents a constant birth (or influx) rate. When this population is in equilibrium, the left-hand sides of the two equations are zero, which gives us the two steady-state populations

$$\bar{H} = \frac{\beta}{\alpha + \mu}, \quad \text{and} \quad \bar{I} = \frac{\alpha \bar{H}}{\nu}.$$

The factor v^{-1} is the (mean) duration of the disease (until death). The *prevalence* P of the disease is the fraction of subjects that have the disease:

$$P = \frac{\bar{I}}{\bar{I} + \bar{H}} = \frac{\alpha}{\alpha + v},$$

which is a proportion. The corresponding odds is given by

$$\frac{P}{1 - P} = \frac{\alpha}{v} = \text{incidence} \times \text{duration}.$$

In particular, this means that if we have two groups of patients and compute the prevalence odds ratio, this is the ratio of the incidences multiplied by the ratio of the durations in the two groups. If we further assume that the duration of the disease is the same in the two groups, the prevalence odds ratio equals the ratio of the disease incidences for the two groups. We must keep in mind that these are statements about populations that are in steady state.

The incidence α is the number of cases per unit time. To estimate it, we start with the following equation for the healthy group:

$$H'(t) = -(\alpha + \mu)H(t), \quad H(0) = H_0,$$

which has the solution $H(t) = H_0 e^{-(\alpha+\mu)t}$. If we count only new cases, $N(t)$, these increase during the observation period at a rate that is proportional to the size of the healthy population. This means that $N'(t) = \alpha H(t)$, from which it follows that

$$N(t) = \frac{\alpha H_0}{\alpha + \mu}(1 - e^{-(\alpha+\mu)t}), \quad 0 < t < T.$$

If we ignore the deaths here and take $\mu = 0$, we have that the *cumulative incidence function* (CIF) $G(t) = N(t)/H_0$ of new cases is given by $G(t) = 1 - e^{-\alpha t}$. This quantity is (approximately) what we observe, and from it we can obtain the estimate for α. If α is small, we have $G(t) \approx \alpha t$, so we estimate α as the observed number per person-year. Better than to ignore deaths is to consider such events as censored; an individual who dies without having developed the disease can only be counted as being at risk before he dies. This means that we observe this individual for only a part of the observation period, which we should take into account when we estimate α. This is further discussed in Chapter 11.

The CIF gives the proportion of individuals observed for whom the event has occurred at a particular point in time. A related question is this: what is the distribution of the total number of events that occur – for example, the number of subjects who will acquire the disease in the population we study during the observation time? The discussion in Box 2.10 shows that it is a reasonable approximation when the events are rare enough (so that there is not much difference between the conditional mean hazard and the overall mean hazard) that this number has a $Po(\alpha T)$ distribution, where T is the total exposure time for the whole group and α is as above. There is an implicit assumption here that the disease is not contagious, since that would make events come in clusters, which we assume is not the case. (Note that the Poisson approximation implies that α is estimated by the number of events divided by observed person-years.)

Box 2.10 The Poisson distribution and Poisson process

Consider a large population in which a particular disease occurs with a subject-specific but constant hazard rate, which therefore defines a stochastic variable θ. Assume also that the events are rare in the population. If we start a clock and count the cumulative number $N(t)$ of disease occurrences up to time t in the population, we get a stochastic variable which, to a good approximation, has a Poisson distribution. Here is a heuristic explanation.

Let $P_k(t) = P(N(t) = k)$ be the probability that exactly k events have occurred before time t and assume that events do not come in clusters. We then have for $s > t$ but very close to t that (approximately, but with an error which we ignore)

$$P_k(s) = P(N(s) = k|N(t) = k)P_k(t) + P(N(s) = k|N(t) = k - 1)P_{k-1}(t).$$

If we differentiate this with respect to s and then let $s \to t$, we find that

$$P_k'(t) = -\lambda(t)P_k(t) + \lambda(t)P_{k-1}(t), k > 0, \quad P_0'(t) = -\lambda(t)P_0(t),$$

where $\lambda(t) = E(\theta|T \geq t)$ is the average hazard rate among those who have not yet had an event at time t. The general solution to this system of differential equations is

$$P_k(t) = \frac{\Lambda(t)^k}{k!}e^{-\Lambda(t)}, \quad \text{where } \Lambda(t) = \int_0^t \lambda(s)ds$$

is the cumulative intensity function in the population. This means that $N(t) \in \text{Po}(\Lambda(t))$, where $\text{Po}(\mu)$ denotes the *Poisson distribution* with mean μ. This is the discrete distribution that takes on the values $k = 0, 1, \ldots$ with probabilities $e^{-\mu}\mu^k/k!$. The Poisson distribution is one of the important distributions in elementary statistics. One of its more important properties is the addition formula that if X is distributed as $\text{Po}(\mu_1)$ and Y as $\text{Po}(\mu_2)$, then $X + Y$ is distributed as $\text{Po}(\mu_1 + \mu_2)$, provided the variables are independent.

2.8 Comments and further reading

Much of the early discussion in this chapter is about epidemiology and case–control studies, to which Rothman (2002) may serve as an introduction. The style of the beginning of this chapter was much inspired by the book by Evans et al. (2006), which is now (as of this writing) out of print, but can be found on the Internet in the *James Lind Library*, a site devoted to explaining and illustrating the development of fair tests of treatment in health care. The particular epidemiological story about lung cancer and smoking is discussed by Richard Doll (1998b) in a volume of *Statistical Methods in Medical Research* which has the statistics of smoking as its theme; the quote from Thomas Lewis at the end of Section 2.2) is taken from Doll (1998a). The original articles referred to in Box 2.3 are Doll and Hill (1950, 1956). The causality question here is interesting in itself, since some eminent statisticians, R.A. Fisher in the UK and Berkson in the US, remained unconvinced about the interpretation that this was evidence for smoking as a causative factor. (For Fisher the argument seems to have been around a common genetic cause for both lung cancer and the smoking habit, or one preventing both.) However, biology has moved on since then, and now the importance of smoking in the development of lung cancer is beyond any reasonable doubt, despite the non-existence of a properly randomized trial to confirm this. In addition to this there are numerous other diseases for which there is clear evidence that smoking is an important risk factor, including cardiovascular disease. The WHI study reference mentioned in Box 2.4 is Writing Group for the Women's Health Initiative Investigators (2002).

The importance of James Lind in combating scurvy is debated in (Wootton, 2007, Chapter 8), where it is argued that the importance of this particular experiment is a modern construction with no support in history. Lind had no influence on subsequent developments, and perhaps did not understand what he had discovered himself. In the same book, Chapter 16 also contains a discussion about what happened after the 1950 result of Doll and Hill on smoking and lung cancer.

We noted above that the odds ratio may be the preferred way to describe an association in a 2×2 table, because it is independent of the study design. The main argument against its use is that the odds ratio is hard to interpret, but there is also an argument that the strength of an association should not be measured in this way at all. A critique of the odds ratio, together with a general discussion about different measures that can be obtained from the four numbers of a 2×2 table, is given by Kraemer (2004).

Simpson's paradox was extensively studied in Simpson (1951), but had been observed and discussed earlier. An in-depth discussion of it, and the causality problem, is given by Pearl (2000). Cornfield's inequality (Cornfield et al., 1959) appeared as a response to Fisher's suggestion of a common genetic cause for smoking and lung cancer. The kidney stone example is taken from Charig et al. (1986) and has been used by others to this end. The low-birthweight paradox is discussed in Wilcox (2001), where it is shown that our quantitative argument applies better as a high-altitude paradox than a low-birthweight paradox. This paper demonstrates that life at high altitudes (as in Norway or Colorado, US) leads to a shift in the birthweight distribution in the same way as maternal smoking does, but has no effect on child mortality. When it comes to smoking, it turns out, not surprisingly, that smoking actually has an effect on mortality – it increases it. In other words, the true effect is the reverse of what the paradox claim.

The discussion in the last section was focused on the relationship between the two concepts incidence and prevalence in epidemiology, but is also the starting point for some statistical models with important application in biostatistics. What we have is a competing risk model, where the acquiring of the disease is in competition with death without having acquired it. This type of model will be revisited in Chapter 11.

References

Charig, C., Webb, D., Payne, S. and Wickham, O. (1986) Comparison of treatment of renal calculi by operative surgery, percutaneous nephrolithotomy, and extracorporeal shock wave lithotripsy. *British Medical Journal*, **292**, 879–882.

Cornfield, J., Haenszel, W., Hammond, E., Lilienfeld, A., Shimkin, M. and Wynder, E. (1959) Smoking and lung cancer: Recent evidence and a discussion of some questions. *Journal of the National Cancer Institute*, **22**, 173–203.

Doll, R. (1998a) Controlled trials: the 1948 watershed. *British Medical Journal*, **317**, 1217–1220.

Doll, R. (1998b) Uncovering the effects of smoking: historical perspective. *Statistical Methods in Medical Research*, **7**(2), 87–117.

Doll, R. and Hill, A.B. (1950) Smoking and carcinoma of the lung. preliminary report. *British Journal of Medicine*, **2**(4682), 739–748.

Doll, R. and Hill, A.B. (1956) Lung cancer and other causes of death in relation to smoking; a second report on the mortality of British doctors. *British Journal of Medicine*, **2**(5001), 1071–1081.

Evans, I., Thornton, H. and Chalmers, I. (2006) *Testing Treatments. Better research for better healthcare*. London: British Library.

Kraemer, H.C. (2004) Reconsidering the odds ratio as a measure of 2×2 association in a population. *Statistics in Medicine*, **23**, 257–270.

Pearl, J. (2000) *Causality: Models, Reasoning, and Inference*. Cambridge: Cambridge University Press.

Rothman, K.J. (2002) *Epidemiology. An introduction*. Oxford: Oxford University Press.

Simpson, E. (1951) The interpretation of interaction in contingency tables. *Journal of the Royal Statistical Society, Series B*, **13**, 238–241.

Wilcox, A.J. (2001) On the importance – and the unimportance – of birthweight. *International Journal of Epidemiology*, **30**, 1233–1241.

Wootton, D. (2007) *Bad Medicine: Doctors Doing Harm since Hippocrates*. Oxford: Oxford University Press.

Writing Group for the Women's Health Initiative Investigators (2002) Risks and benefits of estrogen plus progestin in healthy postmenopausal women. principal results from the Women's Health Initiative Randomized Controlled Trial. *Journal of the American Medical Association*, **288**(3), 321–333.

3

Study design and the bias issue

3.1 Introduction

In our first chapter we discussed the different steps that are involved when evidence is obtained using statistical methods. Not only do we need the experiment to provide us with the data for analysis, we also must ensure that the design allows us to draw the conclusions we want to from this statistical analysis. A small p-value alone does not guarantee the validity of the conclusion. The design elements that were discussed included the importance of avoiding the multiplicity issue. This has become a statistical issue, because of the habit of forcing conclusions from p-values. However, even when we have planned our statistical work properly, and know how to draw conclusions from such information, there remain design problems to be addressed. These are not really statistical issues, but are very often considered to be so by physicians when it comes to medical research. They are about eliminating alternative explanations from the experiment.

In the previous chapter we discussed clinical study design in terms of what kind of studies we can do. Here we will discuss in more detail some aspects that relate to the problem of how to avoid bias. Bias is about having some alternative explanation operating in the background which influences the result, but which is not acknowledged when the conclusions are drawn. Understanding the bias problem has led to the development of the controlled, randomized, double-blind, clinical study as the state-of-the-art study design when it comes to clinical research, provided it can be done. We will address different aspects of bias, pointing out what it is randomization and blinding try to accomplish. Following that we will take a look at the analysis stage. Statisticians, especially in the pharmaceutical industry, have invented ways to undo some of the careful planning that goes into a state-of-the-art clinical study. I call this self-inflicted bias, where the generally accepted term is that a per protocol (PP) analysis is done. It is about selecting data for analysis, and we will see an example which shows how data, which are originally unbiased, become biased despite the best of intentions.

In Section 3.9 we will address the bias involved when we try to obtain knowledge by summarizing the medical literature. Decisions taken from such reviews are what is generally referred to as *evidence-based medicine*. This involves combining results from different studies

Understanding Biostatistics, First Edition. Anders Källén.
© 2011 John Wiley & Sons, Ltd. Published 2011 by John Wiley & Sons, Ltd.

reported in the literature in so-called meta-analysis. The bias referred to is the publication bias which follows from the fact that it is predominantly positive findings that get published.

3.2   What bias is all about

In everyday language bias is about prejudice – about opinions or feelings that tend to favor one side in an argument. In statistics in particular, and science in general, it is any factor or process that tends to drag the results away from the truth. To clarify what this means in a statistical context, consider a situation where we have compared two groups.

- The statistical methodology, including the p-value, tells us what confidence we can have that there is a difference between the *groups* for the outcome measure we have analyzed.

- Our objective is usually not to claim that there is a difference between the groups but to claim that the difference is due to a particular factor, such as a drug treatment, which defines the groups. The question is whether this is a valid conclusion to draw from the statistical analysis.

Bias enters at both these stages; in the first stage because many statistical methods have some intrinsic, but usually rather mild, bias built into the estimation of the effect. In this chapter we ignore this particular kind of bias and concentrate fully on the issue in the second stage: how we certify that we are making the right claim.

The basic point which we must bear in mind at all times is this: if we want to prove that a particular exposure increases the risk of an outcome Y, or the magnitude of an outcome variable Y, we may divide patients into two groups, exposed and non-exposed, and analyze how Y differs between the two groups. The confidence in the conclusion about Y hinges on the extent to which we can guarantee that the two groups do not differ in any other respects than the exposure factor. When we say 'differ', we are not referring to the random differences that always will occur when you sample from two groups, and which are handled by the statistical methodology. What we are referring to are any *systematic* differences, which is what defines bias.

In coming sections we will discuss the following basic variants of bias:

- selection bias, which occurs when our sample is not representative of the population we want to make claims about;

- incomparable groups (apart from the factor(s) under investigation) when we start, and how we use randomization in experimental studies to avoid this;

- information bias, which refers to when the measurements are not taken in identical ways in the two groups – this includes differences in the conduct of the study in the groups of an intervention study.

3.3   The need for a representative sample: on selection bias

The simplest example of bias is when our purpose is to obtain a description of a population, using a sample from it. It could be something as simple as the male/female ratio in the

---

Box 3.1 The rise of the Gallup poll

In 1936 the US presidential election was a contest between the Republican candidate Alfred Landon and the incumbent president, the Democrat Franklin D. Roosevelt. In order to predict the outcome of the election, the magazine *Literary Digest* surveyed over 2 million voters by post, from which they predicted that Landon would win by a wide margin; he was expected to receive 370 electoral votes, with Roosevelt expected to receive 161. The actual outcome of the election was a dramatic reverse: Roosevelt beat Landon in all states but Maine and Vermont, mustering 523 votes versus 8!

The poor prediction can be fully explained by selection bias. The magazine survey represented a sample collected from readers of the magazine, supplemented by records of registered automobile owners and telephone users. This sample included an over-representation of individuals with high income, who were more likely to vote for the Republican candidate.

In contrast, a poll of only 50 000 citizens selected by George Gallup's organization successfully predicted the result, leading to the popularity of the Gallup poll. Note the difference in numbers: 2 million versus 50 000. We cannot overcome bias by sheer numbers.

---

population, or its age distribution, or something more complicated like an exposure–effect relationship. The discussion in this section is primarily focused on observational studies, and we start with a cross-sectional one. In such studies we want a sample that is representative of the whole population.

The best way to get a representative sample of the population is to take a *random sample*, which means that when we select our sample, every individual in the population has the same probability of being chosen; if there are N individuals in the population, every individual should have a probability of $1/N$ of being chosen. Only with a random sample can we expect to get it at least approximately right, when we try to figure out what the whole population looks like from the sample. When we sample in any other way, we have a *selection bias*, and in such a case we are almost certain to get a distorted picture of the full population. A classical example of how non-representative sampling gives the wrong conclusion is outlined in Box 3.1. Occasionally we know what the bias is, and we can then handle it in the analysis. In a nationwide survey we may, for example, deliberately over-sample from minority groups in order to gain precise information about them.

When we analyze a subset that is not taken at random, the problem becomes one of giving convincing arguments for why a study of that group still produces valid conclusions about the general population. Sometimes this simple, and obvious, problem is disguised in the description of the problem. For example, assume you want to investigate the association between a disease and a risk factor. You look for an odds ratio, for which you need a 2×2 table. To get the data you take on the task of checking out the records for all in-patients at a particular hospital and build your table from that. Suppose that you get an odds ratio of 2, with all the associated confidence needed. Is that evidence for a relationship?

It would be if you pose the question appropriately: in patients who are hospitalized, is there an association between the risk factor and the disease? Because that is what you have studied, and the conclusion would be valid if this hospital can be considered representative of

---

Box 3.2 Berkson's paradox in action

A famous example of when Berkson's paradox was most likely operating is in a paper by
R. Pearl on cancer and tuberculosis in 1929 in the *American Journal of Epidemiology*.
It was based on a detailed analysis of the first autopsies conducted at the Johns Hopkins
Hospital in Baltimore, Maryland, on persons who had died with malignant tumors,
together with the same number of controls without evidence of malignancy, matched by
sex, race and (as far as possible) year of death. On these autopsies, active tuberculosis
(TB) was found distributed as follows:

	TB	No TB	Total
Cancer	54	762	816
No cancer	133	683	816
Total	187	1445	1632

The odds ratio in this table is 0.36, with 95% confidence limits (0.26, 0.51). There
is therefore considerable evidence that there is a negative association between active
tuberculosis and malignancies. The author concluded that one should treat cases of
recurrent, biopsy-proven, cancer with extracts (tuberculin) from the mycobacterium
responsible for tuberculosis, and actually initiated such treatment.

Other authors obtained similar results, but also noted that the same type of association was obtained if one replaced cancer with heart disease. They therefore concluded
that 'the only proper control for the association of active tuberculosis and cancer is the
incidence of active tuberculosis in some other disease'. There is no such association in
reality, and the results all come down to what selection procedure is used for who gets
autopsied; for some reason subjects with two diseases were less often selected.

---

all hospitals. It may, however, be a very misleading conclusion to draw for the out-of-hospital
population. If we want to claim a true biological relation between the risk factor and the disease
we need to look more closely at how the data were obtained. Suppose that the risk factor is
another medical condition, such as hypertension, manifesting itself as headache, and that the
disease is a skin affection. Assume that all patients with only one of these conditions have
about the same admission rate to hospital as those who have neither, whereas those who have
both have an increased (say, doubled) tendency to seek medical advice (doubled admission
rate). In such a case, an odds ratio of 1 in the general population will show up as an odds ratio
of 2 in the hospital records, and consequently indicate a (false) association. This observation
has sometimes been considered a paradox, and as such it is referred to as *Berkson's paradox*
(see Box 3.2).

Another problem related to the computation of odds ratios from 2×2 tables is about
classification methods. The question to ask is whether the disease is equally likely to be
diagnosed for patients who have the risk factor as compared to those patients who do not have
it. Do patients with the disease have the same probability of remembering an exposure as those
without it? It is important to keep in mind that we never investigate the relationship between
a disease and a risk factor, but the relationship between the patients who have been diagnosed

with the disease and the risk factors which have been identified. Both of these may be prone to misclassification, and if misclassifications do not occur randomly in the table cells, we may get spurious relationships. For instance, knowledge about a subject's prior exposure to a putative cause may lead to a more thorough search for the disease, and, as a result, exposed subjects will be more likely to have the disease on record. When this occurs in a case–control study, it is an example of selection bias. When it occurs in an prospective cohort study, it is an example of information bias, as will be discussed in Section 3.5.

3.4 Group comparability and randomization

In the simple cohort study we compare groups of subjects. For simplicity, assume we have two groups. One group consists of individuals exposed to some particular risk factor; the other group is a group of non-exposed individuals. By comparing some outcome variables between the two groups, we want to assess the effect of the risk factor on these outcomes. In order for us to be able to draw a valid conclusion, the groups should as far as possible be comparable in all other respects. In other words, potential confounder factors should, at least in an average sense, be evenly distributed between the groups. Only then can we attribute any difference we find between groups to the risk factor.

For an observational study this means that we must identify a non-exposed group which fits with the exposed group, except for the particular risk factor that we are studying. The confidence in the results of the study will depend on how well we can justify that the control group is appropriate, so that no important confounders play a substantial role on the results. To balance potential confounders is not always achievable in such a study, in which case we need to resort to statistical modeling to eliminate the influence of (known) confounders. The optimal way to avoid the controversies around unknown confounders would be to construct the groups in such a way that the construction method guarantees comparability, at least on average. The preferred method to achieve this is *randomization*, which requires an experimental study. This cannot always be done; it may not be ethically acceptable to expose subjects to harmful risk factors in the name of science. However, there are many occasions when it can be done, including the important case when we want to compare the benefit of an intervention, such as a particular drug treatment.

To randomize a study means that we randomly allocate subjects to the predefined groups. Each patient must have the same probability of being allocated to the different groups. In the early twentieth century new treatments were invariably introduced on the basis of the outcome of small series of patients performed by some medical authority in a purely observational manner, with no concurrent control. However, the need for concurrent controls was slowly realized and introduced into medical research in the 1930s, usually in such a way that alternate patients were given alternate treatments. This meant that allocation was made sequentially to the groups with a scheme such as

$$ABABABABAB\ldots$$

If no one is cheating, this should produce comparable groups – the problem is to prove beyond any reasonable doubt that there is no cheating (deliberate or not) involved. There is always a risk of bias when the physician knows which treatment the next patient is allocated to, because, depending on his prejudice, he can decide not to include a particular patient on a particular occasion, because he may not think this patient is appropriate for that particular treatment. Or the reverse.

One alternative to avoid this bias would be to divide the patients into two groups first, and then randomly allocate the treatments to the groups. In this way we ascertain balance between the two groups for known confounders by using some matching procedure. The actual randomization can be done by tossing a coin. There are two drawbacks with this: (1) it is not possible to match for all relevant confounders, in particular those unknown, and (2) there is no way to measure the relevant random error, and therefore to compute summaries like p-values and confidence intervals. These difficulties are overcome if we randomize individual patients instead, which we do by generating a list by some mechanism using random numbers in such a way that patients are allocated to all groups, right through the study. Such a list may look like

$$ABBAABABBAABAABB\ldots$$

When a patient presents himself, the physician must decide to randomize the patient before he learns which intervention he is randomized to. This removes the risk, mentioned above, that he, consciously or unconsciously, allocates one treatment to one particular type of patients and the other to another type of patients, or that a certain group of patients will prefer one treatment whilst another group will prefer the other.

The first clinical trial reported which was based on a randomized allocation of patients, was a study of the antibiotic streptomycin for the treatment of pulmonary tuberculosis, discussed in Box 3.3. However, the concept of randomization was controversial, and its subsequent spread was slow and not without opposition from clinicians. (The streptomycin trial is interesting from another point of view as well. In the trial the drug proved very effective, but when the drug later came to market it was much less effective, since the tubercle bacillus rapidly became resistent to the drug. This discrepancy between what is seen in clinical trials and in real life is an ever-present risk.)

One aspect of a proper randomization is that patients are allocated to all groups during the whole course of the study. This is usually accomplished by generating the randomization list in blocks of reasonably short length, in such a way that within a block there is an equal allocation to all groups (Box 3.4). Different blocks look different, so the whole list is still randomly generated. There is no problem in designing the study so that, for example, treatment A occur twice as often as treatment B. Another reason for using randomization in blocks is when we conduct a multi-center trial. If effects are different at different centers, we gain in efficiency in the analysis if we have (approximately) the same number of patients on each treatment within each center. We may therefore want to randomize each center separately. The design-driven analysis of such a study is as a stratified analysis, where center is the stratum, and we have the same problem with how to weight the response at different centers (according to precision or something else) as was briefly discussed in Section 2.6.

Randomization is done in order to minimize bias and ensure that the patients in each treatment group are as similar as possible with regard to all known and unknown confounders. Therefore any differences found between the groups in the outcome(s) of interest must be due to treatment, and not to other differences between the groups. This means that the groups should be equal at baseline in the sense that their true distributions are the same. It does not mean that there is no imbalance in a particular study, due to random fluctuations. But

Box 3.3 The first randomized clinical trial

Streptomycin is an antibiotic drug that was first isolated in 1943 and immediately showed striking results in inhibiting the tubercle bacilli both *in vitro* and in animal experiments. In order to investigate its usefulness to human pulmonary tuberculosis patients, a clinical study was designed in 1946. The results were published in 1948 in the *British Medical Journal*.

In this study, one group of patients was treated with bed rest alone and another with both streptomycin and bed rest. The patients were to remain in bed for at least 6 months, and the results were assessed at the end of that period. When the results were collated, they were striking. For example, mortality is summarized in the following table:

Treatment	Died	Survived
Streptomycin	4	51
Bed rest alone	14	38

Analysis of this table shows that the relative risk of death is only 0.27, with 95% confidence interval (0.08, 0.70), for the patients getting streptomycin compared to those not getting it ($p = 0.006$). Similar striking results were obtained from radiological judgement of clinical improvement, where 51% of the streptomycin patients improved as compared to 8% of the controls.

However, it is not this landmark result in the fight against a killer disease that makes this trial famous in the history of clinical trials. It is the way in which it was decided who was to be treated with bed rest and streptomycin and who was to be treated with bed rest alone:

Determination whether a patient would be treated by streptomycin and bed-rest (S case) or by bed-rest alone (C case) was made by reference to a statistical series based on random sampling numbers drawn up for each sex at each center by Professor Bradford Hill: the details of the series were unknown to any of the investigators or to the coordinator and were contained in a set of sealed envelopes, each bearing on the outside only the name of the hospital and a number. After acceptance of a patient by the panel, and before admission to the streptomycin centre, the appropriate numbered envelope was opened at the central office; the card inside told if the patient was to be an S or a C case, and this information was then given to the medical officer of the centre. Patients were not told before admission that they were to get special treatment.

The decision to make lottery decide who gets the drug was not an easy one, since patients were expected to receive the best available treatment. The winning argument at the time was that the drug was short in supply so that it was not accessible to everyone anyway.

This study was actually not the first one designed as a randomized study, but it was the first to be reported. Another study, on immunization for whooping cough, was designed earlier, but lasted longer and was reported later.

Box 3.4 The origin of randomized blocks

Randomized blocks were introduced by R.A. Fisher in 1926 when he worked at the Rothamsted agricultural research station of in Harpenden, England. When a trial was carried out in agriculture, fields were divided into plots. Sometimes the field would vary in fertility, and the plots were grouped into blocks of approximately equal fertility. Treatments were then randomized within blocks, so that they were equally well represented in each block. In agriculture we have the blocking structure first, and allocate treatments within these, whereas for clinical trials the treatment structure is defined first and the patients enter as the trial is run.

The agricultural way for a clinical trial would be to identify the patients before the trial and group them according to various prognostic factors into blocks, within which one would randomize. In reality one usually uses only a few factors to define the blocks, the most important of which is the center. If more than one prognostic factor is used for the blocks, we talk of a stratified randomization. We may, for example, want to have equal number of males and females in the treatment groups, which may be achieved by stratifying on sex. (Or equal number of treatments at each center.)

If we want to achieve what Fisher achieved in a clinical trial, we need a dynamic procedure that allocates patients as if we had them collected before the trial. Such methods, in which patients are allocated depending on how the study is balanced with respect to defined prognostic variables when that particular patient is randomized, do exist. They involve some automated telephone or web-based system for treatment allocation. Such methods have their proponents, but their value remain unclear since the analysis often can take care of the known prognostic factors that form the basis of such schemes.

From a Fisherian perspective, the basic reason for randomization is really that it provides us with an unbiased estimate of the variance; that it balances prognostic factors is a lucky side-effect. Methods that primarily aim to balance such factors have a tendency to underestimate the variance, which affects the validity of the statistical conclusions.

---

such differences, in contrast to systematic bias, are handled in the statistical analysis and expressed in the *p*-value and other confidence statements. However, for each particular study there will be some imbalance at baseline, and that imbalance may propagate to an effect in the *estimate* of the treatment effect. But this can be handled by the appropriate statistical analysis.

However, randomization does not guarantee that the result we obtain in the study applies to the whole target population. In fact, quite the contrary. In order for us to deduce that the results apply to the whole population we need to ensure that we have a representative sample. This is almost never the case in clinical trials. Not all patients want to participate in clinical studies, and the investigators may have their prejudices against some patients. They may consider some of them unreliable and unable to follow the specifications of the study protocol. Moreover, protocols typically put restrictions on which patients should be enrolled into the study. What the results of the randomized study actually tell us is that for the patients we have studied, if we randomly assign them to the groups, we should find a difference between the groups. The claim we can make is only about these patients, because what is random in the study is the allocation of the patients to the two groups. If we want to generalize to the whole population, and we do, we need to use other, non-statistical, arguments. But there is really no alternative method available at present.

Before we end this section, let us point out that it serves no purpose to perform hypothesis testing on baseline data between randomized groups. Irrespective of what p-value such a test provides, this is a situation in which we know the answer: the true distributions for the baseline data for the randomized groups are the same. We know this, so also if we get a small p-value, the only explanation available is that an unlikely event has happened. The groups will probably differ numerically, because there is no way we can divide a group randomly into two identical subgroups. However, the random noise that is introduced is accounted for in the statistical analysis. The practice of baseline testing is unfortunately in common use, and seems to be used as a way to judge the adequacy of a given allocation. This is wrong, since what the test addresses is the randomization process as such. A statistically significant result, if not a chance finding, would imply that the method of randomization is not appropriate. In most cases we trust our randomization process and this leaves us with the only alternative explanation: that a rare event has occurred. On the other hand, it may be proper to adjust for differences in background variables in the statistical analysis for reasons to be discussed in more detail later.

3.5 Information bias in a cohort study

Even when there is complete balance with respect to patient characteristics in the groups in a cohort study, there are still potential bias problems. The problem to which we now turn is whether the measurements are taken in the same way in the groups. The collective term for this type of bias is information bias, or ascertainment bias. It consists of a variety of different types of bias.

With an exposed group and a control group, and looking for an effect of the exposure, there is always a risk that we look for the effect with different degree of scrutiny in the two groups, referred to as surveillance bias. This occurs when individuals under frequent or close surveillance are more likely to have a disease diagnosed. For example, postmenopausal women taking estrogen may be more likely to have breast cancer diagnosed because they visit their physician more frequently. Another example is when an exposure, rather than causing the disease, causes symptoms that lead to a search for the disease, so that the disease is discovered earlier in that group. This means that a higher proportion of patients with disease would be found in the exposed group than in the control group, when in reality there is no such difference (similar to Berkson's paradox). An early report on postmenopausal estrogen replacement therapy and endometrial cancer was criticized on these grounds. It was suggested that subclinical cancers were being diagnosed more frequently in exposed women because estrogen use could cause patients with no other symptoms to bleed, which, in turn, led to more thorough investigations into the causes of this bleeding, with earlier detection of a cancer as a consequence.

A related example is stage migration with screening activities mentioned in the previous chapter (screening will also be discussed in more detail in Section 4.3.2). We may want to study the usefulness of screening for a particular cancer, by assessing whether it has led to an improved survival time. Survival time is defined as the time from diagnosis to death, and if the screening activity has led to earlier detection, this will increase – not because anyone lives longer, but because one has known about the cancer for a longer time. A comparison between a screened population and a reference population would therefore falsely conclude improved survival.

In experimental studies we try to avoid bias in the collection of data by informing neither the investigator nor the patient about what group the latter belongs to. This process is called *blinding*, and its usefulness is that it ensures that groups continue to be comparable, in terms of how measurements are made, after the allocation to treatment has been done. To what extent this can be done depends on the experimental setup, but it is standard procedure in drug development. It is more problematic if you want to compare an invasive treatment, such as surgery, to a non-invasive treatment, such as a drug treatment. For example, if we want to compare surgery to antibiotics alone as treatment for appendicitis.

Clinical studies are labeled according to what degree of blinding they have: open (no blinding), single-blind, double-blind, etc. Usually a double-blind study is one in which all handling of patient-specific data is done by personnel with no knowledge of which treatment a particular patient is given. This includes not only the obvious study measurement taken on the patient in the clinic, but also the interpretation of some data, for example reading X-rays or staging biopsy data. The treatment code is broken only after the data have been collected and verified. In a single-blind study one party, usually the patient, is not informed about the treatment allocation, whereas others, such as the physician and other study personnel, may know. The importance of doing investigations in a blinded manner have been demonstrated in cases where both a blinded and an unblinded doctor have performed the same investigation and where the latter tended to see treatment effects whereas the former did not.

The first to ask for a blinded experiment was probably Louis XVI, king of France, when he called for an investigation into Franz Mesmer's claims about the benefits of what he called animal magnetism (see Box 3.5). The king wanted to know whether the effects were due to a 'real' force or only an 'illusion of the mind'. Blindfolded people were told – sometimes misleadingly – either that they were or were not receiving the treatment. As it turned out, people who were tested only felt effects when they had been told that they were receiving the treatment.

But blinding may not help. It would not help with the endometrial cancer example above, since blinding would not remove the symptoms that lead to the search for the disease. A related situation occurs when the treatment affects how concomitant medications are taken, medication which in turn affect the outcome variable. As an example of this, assume we are doing a clinical study on pain reduction in patients with rheumatoid arthritis in which we want to compare no treatment (i.e., placebo) with a pain-reducing treatment. During the course of the study some patients may take another pain reliever, called a rescue medication, when they need to. If the outcome variable in the study is a pain score reported by the patient, we have two factors that affect the score: the product under investigation and the rescue medication. The difference we measure in the pain score between the two groups is dependent on how much rescue medication the patients used in the two groups. Other examples of studies with the same type of complication are studies in asthma, in which bronchodilators take the role of pain relievers. Carrying out such trials without access to the rescue medication would not be ethical, and probably not possible.

The question now becomes how we should handle data obtained after intake of rescue medication. Assume that our study is such that the patient is asked to record daily the degree of pain they experience. We want that pain score to be a marker of disease intensity, and therefore free of influence from any rescue medication. But the disease fluctuates in severity: on some days, the bad days, the patient will take a pain reliever, whereas on other

Box 3.5 About the word 'mesmerized'

The transfixed state in which one is unaware of one's surroundings, or *mesmerized*, was named for the Austrian physician Franz Mesmer. In 1774, at a time when magnetism became fashionable, he treated a young, mentally ill, woman with magnets (she swallowed a preparation containing iron, whereupon Mesmer attached magnets to various parts of her body). He then discovered he could do as well without the magnets, and came to the conclusion that there was some universal fluid between him and his patients. By manipulating that fluid, he launched the concept of *animal*, as opposed to mineral, *magnetism* and claimed to cure all diseases. After failing to cure the blindness of an 18-year-old woman in 1777, he was forced out of Vienna by the medical establishment, and moved to Paris. Here, his reputation preceding him, he was able to set up a lucrative practice at Place Vendôme. He also wrote a book about his method and ideas.

His claim for his method, which looked much like hypnosis, was that he enabled a free flow of vital energy through the body. He was so much in demand that he scaled things up by moving into group therapy. In the process he wanted the endorsement of the Faculté de Médicine and suggested in 1780 that the *effect* of his method should be tested in a fair test, in which patients were to be assigned to treatment groups by drawing lots. Had this study been run, it would have been the first randomized trial. However, the Faculté refused to participate.

A few years later, mesmerism was considered to be a threat, potentially deleterious to both mind and body. King Louis XVI was therefore prompted in 1784 to set up a Royal Commission, which included, among others, Benjamin Franklin and the chemist Lavoisier, to investigate the legitimacy of Mesmer's claim to cure. A series of experiments were conducted to determine whether he had discovered a new physical fluid. They found no such evidence and concluded that 'the practice of magnetism is the art of increasing the imagination by degrees'. As a result Mesmer became the subject of ridicule, left Paris and returned to Austria, having retired. He died in 1815, at the age of 85.

---

days, the good days, the rescue medication will not be needed. The number of good and bad days may differ between the groups, depending on the efficacy of the drug under investigation. What are we to do with the data that are influenced by rescue medication? If we ignore the scores from these bad days, that is, consider those measurements missing, we will only look at the pain scores on the patient's good days. Ignoring bad days must produce a bias. The alternative is also biased. If we ignore the fact that the patient has taken rescue medication and take the pain scores at face value, we will get the impression that he is in less pain than he actually is. In terms of a group comparison, if the placebo patients use more rescue medication, the true score difference will be biased toward no difference between the groups.

To address this problem we must come up with a scoring system that in some way uses both the pain score and the use of rescue medication. For rheumatoid arthritis there are such tools available. In asthma, however, it is about lung function and rescue medication, and it is less clear how to get a valid and clinically meaningful summary of these.

Box 3.6 Ethics and the use of placebo

The use of placebo in clinical research is controversial. In the 2000 version of the Declaration of Helsinki the first sentence in paragraph 29 reads: 'The benefits, risks, burdens and effectiveness of a new method should be tested against those of the best current prophylactic, diagnostic, and therapeutic methods', a statement amended only slightly from a version present in the Declaration since it was first adopted in 1964. Since this declaration sets the moral standard for clinical research, this statement has been a powerful argument against the use of pure placebo arms in clinical trials.

However, absolute efficacy of an intervention cannot be assessed without a proper comparison to no treatment (i.e., placebo) in a randomized, double-blind trial, and the above-mentioned paragraph is therefore to some extent at odds with the evidence-based medicine movement, which expects interventions to have proven their worth before being used on a large scale. This has led to the development of indirect methods to prove efficacy, in particular the non-inferiority argument discussed in Box 1.4. However, the US regulatory agency, the FDA, seems to have all but ignored this paragraph – or at least interpreted it to mean that placebo treatment is acceptable as long as it does not put patients at serious risk.

In the latest version (2008) of the Declaration of Helsinki the above statement has been modified in the direction of how FDA has operated on it. It now allows an exception to the rule: 'Where for compelling and scientifically sound methodological reasons the use of placebo is necessary to determine the efficacy or safety of an intervention and the patients who receive placebo or no treatment will not be subject to any risk of serious or irreversible harm'. To what extent this changed formulation will allow for more placebo controlled trials remains to be seen.

3.6   The study, or placebo, effect

It is a general rule that one should not compare post-treatment data to pre-treatment data alone, in order to evaluate the effect of a treatment in a single, treated, group. This is because in any experimental study one expects to see effects that are unrelated to the treatment. It is helpful in this context to make a distinction between an effect on the outcome of a disease and an effect on the experience of illness. Many serious diseases do not show symptoms for a long time, while feeling unwell is not always an expression of a disease. The outcome measures in clinical studies are often related to the symptoms and signs of the disease, especially when more objective measures are hard to find. Participation in a clinical study may make patients feel better without actually affecting the disease. The popular term for this effect is the placebo effect.

The word *placebo* is Latin for 'I will please'. Originally it referred to something a doctor would give a patient when asked for a remedy and none was available. It has been described as 'an epithet given to any medicine adapted more to please than to benefit the patient'. In other words, the purpose of the placebo was to make people to feel better simply because they expected to feel better. (When the patient's expectation is that he will be hurt by the intervention we talk about a *nocebo*, meaning 'I will harm').

What is involved with the placebo effect can be, and is, disputed.

- If you are medically inclined you may say that by taking care, you start physiological and psychological effects in the patient which improve the condition. Patient expectation may, for example, trigger the production of certain painkillers (endorphins, enkephalins) in the brain.

- If you are less medically, and instead more statistically, inclined you may say that what is observed is largely the natural course of the disease; that what is involved is the general phenomenon of *regression to the mean*, which we will discuss in more detail in Section 7.5. With a fluctuating condition, if you assess it when it is in a bad state, you expect it to improve. On the other hand, if you assess it in a good state, you expect a deterioration to follow. Patients more often seek medical advice when in the former state than when in the latter.

In a clinical study there is a protocol which defines which patients to include, and in many cases these inclusion criteria are such that patients are enrolled into the study when they are symptomatic. Once in the study, the study procedures and the interest the patients generate with the doctors and nurses may make them improve. As a consequence of participating in a study they may also, at least for a while, change their lifestyle in a way that has a beneficial effect on their underlying condition. Each study protocol therefore implies some kind of study effect. This is why it is usually not appropriate to assess the benefit of a particular treatment by comparing the pre-treatment state to the post-treatment state alone. It is, however, appropriate to compare the change in the outcome variable from pre-treatment to post-treatment between the two groups. But there must be something in the design that provides information on the non-treatment related effects of participation in the particular study. This is typically achieved by having a control group, and such studies are said to be controlled studies.

If we want to assess the absolute efficacy of a drug, the control arm should only contain a placebo treatment, where placebo refers to an inert substance, a 'dummy pill', that is intended to be indistinguishable from the active drug. To be fully successful it should be identical in all aspects, except for the active ingredient. A study which investigates the effect of a drug can often accomplish a placebo treatment, whereas other treatments such as counseling, physiotherapy, or acupuncture may well struggle to do so, with consequential uncertainty about the interpretation of the results.

That a study is placebo-controlled does not necessarily mean that there is a group that gets no treatment other than a sugar pill. What it means is that there is a group which receives a placebo corresponding to the drug given in the other treatment group. All patients in the study may be on some other background medication, either protocol specified or the treatment the patient had at enrollment. However, formally, the meaning of the drug's absolute effect is what the drug accomplishes compared to a situation when no drug is given. In the presence of a background medication in the study, what is measured is not the full effect of the drug, but the add-on effect over and above this background medication.

Blinding and placebo control can be more or less complicated, and it may be that the only way to accomplish this will in some ways compromise the ultimate purpose of the study. On occasion a drug developer may want to compare a formulation of a drug that can be given once daily to one that needs to be given twice daily. The main argument for why a once daily formulation may be better is that it may improve compliance, since the patient may miss fewer doses. However, if we design a blind placebo controlled study to study this,

---

Box 3.7 Quality scoring of clinical trials: the Jadad scale

The Jadad scale is a (not uncontroversial) procedure to independently assess the methodological quality of a clinical trial. The system was designed by the Colombian physician Alejandro Jadad-Bechara with the aim of assigning a score between 0 (very poor) and 5 (rigorous) to how well potential methodological errors have been take account of.

To form the Jadad score for a given paper one answers the following three questions on its content:

1. Was the study described as randomized? If yes, score 1 point. If, in addition, the method of randomization was described in the paper, subtract 1 point if the method was inappropriate, but add 1 point if it was appropriate.

2. Was the study described as double-blind? If yes, score 1 point. If, in addition, the method of blinding was described in the paper and it was appropriate you add 1 point; if it was inappropriate you subtract 1 point instead.

3. Was there a description of withdrawals and dropouts? If yes, score 1 point.

The result is a score between 0 and 5.

---

we need the patients randomized to the once daily arm to also receive a placebo at the other occasion when the twice daily arm should take medicine. That would take away (some of) the advantage of the once daily administration. One remedy for this would be to introduce two placebo groups, one with a once daily treatment and one with a twice daily treatment. The patients are then randomized to one of four treatments, but they are only pairwise blinded: the once daily to its placebo and the twice daily to its placebo. The comparison between the once daily and the twice daily administration then goes via comparison of active to its placebo. In this context we should also mention the double-dummy technique. If we want to compare two drugs, which come in forms that are distinguishable, for example tablets that look or taste differently, we need to produce a placebo for each of them. In order to be blinded each patient needs to take two tablets, one of which contains an active ingredient and the other does not.

In summary, the state-of-the-art clinical study design is the controlled, double-blind, randomized study. In fact, one way to assess the quality of a report of a clinical trial is to score how these key elements are described in the publication, as exemplified by the Jadad scale described in Box 3.7. Such scoring may be useful when one wants to perform meta-analysis of many small trials in order to provide 'evidence-based medical proof'. But they have many weaknesses as well; the Jadad scale says nothing about the quality of the conduct or the statistical analysis of the study. It is primarily about the design elements.

3.7   The curse of missing values

The state-of-the-art clinical study design is the controlled, double-blind, randomized study. But even the best-designed study will need to face reality, and in real life, as any military commander will tell you, the best laid plans will fail to some extent.

Ideally, all participants in a trial should follow the protocol, complete the study, and provide data at all time-points they are measured on all the outcomes of interest. In reality, we are likely to end up with some missing data in the data sets that form the basis for the statistical analysis. Data can be missing because patients withdraw early from the study for one reason or another, or because some assessments were simply missed, either deliberately or accidentally. Alternatively, measurements may have failed for some, possibly technical, reason. The actual reason why a particular piece of data is missing has consequences for what bias its omission may entail. In this section we will focus primarily on data that are missing because patients prematurely discontinued the study. This means that data are missing after the last observed data value. No other data are missing.

There are many reasons why patients may choose to withdraw from a study they have agreed to participate in. Some of the reasons may be unrelated to the treatment: the patient may have a new job in another town, or have been offered a last-minute trip to some holiday resort. It may even be that the patient dislikes the investigator. But more commonly, patients discontinue either because they experience some adverse event during the study, or because they do not feel the treatment is making them any better and therefore want to return to standard care. On occasion they may discontinue because they actually have recovered. Whatever the reason for discontinuing the study, the data the patient should have produced after the point of discontinuation are missing.

In most clinical studies the data analysis is done on the outcomes recorded at the end of the treatment period. Randomization guarantees that we have comparable groups when we start the experiment, but as patients withdraw the groups shrink to subgroups of the groups that started the study. This means that there has been a selection process operating on the patients during the course of the study. This selection mechanism can differ between the groups, as a result of which they may no longer be comparable at the end of the study – when we analyze the data.

To take a simple example, assume the new drug is effective in improving some symptoms of a disease, and that when patients feel that their condition is worsening, they tend to withdraw from the study. Assume that the disease we study varies in severity between patients, and that patients with severe symptoms are more likely to withdraw from the study. At the end of the study we then find an over-representation of patients with a more severe disease (as measured at baseline) in the active group than in the placebo group. The effect we measure may therefore be affected by selection bias. Perhaps we have comparable groups in terms of disease severity at the end of the treatment period, taking treatment effect into account, but that is not what we are interested in.

To illustrate this graphically, consider Figure 3.1 which is a simplified version of the discussion. There are two curves for the placebo group and two curves for the active group. One curve is dashed, one is solid. The two dashed curves show the true distributions of the response (change from baseline in some outcome variable) for the groups. The difference between the dashed curves is therefore the true effect of the drug, which corresponds to a one-unit horizontal shift.

The solid curves are what we actually see, given some assumptions that we now specify. We assume that, sometime during the course of the study, patients for whom the end result would have been a decrease of one unit experienced symptoms to such a degree that they decided to withdraw. As a result we have no end-of-treatment measurement for them. The response distributions for the patients who completed the study (solid curves) show what an analysis confined to completers will estimate. The solid curves differ less between the groups

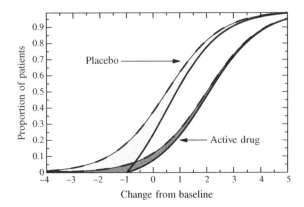

Figure 3.1   Illustration of how the distribution of an effect (dashed lines) changes because of differential drop-out (solid lines).

than the dashed curves do. In fact, the shaded areas illustrate the difference within group between all patients (dashed curves) and completers only (solid curves). This area is larger in the placebo group than in the active group, so the bias within group is larger in the placebo group. In other words, we overestimate the effect more in the placebo group, and the group comparison will therefore be biased toward no effect.

If there are differential withdrawal rates in the treatment groups as above, the final analysis will most likely be biased in some way. At least if a proportion of these discontinuations occurred because of lack of efficacy (or because of recovery). The differential withdrawal rate is in such a case an important finding in itself and a marker for the efficacy of the drug. It may be of interest in such a situation to compare baseline data on completers, in order to see how the randomization has been broken during the study.

When patients withdraw for reasons unrelated to the study (how we know this to be the case is another question) they do not introduce this kind of bias. We can see this from a data perspective: at randomization we have defined two groups for which we are to collect data. When we look at the groups at the end of treatment we find some of these data missing. If missing data occur in a completely random way, it does not introduce any bias, it only decreases the volume of data to analyze – and hence the power of the study. However, if there is a pattern in this, detectable or not, there is bias. We say that such missing data is informative, though we might not necessarily understand what it is trying to inform us about.

How do we handle the missing data in our analysis, in order to avoid the bias we expect from analyzing the subgroup of completers? We need to obtain data for all patients in the study, which may mean that we need to impute data where they are missing. There are two basic ways we can do this, each of which means that we need to redefine our objective (or claim).

1. The first is to continue to collect data also from patients who discontinue treatment with the product under investigation. This way we get data for all patients, but some data are not on the randomized treatment, and some data may be obtained with additional treatment. What we then measure at the end of the study is the response of patients whom we *intended to treat* for a specified period of time with the drug, but

permitting other treatments if it failed. This may mean that we are not doing the study we want, but instead one that permits some background therapy to be used under certain circumstances. Such background medication can be expected to narrow the efficacy 'gap' between an active drug and placebo.

2. The alternative is to redefine the effect variable (and therefore the objective of the study) so that we measure the effect at the end of treatment, whether it is at the end of the study or, for discontinued patients, at the time of the withdrawal. We then build into the protocol that if a patient discontinues the study, a final effect measurement should be taken. Alternatively, we take serial measurements during the course of the study, and use the last one obtained before discontinuation as the outcome measurement for such a patient.

The first of these suggestions is called the 'intention-to-treat' (ITT) approach to analysis in a pharmaco-statistical environment. As a measure of absolute effect it will introduce bias that is expected to operate in the no-effect direction. It is conservative in the sense that the bias will not increase the risk of spurious findings. It is, however, doubtful how relevant the principle is for a placebo-treated group, since it is never our intention to treat the patient with a placebo drug if there are useful alternatives available. Some people refer to the result of the ITT analysis as the *effectiveness* of the treatment, in contrast to what the second suggestion would measure, which is then called *efficacy*. Whether this is a helpful terminology or not is left to the reader to judge.

The alternative has the drawback that if we interpret the difference as the difference at the end of the study, we have actually made a data imputation for the subjects who discontinued the study in that we use the last observation recorded. Considered as an imputation method it is referred to as the 'last observation carried forward' (LOCF) approach. It is simple but makes the strong assumption that the effect is unchanged after the last observation was made, which is most often not a realistic assumption. If we assume that withdrawn patients should continue to get worse, it is reasonable to assume that the true effect is underestimated. It may therefore be useful for many trials, but not for non-inferiority trials, or equivalence trials, in which we look for short confidence intervals around a zero mean difference. For such studies the bias of this imputation method is not conservative.

Example 3.1 Figure 3.2 shows the means of FEV_1 in a crossover study in which a bronchodilator treatment was compared to a placebo treatment. FEV_1 was followed for 26 hours after a single dose administration and the solid curve connects the observed means for the two groups (active above, placebo below). The dots show occasions when a patient took rescue medication (a bronchodilator). As expected, there is more rescue medication taken on placebo treatment than on active treatment.

The placebo curve is a reflection of what the diurnal variation in lung function looks like, intertwined with the effect of the study procedures, such as food intake and sleeping pattern. The effect of the bronchodilator must be assessed by comparison with the placebo treatment. In this case the objective was to see if we can still claim an effect after 24 hours for the drug. The solid curves represent the data as provided, which means that for some patients some of these measurement are influenced not only by the product under investigation, but also by rescue medication. Since there is more rescue taken when on placebo, we expect the assessment at 24 hours to be biased in favor of placebo.

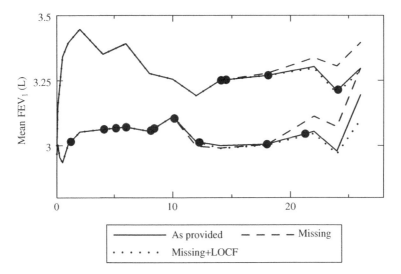

Figure 3.2   The effect of different methods of handling lung function measurements that are influenced by earlier rescue medication. For details, see Example 3.1.

However, it is much less biased than to argue that data after the first intake of rescue should be considered invalid. This would imply that after intake of rescue medication we consider all measurements to be missing and compute the means on what remains; a diminishing number of subjects when we consider later and later time points. This approach is illustrated as the dashed curves in the graph, and we see how they increase at the end of the observation period, when means are computed on a selection of patients. An alternative strategy keeps groups intact by imputing values for patients who took rescue medication. If we do this using the LOCF principle, we get the dotted curves in the graph. It is a non-trivial task to decide which result most accurately describes what it is we want to know: the dashed lines are affected by selection bias, the solid lines by concomitant medication, and the dotted lines by assuming no further change after first rescue medication.

There are other imputation methods for missing data available, one of which is based on applying a linear mixed effects model. We will not discuss such methods here, but wish to mention an old alternative by Larry Gould from 1980, which assumes that we are doing a non-parametric statistical analysis. In non-parametric statistics we rank observations, and Gould's suggestion is to rank missing data according to why they are missing:

- First, remove all discontinued patients from the data set if the reason for withdrawal is unrelated to the effect of the drug.

- Then rank the remaining discontinued patients according to how long they stayed in the study.

- Finally, rank the completers according to their values, but give them all higher ranks than the discontinued patients.

The drawback with this is that, although it provides us with a way to carry out hypothesis testing, it is less helpful in estimating treatment effects (in terms of location measures).

3.8   Approaches to data analysis: avoiding self-inflicted bias

Not only may missing data provide us with biased results, but there are many instances when the analyst manages to introduce bias all by himself, by manipulating collected data in such a way that bias is introduced. The overriding reason why people do this is an obsession with analyzing only valid data. However, if you do not have valid data, the alternative to using non-valid data is to consider such data missing, and we already know that missing data have effects on the comparability of the groups. The following example illustrates this.

Example 3.2   One part of the investigation into the pharmacokinetic properties of a drug is to compute its clearance, which is proportional to the inverse of the area under the curve (AUC) of the plasma concentration profile (provided we control the dose that enters the bloodstream, which we can do by giving the drug intravenously). To estimate this area, we take serial measurements of plasma concentration for some time, say 24 hours, and use a mathematical formula, such as the trapeze formula. This is illustrated in Figure 3.3 as the light gray area. However, we also need to estimate the residual area, the area under the curve from the last measurement onwards, all along to infinity. This we do by fitting a mono-exponential function to the concentration data during the later part of the 24-hour surveillance period, and assume that this provides a good approximation to the plasma concentration thereafter. We compute the residual area using this function (dark gray in the graph) and add it to the previous AUC, in order to get the total AUC.

If the residual area is a large fraction of the total area, many pharmacokineticists consider the estimate of the area under the curve to be poor, and therefore not a valid estimate. For argument's sake, assume they set the limit at 30%. If, in a study with 12 subjects, four have

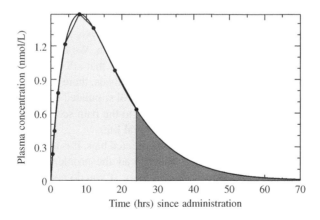

Figure 3.3   How to estimate areas in pharmacokinetics. The darker shaded area is estimated from a mono-exponential function whose rate is determined toward the end of the period that defines the lighter shaded area.

their residual area larger than 30%, how do we handle these? Some take the view that they should be ignored. However, the consequence of that is most likely bias. A large residual area can be the consequence of a high exposure, which means that you discard from the analysis patients with large AUCs, underestimating the true mean AUC, and subsequently overestimating the clearance.

If you instead use these non-valid measurements (i.e., poorly estimated AUCs), you include data that are less precise, but there is no inherent bias. This means that your group averages should be unbiased. The effect of the poor estimates of some individual AUC values is a more uncertain estimate of the mean, but not bias.

As the example tells us, it may be better to use data without the proper quality stamp, because the alternative is to consider such data as missing, and missingness has its own problems with bias. Of course, there may be situations when this is not the case, but when a measurement is of such poor quality, it should have been discarded from the start and never entered into the database. If you measure blood pressure to be 12/8 mmHg in an alive subject, this measurement is better considered due to a faulty machine (or data entry) and should be considered missing from the database. But subjective judgement of data quality should be avoided as far as possible; only cases beyond any reasonable doubt should be rejected.

A common feature of many measurements that statisticians and others want to consider non-valid, is that they are measured in a situation when there is poor compliance. Compliance is the term used for how well the patient, or investigator, follows the protocol that defines the study. It is about taking the drug as prescribed and about taking the measurements as stipulated by the protocol. Deviance from the protocol implies loss of compliance, and it leads to questions about the validity of data (and the results).

The reluctance to analyze data that do not have a proper quality stamp has led statisticians in the pharmaceutical industry to introduce the concept of the per protocol (PP) analysis. This approach to the analysis of data means that one first examines the validity of the data, and the statistical analysis is done only on the subset of what is considered valid data. There is a serious risk of selection bias in this procedure if appropriate care is not taken to avoid it. In an intervention study there is no problem in selecting only a subset of patients to analyze, if the selection is made only on data obtained before randomization. Such selection will not introduce bias, but will decrease the power of the study, since fewer subjects are included in the analysis. The serious problem with PP analysis is when the selection of data is based also on information obtained after randomization. We have already discussed one such problem, namely how to handle rescue medication that affects the outcome measures. In that case, whichever way we choose to make the analysis, there is probably bias. This will be an example of a compliance problem, if the protocol stipulated that the rescue medication should not be taken in a certain time window prior to the pain scoring. This illustrates how hard it may be to follow a protocol to the letter in real life.

The following example is a case story of self-inflicted bias. It is not from a pharmaceutical company study, and thus illustrates how widespread the problem is. It can serve as an illustration of how the best of intentions may go awry, if you do not carefully consider what the procedure you implement would lead to in the absence of any effect of the intervention.

Example 3.3   In some countries, including Sweden, there is routine ultrasound scanning during pregnancy in order to ensure that it is progressing well and that the baby does not have any serious malformation. To address a concern that ultrasound may damage human tissue, in particular brain tissue, a randomized clinical study was designed in which the prevalence of left-handedness (actually non-right-handedness) among boys was studied. The idea was that in some of these children, the cause of the left-handedness would be damage to the left brain brought about by the scan.

The study was performed in 1985–1987 and invited all women who booked for antenatal care in 19 clinics in Sweden to participate. In all 4997 women were randomized into two groups of which one was to have a routine ultrasound scan about 15 weeks into the pregnancy, and the other group was not. A decade later the mothers were sent a questionnaire asking about the handedness of their child. Based on this the children were classified as right-handed or left-handed. Because of a number of factors (miscarriages, failure to fill in the form, etc.) only a subset of all mothers who were randomized actually provided data for analysis. The table for boys only was as follows:

	Right-handed	Non-right-handed
Screened group	636	156
Non-screened group	648	134

The odds ratio for this table is 1.19 with 95% confidence interval (0.92, 1.53) and the p-value is 0.19. There is therefore not sufficient evidence to conclude that an ultrasound scan affects the handedness of the boy.

However, the investigators noted the following. In the non-screened group there were 103 cases where an ultrasound examination had actually been done before week 19 of the pregnancy. Out of these, 30 were included in the table above. These data are obviously invalid in some sense. How should that be handled? If we move them to the other group, and relabel the groups to capture the difference in tables, we get the following table:

	Right-handed	Non-right-handed
Exposed group	655	167
Non-exposed group	629	123

For this table the odds ratio is estimated at 1.30 with 95% confidence interval (1.01, 1.69) and the p-value is 0.04. It therefore appears that we have sufficient evidence to claim that the true odds ratio is larger than one. But is this a basis for claiming that ultrasound screening affects the handedness of the baby? Those who answer yes to this question argue that if we want to draw the conclusion that exposure to something has an effect, we must compare the exposed group with the non-exposed group. Elementary, but is it correct?

The analysis comprises 1284 randomized women and within this group of pregnancies there will be a subgroup for whom there is some kind of problem with the pregnancy such as growth retardation, which medically indicates a closer look, which in most cases means an ultrasound scan. This group of risk pregnancies is expected to be evenly distributed between the two original groups, because of the randomization. When we construct our new groups we move some of these risk pregnancies from the non-screened group to the screened group. The two groups are therefore no longer comparable since the risk pregnancies are not equally distributed between them. It was to ensure an even distribution of the risk pregnancies that we performed the randomization in the first place, but now we have broken it.

In short, by comparing exposed to non-exposed instead of randomized groups, we also expect to find a difference if ultrasound scanning is completely harmless. But, given this, why do we not deal with the 30 boys by just removing them from the analysis? Would that not help? No, it would not, because the basic imbalance that we have between the exposed and non-exposed groups will prevail, though only to half the magnitude.

We next consider a related example, applicable to drug intervention studies. In an open clinical study the patients may not accept the randomized treatment. They may simply refuse it, and demand the alternative. If the investigator yields, because he wants to fill his quota of patients, this will mean that some patients who were randomized to one treatment will actually receive the alternative. How should we handle this in the analysis?

1. If we analyze the data as per the treatment the patients actually received, we have no method to ensure comparable groups at the start of treatment, and whatever conclusion we draw about the treatment could therefore be open to debate.

2. We will get comparable groups at baseline if we instead analyze the data according to the treatment the patient was randomized to. That is hard to swallow for many because then the B group will contain a number of patients treated by A, and why should we want to compare those with other patients who are also treated with A? The answer is, in order to ensure group comparability on all confounders. The price is that we must change the formulation of the claim somewhat. Instead of saying something about a comparison of treatments A and B, you must say something about what happens when you *intend* to treat with A and B, respectively. Which brings us back to the intention-to-treat concept, which may be particularly relevant in these situations.

Another problem related to the present discussion is that of outliers. Outliers are observations that have a different pattern from the majority of the observations in the data set. In its simplest form we may have a single observation which is very large when the majority of observations are confined to another region of the measurement scale. Such data cannot be ignored, but they do pose a problem for many statistical procedures, in particular those that are based on means and standard deviations, since these parameters are rather sensitive to outliers. This fact is often used to promote non-parametric methods. The question is what problem that solves, since a major purpose of a study often is to quantify effects. If there are outliers, the first thing one should ask is what is the appropriate scale for measurement. Perhaps we should log-transform the data (see Box 6.3). Also it is important to note that outliers may try to tell us something. New discoveries are sometimes the result of unexpected observations in data. But that does not take away the basic problem that they sometimes make statistical analysis less precise. It is a judgement call how to handle outlier data, but it is important not to hide them. If it can be proved that they are faulty measurements, they can be ignored, but that presupposes one can have agreed standards for such conclusions. Ultimately it becomes a question of credibility.

3.9   On meta-analysis and publication bias

When a medical issue has been addressed by a number of independent trials it is often appropriate to do an overview of the knowledge obtained so far on the issue. Such overviews can be made in different ways. One obvious way is to ask an expert to write down his interpretation of the collective data, to carry out a systematic review of the literature, discussing the merits of different trials and the consistency of the conclusions. Such reports are often referred to as 'expert reports'. It is natural that if the author is biased, this will show itself in the report. An alternative way to compile overall knowledge from a series of individual studies is to subject them to a meta-analysis. It is not the purpose of this section to discuss the many controversial aspects of meta-analysis, but the reason why we want them performed is related to the concept of evidence-based medicine. This leads us to a discussion of a particular form of bias that occurs when we perform an analysis on published data only.

The Cochrane Collaboration, an international non-profit organization of academics, is considered by some to be one of the more important institutional innovations that took place late in the twentieth century. Its main task is to produce meta-analyses of published and peer-reviewed literature on health-care research. The logo of the Cochrane Collaboration features a simplified graph of a particular meta-analysis. This analysis looked at the usefulness of giving inexpensive corticosteroids to mothers about to give birth prematurely in order to reduce the risk of their babies dying from complications due to immaturity. This medical issue had been investigated in seven small trials in New Zealand in the period 1972–1981. Two of them indicated some benefit from the steroids, but the remaining five failed to do so. A meta-

THE COCHRANE COLLABORATION®

analysis of these trials was then performed in 1989, from which the medical community deduced that there was compelling evidence that steroids did reduce the risk to the prematurely born baby. Two years later seven more trials had been entered, strengthening the evidence even further. In short, stories of this type are what defines evidence-based medicine and justifies that the Cochrane Collaboration (and others) producing meta-analysis on an industrialized scale. The argument is that tens of thousands of children had suffered and died, even when there was enough information available to know what would save them, information that had not been synthesized together.

However, systematic reviews of the literature with subsequent meta-analysis is an industry not without its problems. The basic problem is the same as the problem that meets the researcher who wants to find out what effect a particular treatment may have on a specific biomarker, scans the literature and finds one publication from a small study with the positive effect he was looking for. Encouraged, he then repeats the experiment and gets a disappointing result. In a meta-analytic context, consider the two examples of treating acute myocardial infarction with streptokinase (a protein produced by streptococcal bacteria) and with intravenous magnesium, discussed in an editorial in the *British Medical Journal*. For streptokinase the collective evidence from 15 trials was in 1977 considered sufficient for the medical community to conduct two mega-trials (named GISSI-1 and ISIS-2), which subsequently confirmed the benefit of the drug. Today streptokinase is used as an effective and inexpensive clot-dissolving medication in some cases of myocardial infarction. The story with magnesium is different. In this case a meta-analysis reported in 1993 found the collective evidence of seven trials strong enough to argue that magnesium treatment represented an 'effective, safe, simple and inexpensive' intervention that should be introduced into clinical practice without further delay. When this was put to the test in a mega-trial (called ISIS-4), magnesium appeared completely ineffective as a treatment for myocardial infarction.

The problem here is one faced by anyone scanning the literature for evidence: publication bias. In order to understand what publication bias may mean, consider the following simple situation. Suppose that a new intervention is developed, which is studied at a number of different medical centers. Each of these wants to publish their result (the ultimate drive for a researcher is after all to get as much published as possible), but only those will succeed whose investigation shows that the new method is better than the old one. If you do not get a statistically significant improvement, the journals do not accept you paper (so you may not even bother to write it up), we assume. What will the world look like from a systematic reviewer's point of view? The answer should be obvious: he does not see the full distribution of results, but only a conditional distribution, based on those studies in which the observed effect size is large enough to produce a statistically significant result. It is left to those readers who are sufficiently interested in mathematics to quantify the magnitude of the bias under different assumptions. Such a calculation shows that when there is no effect, the bias is proportional to $1/\sqrt{n}$ and is therefore only slowly decreasing as we increase the size n of the trial.

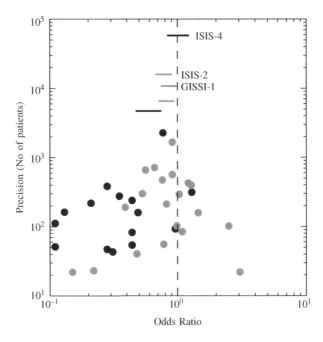

Figure 3.4 Funnel plots illustrating publication bias for two different treatments after myocardial infarction. The black dots illustrate published studies on intravenous magnesium and the gray dots studies on streptokinase. Also shown are the corresponding meta-analyses (unlabeled lines) and the result of large, randomized studies (lines with study labels).

The lesson from all this is that small studies that get published have large treatment effects, whereas large studies do not require effect sizes of similar magnitude. This observation allows us a way to actually assess whether there is a serious risk of publication bias. In fact, if we plot the effect size versus its precision (i.e., the inverse of the standard error, or, more simply, the size of the trial) for each trial, we get something that is called a funnel plot. If there is no publication bias this plot should resemble a symmetric funnel with the results of smaller studies being more widely scattered than those of larger studies. Figure 3.4 shows the funnel plots for the magnesium and streptokinase trials that appeared before the relevant mega-trials, with the mega-trials added. The plot for streptokinase is reasonably symmetric, whereas this is clearly not the case for magnesium. In fact, there is a gap indicating the absence of negative small trials.

This illustrates the non-trivial nature of bias. Any one of the studies included in the meta-analysis is, for all we know, unbiased. It is our sample from the population of unbiased trials that produces the bias, since we rely on published data. The situation is different for a pharmaceutical company that wants to perform a meta-analysis of its studies, since it will not miss any. A meta-analysis of published data is therefore a non-trivial exercise, in which a detailed description in a protocol is needed before the exercise, describing how the studies selected for inclusion were looked for and handled.

3.10   Comments and further reading

This chapter is about issues of bias in clinical studies. Many of these issues are of particular importance in epidemiological research, where they cannot really be avoided by design, but must be explicitly addressed in a discussion about the results of a study. These bias issues are therefore extensively discussed in most books on epidemiology. However, bias is not a non-issue for experimental studies because of missing values and concerns about valid measurements. In this chapter we have looked at the problem from the bird's-eye perspective of following a study population from start to end, to ensure that the game is fair not only at the start, but all the way down to the analysis.

Berkson's fallacy was a theoretical construction when it appeared in 1946, and not seriously considered by most epidemiologists until its real-life relevance was verified empirically (Roberts et al., 1978). Concerning randomization as a guarantee for valid inference from clinical trials there is still some controversy, as discussed by Senn (1994, 2004). The history of the introduction of randomization into clinical research was taken from an educational overview (Doll, 1998) in an issue of the *British Medical Journal* which also reproduces the original article (Medical Research Council Streptomycin in Turberculosis Trials Committee, 1948) on the streptomycin trial. The data in Box 3.2 were taken from Comstock (1999).

Missing data are a part of statistical everyday life. The whole subject of missing values is complex and is discussed in a number of different articles and books; see Senn (2007,Chapter 11) for an overview of the key issues and further references. Our discussion has focused on explaining why they are problematic, and also to explain what the simplest remedy, the LOCF principle, really amounts to. This method is mainly advocated because of its simplicity but is not always appropriate, the classic example being when the effect of a drug is to arrest a deterioration. A much advocated alternative (Mallinckrodt et al., 2008) builds on analyzing longitudinal data in a multivariate way, using a mixed effects model for repeated data. There are also more sophisticated imputation methods (Schafer, 1997), which take covariate information into account, all of which are outside the scope of this book.

Of particular interest in drug development are the different populations that are often discussed in study reports from the pharmaceutical industry, the ITT and PP populations. These are not the only populations that can be discussed in such a context (Gillings and Koch, 1991). It is common jargon to refer to these statistical approaches as the analysis of different populations. This is wrong; there is only one population, the study population. What we have are different approaches to the analysis of available data. It is often claimed that the ITT and PP approaches answer different questions. The ITT is claimed to answer the real-life question of what the total effect of the treatment is, allowing for inadequate compliance with patients not taking the drug. This is also wrong; clinical trials are much more controlled than the real-life situation when the drug is licensed and the results obtained this way have no such relevance. The PP analysis, on the other hand, is claimed to answer the question about the true effect of the drug with full compliance. Hopefully the main text has explained why this is wrong: to justify the claim we must assume that compliance is totally independent of effect. Which seldom seems to be reasonable. I beg the forgiveness of Kieler et al. (1997) for using their study the way I do, but it serves my purpose much better than any other example I know of.

The *British Medical Journal* editorial on meta-analysis is Egger and Smith (1995). Meta-analyses are mentioned now and then in this book, but nowhere discussed in a holistic manner. Usually the term is used when different studies are combined, but there is no essential difference between doing that and analyzing, for example, a multi-center study (randomized within center). The statistical methods are similar (Senn, 2000), their differences lie in the ability to get control of all the data. The particular problem of publication bias should not be used as an argument for not performing systematic reviews, but it is important to discuss potential consequences of it. It may also be possible to use available data to model the amount of 'missing' publications (Sutton et al., 2000). For an overview, with many references, of recent developments in meta-analysis, see Sutton and Higgins (2008).

Finally, the Jadad score was introduced in Jadad et al. (1996) and Larry Gould's suggestion on how to handle missing data, mentioned on page 74, is found in Gould (1980).

References

Comstock, G.W. (1999) Snippets from the past: 70 years ago in the journal. *American Journal of Epidemiology*, **150**(2), 1263–1265.

Doll, R. (1998) Controlled trials: the 1948 watershed. *British Medical Journal*, **317**, 1217–1220.

Egger, M. and Smith, G.D. (1995) Misleading meta-analysis. *British Medical Journal*, **310**, 752–754.

Gillings, D. and Koch, G. (1991) The application of the principle of intention-to-treat to the analysis of clinical trials. *Drug Information Journal*, **25**, 411–424.

Gould, L. (1980) A new approach to the analysis of clinical drug trials with withdrawals. *Biometrics*, **36**, 721–727.

Jadad, A.R., Moore, R.A., Carroll, D., Jenkinson, C., Reynolds, D.J., Gavaghan, D.J. and McQuay, H.J. (1996) Assessing the quality of reports of randomized clinical trials: Is blinding necessary?. *Controlled Clinical Trials*, **17**(1), 1–12.

Kieler, H., Axelsson, O., Haglund, B., Nilsson, S. and Salvesen, K. (1997) Routine ultrasound screening in pregnancy and the children's subsequent handedness. *Early Human Development*, **50**(2), 233–245.

Mallinckrodt, C.H., Lane, P.W., Schnell, D., Peng, Y. and Mascuso, J.P. (2008) Recommendations for the primary analysis of continuous endpoints in longitudinal clinical trials. *Drug Information Journal*, **42**, 303–319.

Medical Research Council Streptomycin in Turberculosis Trials Committee (1948) Streptomycin treatment of pulmonary tuberculosis. *British Medical Journal*, **ii**, 769–782.

Roberts, R.S., Spitzer, W.O., Delmore, T. and Sackett, D.L. (1978) An empirical demonstration of Berkson's bias. *Journal of Chronic Diseases*, **31**, 119–128.

Schafer, J.L. (1997) *Analysis of Incomplete Multivariate Data,* vol. 72 of *Monographs on Statistics and Applied Probability*. London: Chapman & Hall.

Senn, S. (1994) Fisher's game with the devil. *Statistics in Medicine*, **13**(3), 217–230.

Senn, S. (2000) The many modes of meta. *Drug Information Journal*, **34**(2), 535–549.

Senn, S. (2004) Added values. Controversies concerning randomization and additivity in clinical trials. *Statistics in Medicine*, **23**, 3729–3753.

Senn, S. (2007) *Statistical Issues in Drug Development*. Chichester: John Wiley & Sons, Ltd.

Sutton, A.J. and Higgins, J.P.T. (2008) Recent developments in meta-analysis. *Statistics in Medicine*, **27**(5), 625–650.

Sutton, A.J., Song, F., Gilbody, S.M. and Abrams, K.R. (2000) Modelling publication bias in meta-analysis: a review. *Statistical Methods in Medical Research*, **9**(5), 421–445.

4

The anatomy of a statistical test

4.1 Introduction

This chapter is an introduction to the nature of statistical tests, the tests that produce the p-values discussed in Chapter 1. Statistical tests are not much different from other kinds of tests, except that they involve numbers instead of, say, chemicals. In this chapter, in order to take away some of the mysticism around them, we will first discuss how they are related to tests that are not seen with similar skepticism, tests that are used in order to make a proper medical diagnosis.

After having explored such analogies, we will take a closer look at how statistical tests work. In this chapter we will consider special cases and defer a more general discussion to the next chapter. There are two aspects of (inferential) statistics: hypothesis testing and parameter estimation. We will discuss the former in an example and the latter in the simple case of estimating a binomial parameter. In the latter case we also take the opportunity to compare the two most important approaches to statistics, that of the frequentist and that of the Bayesian. Recall that these approaches are in some way reflections of the two philosophical approaches to science, falsification and induction, respectively.

Finally, we will introduce the best-known distribution in biostatistics, the Gaussian distribution. We do this from a historical perspective, with the emphasis on trying to explain what makes it so important: the central limit theorem in probability theory.

4.2 Statistical tests, medical diagnosis and Roman law

There is a rule in soccer called the offside rule. When such a situation arises in a game, the referee has to decide whether what he saw was offside or not. The rule is clear enough that in each situation the player is either offside or not, so there exists a factual state of offside or not offside. However, the referee can only use his senses and assistants on the lines to make a judgement as to whether it was offside or not. His conclusion from this information is his observation, and this can be right or wrong. It can be wrong in two ways:

Understanding Biostatistics, First Edition. Anders Källén.
© 2011 John Wiley & Sons, Ltd. Published 2011 by John Wiley & Sons, Ltd.

he can rule for offside when there was no offside, or he may fail to do so when in fact there was an offside.

A similar problem arises whenever we make decisions based on information. This includes the case where a physician has to decide on a diagnosis based on various signs and symptoms. Whether the patient has a particular disease or not is one thing, whether the physician finds it is another. The physician can make the same kind of errors as the referee above: he may miss the disease or he may say that the patient has it when he does not.

Decision making based on statistical tests is subject to the same problems as decision making in general: you may get it wrong. However, the fundamental and key aspect of statistical testing, which is the reason why it is so important in medical research, is that statistics tries to quantify the risk of such errors.

In this section we will explore the analogy between statistical testing and medical diagnostics, and also what these have in common with the judicial problem of finding someone guilty of a crime. To simplify the description of the statistical test, we often assume that it is used to compare a new drug to placebo, in order to prove efficacy of the former. There is nothing restrictive about this case, any other particular test would do, but being specific sometimes clarifies the discussion.

All these situations, statistical testing, medical diagnostics and courts of law, are part of a more general problem, namely the signal detection problem: a signal is either present or absent, and we want to determine which it is. The signals in our cases are as follows:

Medical diagnosis. A particular disease, which the patient may have or not have.

Justice system. The guilt of the person accused, which is absent if he is innocent.

Statistical hypothesis. The drug is effective (in general, the null hypothesis is false), or it has no effect.

The test we employ for signal detection has built-in uncertainties, so we cannot with 100% certainty say whether the signal was present or not from the test. The test therefore gives an answer which may, or may not, be correct. We can summarize this in a 2×2 table:

		Signal	
		Present	Absent
Test	Yes	Hit	False alarm (Type I error)
	No	Miss (Type II error)	Correct rejection

An alternative way of describing the cell entries is according to whether they are true or false, and positive or negative. With such a description a hit corresponds to a true positive, a correct rejection to true negative, whereas a Type I error is a false positive and a Type II error a false negative outcome.

All test procedures start with the assumption that there is no signal, and gather evidence of a signal. The final judgement call is to rule whether sufficient evidence has been collected so that we can decide that there is a signal. The whole setup is such that the effort is directed toward rejecting the assumption of no signal (you are innocent until proven guilty). The tests in our three situations are as follows:

Medical diagnosis. The physician carries out one or more examinations based on which a conclusion is to be drawn about the patient's condition. This can be as simple as a single laboratory test, or it can be a subjective interpretation of a multitude of information from symptoms, laboratory tests, CT scans, etc.

Justice system. The court (or a jury) weighs the information that has been presented by the prosecutor and the defense.

Statistical hypothesis. A statistical test is performed by calculating the probability of an observation as extreme as what we found, under the assumption that the null hypothesis of no signal is true.

All of these tests need a standard of judgement, which corresponds to a cut-off point, which is such that if our observation lies on one side of it, we consider the test to be positive, and if it lies on the other side of it, we consider it to be negative. In complex medical diagnosis settings, as well as for the courtroom, this standard of judgement may not be easily quantifiable, but that does not really change the picture. The cut-off point determines the properties of the test and the ultimate objective of its selection is to minimize the size of the Type I and Type II errors:

Medical diagnosis. The Type I error is that the physician tells the patient that he has the disease, when in fact he does not have it. The Type II error is that the physician declares the patient healthy when he actually has the disease.

Justice system. The Type I error is that an innocent person is sentenced, whereas the Type II error is that a guilty person is set free.

Statistical hypothesis. The Type I error is that we declare a drug effective when it is useless (in general, that we reject the null hypothesis when it is true). The Type II error is that we fail to show that the drug is effective, when in fact it is (in general, that we do not reject the null hypothesis despite it being wrong).[1]

However, we minimize one type of error at the expense of the other. For example, if we want to make absolutely certain that an innocent person does not go to jail, we may need a standard of judgement in the court such that a fair number of true criminals go free as well – the benefit of the doubt. If we want to make certain that all criminals are convicted, we need to set standards so low that some innocent people will also be put in jail.

4.3 The risks with medical diagnosis

4.3.1 Medical diagnosis based on a single test

To understand the risks associated with making a medical diagnosis from a patient's signs and symptoms, we use a test for alcoholism, much discussed in the 1980s. It uses a particular enzyme, gamma-glutamyl transpeptidase (GT), which exhibits elevated activity in serum in

---

[1]This is actually where the confusion between the Fisher and Neyman–Pearson approaches comes into play. In the pure Neyman–Pearson setting one has a choice between a null hypothesis and an alternative hypothesis, and should choose the one that the data point to. That means that if we fail to reject the null hypothesis, we should accept it. However, that is a decision-theoretic construction, which is not necessarily relevant to science.

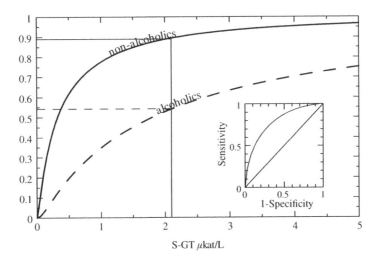

Figure 4.1 The main graph shows the CDFs for serum GT among alcoholics and non-alcoholics, respectively (hypothetical data). The inserted graph shows the CDFs versus each other, which defines the ROC curve (see text).

liver diseases. In fact, serum GT increases are mainly associated with certain types of drug treatment and with excessive intake of alcohol. Before we proceed, let us recall that a laboratory measurement is accompanied by a reference range. This range is assumed to be constructed so that 95% of observations from healthy subjects are within its limits. A value within this range may then be considered a normal value, whereas a value outside would indicate an unlikely observation in a healthy subject. This defines a test for alcoholism, for which we have control over the Type I error. If the reference range is 0–3.5 μkat/L, with this logic a value above 3.5 μkat/L would support a suspicion of alcoholism.

The serum GT reference limits correspond to a one-sided test (since negative measurements are not possible); only high values matter. For other laboratory measurements the reference limits allow for both small and large values to be abnormal. Such a range defines a two-sided test. This difference between one-sided and two-sided tests is important in the context of statistical tests.

Using the reference range as a test for alcoholism puts all focus on the Type I error. In order to see if a better test can be constructed, we need an understanding of the distributions of serum GT values among alcoholics and non-alcoholics, respectively. Tentative such distributions are shown (not based on real data) in Figure 4.1 as CDFs (see page 42).

When we design a test for alcoholism based on the level of serum GT, we decide on a cut-off limit (standard of judgement) such that if the value is above this limit, we diagnose the subject as an alcoholic. If we choose the cut-off limit to be 2.1 μkat/L, Figure 4.1 shows that 90% of the reference population has a serum GT value less than this value, whereas 45% of the alcoholics have a serum GT value that is higher than this value. The test defined by this cut-off limit therefore has specificity 0.90 and sensitivity 0.45 according to the following definitions:

Specificity. This is the probability that the test is negative when there is no signal. Thus 1 minus specificity is the magnitude of the Type I error.

Sensitivity. This is the probability that the test is positive when the signal is present. Thus 1 minus sensitivity is the magnitude of the Type II error.

We can compare this to the situation where we use the reference limits as our guide, using the cut-off limit 3.5 μkat/L. It has specificity 0.95 (because of its construction) and sensitivity 0.33 (read off from the dashed curve). This test therefore misses quite a few alcoholics.

For a more general discussion, let $F(x)$ denote the distribution function for the healthy population and $G(x)$ the distribution function for the population of alcoholics. If we define our test by using the cut-off limit x, we have that $F(x)$ is the specificity, and $1 - G(x)$ is the sensitivity. If we plot the sensitivity as a function of 1 minus specificity, $F(x)$ versus $G(x)$ for different x, we get the curve shown in the inset graph of Figure 4.1, which is called the receiver operating characteristic (ROC) curve of the test. (ROCs were developed during World War II for the analysis of radar signals, which explains the name.) The choice of cut-off point for the test is often better the closer the point it defines on the ROC curve is to the point (0, 1) (which is the test with no uncertainties) and the graph illustrates the trade-off between sensitivity and specificity that needs to be addressed for this decision. We see that in order to get as close as possible to the optimal test we need to have both the sensitivity and the specificity around 0.7, which would correspond to a cut-off point of about 0.8 μkat/L. However, that does not mean that this is the best test in any other sense, and it may not correspond to how we want to weight the relative importance of specificity and sensitivity. Each decision is associated with a cost, not only monetary (see Box 4.1), which depends on how useful it is to know that one has the disease (can it be treated?).

It may be worth noting that because diagnostic methods are not error-free there is a discrepancy between the true proportion P of subjects with a particular disease, and the perceived proportion P^* from using a diagnostic test:

$$P^* = P \times \text{sensitivity} + (1 - P) \times (1 - \text{specificity}).$$

If we actually know the sensitivity and specificity we can solve for P in this expression.

It is a widely held belief that the specificity and sensitivity are independent characteristics of the test, and can be transferred unchanged from one application of the test to another. This is not the case, since these concepts depend on two well-defined populations from which we can derive them. In our case, we need a definition of who is an alcoholic in order to be able to derive the CDFs in Figure 4.1. Such a definition will depend on the amount of alcohol intake and various behaviors, but different definitions will produce different CDFs and therefore different values of specificity and sensitivity. To take data from one context to another, where the disease definitions differ, may therefore change the values of these test characteristics, and different definitions imply different prevalences of the disease.

4.3.2   Bayes' theorem and the use and misuse of screening tests

Medical screening is a strategy for early identification of diseases in a population. This allows for earlier intervention and management, hopefully leading to a reduction in both mortality and suffering. Several disorders, especially cancers, are the target of many national screening programs, and numerous private clinics promote regular health check-ups – essentially a battery of screening tests – claiming that these will help their clients to stay healthy. Well-known examples of screening programs are the use of prostate-specific antigen for screening for prostate cancer in men, and mammography for breast cancer in women. Although screening

---

Box 4.1 Calibrating a screening test based on costs

How should we choose the cut-off point for a diagnostic test? One approach is to use the expected cost C of the test, which can be computed from

$$C = c_0 + c_{TS} P(TS) + c_{T^c S} P(T^c S) + c_{TS^c} P(TS^c) + c_{T^c S^c} P(T^c S^c),$$

where T denotes a positive test, S means that there is a signal, c_0 is some overhead cost, and the remaining coefficients c_{XY} represent the costs for the events XY. The cost C should not only be monetary, but also include patient discomfort and other aspects, though they need to be given a price tag.

Introducing conditional probabilities and after some algebra, we can rewrite C as

$$C = c_0 + \Delta c_{TS} P(T|S) P(S) + c_{T^c S} P(S) - \Delta c_{T^c S^c} P(T|S^c) P(S^c) + c_{T^c S^c} P(S^c),$$

where we have introduced the cost differences

$$\Delta c_{TS} = c_{TS} - c_{T^c S}, \quad \Delta c_{T^c S^c} = c_{T^c S^c} - c_{TS^c}.$$

The ROC curve is such that if $x = P(T|S^c)$, then $ROC(x) = P(T|S)$. We therefore have the cost as a function of x:

$$C(x) = c_0 + (\Delta c_{TS} ROC(x) + c_{T^c S}) P(S) - \Delta c_{T^c S^c} x + c_{T^c S^c}) P(S^c).$$

The best test should be the one when this is at a minimum, which occurs when $C'(x) = 0$. The condition for this is that

$$ROC'(x) = \frac{\Delta c_{T^c S^c}}{\Delta c_{TS}} \frac{P(S^c)}{P(S)},$$

which means that we should find the point on the ROC curve for which the slope equals what is at the right-hand side. We see that the right-hand side is inversely proportional to the odds for the disease, so for a rare disease we need a point with a steep gradient, which means being close to the origin in Figure 4.1. We must minimize our false positives, even at the expense of missing true positives.

---

may lead to an earlier diagnosis, not all screening tests have been shown to benefit the person being screened; over-diagnosis, misdiagnosis, and creation of a false sense of security are some potential adverse effects of screening. With respect to prostate cancer screening, it has been observed in autopsy studies that a high proportion of men, who have died for other reasons, have prostate cancer when the prostate is examined under a microscope. This has led some to conclude that 'most men die with prostate cancer, not from it', though the jury is still out. Therefore screening is not as uncontroversial as it may appear at first; while some types of screening are helpful – measuring blood pressure may be an example – others may be harmful. What are the issues involved?

Once the diagnostic test is defined, the population will be divided into four groups which can be described in a 2×2 table, based on whether they have the disease or not, and whether the test is positive or negative. We denote a positive test by T (so a negative test is T^c), and the situation where the subject actually has the disease by S (so absence of the disease is denoted by S^c). The terminology introduced in the previous section is that the probability $P(T|S)$ (the test is positive if the subject has the disease) is the test sensitivity, whereas the probability $P(T^c|S^c)$ (the test is negative in a subject without the disease) is the test specificity.

Specificity and sensitivity may be interesting properties of the test, but what matters to the individual is how likely it is that he has the disease, when the test is positive (in fact, also if the test is negative). This probability, $P(S|T)$, is called the *predictive value* of the test. The test does good for the subject if it correctly predicts that he has the disease (provided that this information is useful to have). The test is bad for the subject when it wrongly says that he has the disease. In that case he or she may undergo some risky surgery or treatment unnecessarily. From the perspective of the individual, a large predictive value is the key ingredient of a screening test.

However, the predictive value of a screening test is not computable from knowledge of the sensitivity and specificity alone. It also requires knowledge about the disease prevalence; the proportion in the population with the disease. Let us continue the example with the serum GT test using a cut-off limit of 2.1 μkat/L as a screening test for alcoholism. Recall that the test had a sensitivity of 0.45, meaning that 45% of all alcoholics have a positive test, and a specificity of 0.90, meaning that only 10% of non-alcoholics end up with a positive test. Now assume that we have 10% alcoholics in the population. To compute the predictive value, consider a sample of 1000 people from the population. We expect 100 of them to be alcoholics, contributing 45 positive tests. At the same time we have 900 non-alcoholics, contributing 90 positive tests. It follows that only one third of the positive tests come from the alcoholics. From the perspective of the screened individual, there is a fair risk that the wrong conclusion is drawn.

Under the same assumption, we can compute the probability that the subject is not an alcoholic if the test is negative. Arguing as above, our 1000 subjects produce 55 negative tests among the alcoholics and $0.9 \cdot 900 = 810$ among the non-alcoholics, so the likelihood that a negative test comes from someone who is not alcoholic is $810/(810 + 55) = 0.94$. In other words, among the people with a negative test we find only 6% alcoholics, those who were missed by the test.

For these reasons a test used in a screening program, especially for a disease with low incidence, must have good specificity in addition to acceptable sensitivity. But even that may not be sufficient. Suppose that we have a test for a particular drug, say a narcotic, that has both a specificity and a sensitivity of 99%. If a big company starts to test all its employees for this drug, a drug that is actually used by only 1% of them, we see that for a subject with a positive test, there is only a 50–50 chance he actually is a drug user. Is that an acceptable false-positive rate? The calculations performed above are formalized in mathematical terms in what is called Bayes' theorem (see Box 4.2), and which forms the foundation of Bayesian statistics (to be discussed later in this chapter).

4.4   The law: a non-quantitative analogue

What is the counterpart of all this for the legal problem? In the law setting the test diagnostic is some measure of 'appearance of guilt'. If this could be quantified, it would have a distribution in the population. Innocent people may appear guilty to some degree; they may just happen to be in the wrong place at the wrong time, or they can have a history of similar crimes. The amount of guilt appearance therefore varies between individuals and defines a distribution

---

Box 4.2 Bayes' theorem

In 1763 an essay with the title *An Essay towards Solving a Problem in the Doctrine of Chances* was posthumously published. The author was Thomas Bayes, a Presbyterian minister in England, and the essay contained a theorem which is named for him. Actually, he never wrote down the theorem explicitly, only the ideas. The actual theorem was first formulated by Laplace some forty years later and is therefore sometimes called the Bayes–Laplace theorem. It lies at the heart, mathematically, of Bayesian statistics.

The theorem is obtained by using the symmetry in A and C in the left-hand side of the formula $P(AC) = P(A|C)P(C)$, to derive the expression

$$P(C|A) = \frac{P(AC)}{P(A)} = \frac{P(A|C)P(C)}{P(A)}.$$

This is combined with the equally simple observation that if A has occurred, it must have done so together with one of the events C and C^c, so that $P(A) = P(AC) + P(AC^c) = P(A|C)P(C) + P(A|C^c)P(C^c)$. Combining these observations gives us Bayes' theorem,

$$P(C|A) = \frac{P(A|C)P(C)}{P(A|C)P(C) + P(A|C^c)(1 - P(C))},$$

which relates the conditional probability $P(C|A)$ to that of the inverse conditional probabilities $P(A|C)$ and $P(A|C^c)$. But to carry out the computations we need to know $P(C)$.

In Bayesian statistics the theorem is applied in situations where we want to make statements about a parameter θ from an observation x. The A above then refers to probability statements about the observation and C to probability statements about θ. We then get an expression of the form

$$dP(\theta|x) = \frac{dF(x|\theta)dQ(\theta)}{\int dF(x|\theta)dQ(\theta)},$$

which will be further explained in Section 4.6.2.

---

which corresponds to the distribution for the non-alcoholics in Figure 4.1. The criminals, those who commit the crimes, also differ in their appearance of guilt, which corresponds to the distribution for the alcoholics in Figure 4.1. The standard of judgement in the court corresponds to the cut-off point to be used when handing down the verdict of guilty or not guilty.

The main problem in the law situation is that it is not possible to identify any of the distribution functions involved, so we cannot produce Figure 4.1 and the whole problem becomes one of an overall judgement. It is, however, instructive to see the connection with the diagnostic tests above, since the implications of the standard of judgement in the courtroom are readily understood. In fact, in many judicial systems there are three different requirements on the evidence for guilt, constituting different levels of the *legal burden of proof*: (1) 'probable cause' is a relatively low standard of the evidence which determines if an arrest can be made or a case tried at court; (2) 'clear and convincing evidence' is a level employed in civil courts and for criminal cases with less serious punishments; while (3) evidence 'beyond reasonable doubt' is required for more serious punishments, including capital ones in some countries. Definitions differ, but the message is that more drastic actions require more confidence that the action to be taken is the right one.

A particular piece of evidence, denoted A, is presented in court, corresponding to a particular level of appearance of guilt. Let G be the event that a particular individual is guilty. The argument in court may then be that the probability of A for an innocent person, the probability $P(A|G^c)$, is so low that the person must be guilty. This is an indirect argument, similar to how we use p-values in statistics. There is a difficulty in finding the correct probability model for a court case, but that has not always stopped prosecutors and lawyers from using them, sometimes with serious effects, as the infamous case described in Box 4.3 shows.

However, the main point of confusion here is perhaps that the indirect evidence probability $P(A|G^c)$ is sometimes taken to be the probability that the person in question is guilty given the evidence. In other words, it is mistaken for the probability $P(G|A)$. In this context this confusion is called the *prosecutor's fallacy*. From Bayes' theorem we know that in order to be able to assess this probability we need to know the probability of guilt $P(G)$ in the absence of this particular piece of evidence; the *a priori* probability of guilt. (We also need to know $P(A|G)$, but that should be close to one if A is to be taken as evidence of guilt.) For more on this, see Section 4.8.

4.5 Risks in statistical testing

4.5.1 Does tonsillectomy increase the risk of Hodgkin's lymphoma?

In statistics the situation is similar to the use of reference ranges for medical diagnostics in that there is a focus on control of the Type I risk. The Type II risk is addressed when we plan the experiment and guides us in designing it (in the choice of sample size). To introduce statistical testing, we first use an example which investigates the relationship of an exposure E to an outcome C, with data that can be summarized in a 2×2 table as follows:

	E	E^c	Total
C	x_{11}	x_{12}	n_{1+}
C^c	x_{21}	x_{22}	n_{2+}
Total	n_{+1}	n_{+2}	n

Here the notation in the margins represents the appropriate sum of x_{ij}s. We want to describe any potential association between the exposure and the outcome.

In a cross-sectional study, data are obtained by picking n subjects and classifying them into the groups $CE, C^c E, CE^c$ and $C^c E^c$. In such a case, we have an observation of a multinomial distribution

$$(X_{11}, X_{12}, X_{21}, X_{22}) \in \text{Mult}_4(n, p_{11}, p_{12}, p_{21}, p_{22}).$$

For the definition of this distribution, see Box 4.4. Here p_{ij} denotes the probability that a randomly sampled individual belongs to cell ij ($p_{11} = P(CE)$, etc.) which is the fraction of patients in the whole population that belong to this group. The sum over all the p_{ij} is one.

Box 4.3 The sad case of Sally Clark

In 1999 the solicitor Sally Clark stood trial for, and was convicted of, the murder of her two children. The accusation was that she killed her first son when he was 11 weeks old, and then killed her second son when he was 8 weeks old. The defence claimed that these were two cases of sudden infant death syndrome (SIDS). What decided the case for the court was the testimony of expert witness Sir Roy Meadow, a pediatrician, who claimed that the probability of two children in the same family dying from SIDS is as small as 1 in 73 million. Since this is so small, the court argued, the hypothesis of SIDS deaths must be rejected, and from this it follows that it is murder (no other alternatives were mentioned). Even though a higher court quashed Mrs Clark's conviction in January 2003, she never recovered from the court case and died of alcoholism in her home in March 2007, at the age of 42.

The are two flaws in the argument put to the court.

1. The figure of 1 in 73 million was derived as $(1/8543)^2$, where Sir Roy drew on published studies to obtain the frequency of one SIDS death out of 8543 births in families sharing some of the characteristics of the family on trial. However, if we let A_1 denote the event that one child dies of SIDS and A_2 the event that two children in the same family die of SIDS, then we have that $P(A_2) = P(A_2|A_1)P(A_1)$, but the assumption that the conditional probability $P(A_2|A_1)$ of suffering another death in SIDS when you have already had one is the same as the probability of a single case, $P(A_1)$, may not be true. This is an assumption of independence, but there may well be environmental or genetic factors in play, so that in fact the probability of two SIDS deaths is much larger than the probability of one squared. However, even though this disputes the estimate, it is not the key error made by the court.

2. The problem is that the probability $P(A_2)$, whatever it is, is not relevant at all, because rare events do occur. The relevant probability is the conditional probability that the mother murdered her children, when we know they have died. Since we have two alternatives, SIDS or murder, given that two deaths have occurred, which is the most likely alternative?

Before the event, double SIDS and double murder are both very unlikely events. But that is irrelevant, since we are faced with the fact that two babies have died. Which is then the most likely, SIDS or murder? We do not know. But the argument put forward in court is not valid.

Note that for the cross-sectional study, p_{ij} is naturally estimated by x_{ij}/n. The same 2×2 table can be obtained in other study designs as well:

1. For a cohort study, including the randomized parallel group experiment, we fix the number of exposed at n_{+1} and number of non-exposed at n_{+2}.

2. For a case–control study we fix the margins n_{1+} and n_{2+} instead.

Because the method used to sample subjects is different for cohort and case–control studies than for cross-sectional ones, the distributions involved are different. For both study types the data are from two independent binomial distributions, but the probabilities in the cells are estimated in a different way: in the cohort case the probability p_{11} is estimated by x_{11}/n_{+1}, whereas in the case–control case it is estimated by x_{11}/n_{1+}. But they have changed meaning as well: in the cohort case p_{11} is the conditional probability $P(C|E)$, whereas in the case–control case it is the conditional probability $P(E|C)$. Note that in neither of these cases do the four probabilities p_{ij} sum to one.

If the two events E and C are not associated, they are independent. We can test for this independence using a conditional test, called Fisher's exact test, which is based on the fact that if there is independence, X_{11} has a hypergeometric distribution (see Box 4.4). To exemplify this we consider a specific association, between the cancer known as Hodgkin's lymphoma and the medical practice of removing tonsils.

Lymph tissues constitute an important part of our immunological defense system and include the tonsils. Sometimes the tonsils are removed in children with repeated throat infections. Hodgkin's disease is a cancer in the lymph glands, and one reasonable question to ask is whether the removal of this lymphatic barrier increases the risk of Hodgkin's disease.

A case–control study was carried out to investigate this. A group of 109 Hodgkin's lymphoma patients were matched with a control group of 109 patients on characteristics such as age, sex, general disease history, and occupation, as well as other variables that could be prognostic for the disease. Then hospital records for each of these patients were inspected in order to find out whether a tonsillectomy had been done or not. This was not possible for 8 patients with Hodgkin's disease and 2 patients in the control group, so the investigators arrived at the following table in which E denotes tonsillectomy and C Hodgkin's lymphoma:

	E	E^c	Total
C	67	34	101
C^c	43	64	107
Total	110	98	208

The hypothesis of interest is that E and C are independent events, an assumption we can express as the null hypothesis $P(C|E) = P(C|E^c)$. In words: the risk of acquiring Hodgkin's disease is the same whether or not the tonsils have been removed.

In Box 4.4 we see that if we assume the null hypothesis and that all margins are given, the conditional distribution of the observation in the upper left-hand corner is an observation of a hypergeometric distribution, namely Hyp(101, 107, 110). The CDF of this discrete distribution is illustrated in Figure 4.2. We see that the probability that $x_{11} \leq 59$ is 95% – which means that if E and C are independent events, the probability of finding a number 60 or more is as small as 5%. (Actually, since the distribution is discrete we cannot solve exactly for the probability 95%. The first probability above is actually 95.5%, so the second is 4.5%.) We found an even higher number, 67, for which the probability is 0.00013. Because such a large observation is unlikely if E and C are independent, we conclude that there is an association between Hodgkin's lymphoma and tonsillectomy. Note that the p-values referred to above are one-sided.

Box 4.4 The multinomial distribution and its descendants

By definition, $X = (X_1, \ldots, X_k) \in \text{Mult}_k(n, p_1, \ldots, p_k)$, where $\sum_j p_j = 1$, if its probability function is

$$p(x_1, \ldots, x_k) = \binom{n}{x_1 \ldots x_k} p_1^{x_1} \ldots p_k^{x_k},$$

where $\sum_j x_j = n$. The two-dimensional distribution $\text{Mult}_2(n, p, 1 - p)$ is equivalent to the $\text{Bin}(n, p)$ distribution for X_1. The coefficient $\binom{n}{x_1 \ldots x_k}$ denotes the number of ways we can divide n identical elements into k subsets with x_i elements in subset i.

The conditional distribution of the quadrinomial $X = (X_1, X_2, X_3, X_4)$, given the two conditions $X_1 + X_2 = n_1$ and $X_3 + X_4 = n_2$, consists of two independent binomial distributions, $X_1 \in \text{Bin}(n_1, p_1/(p_1 + p_2))$ and $X_3 \in \text{Bin}(n_2, p_3/(p_3 + p_4))$. If, in addition, we assume that $X_1 + X_3 = r$, the probability function of the conditional distribution of X_1 is given by

$$\frac{P(X_1 = j)P(X_3 = r - j)}{P(X_1 + X_3 = r)} = \frac{\binom{n_1}{j}\binom{n_2}{r-j}\theta^j}{\sum_{k=0}^{n_1} \binom{n_1}{k}\binom{n_2}{r-k}\theta^k},$$

where $\theta = p_1 p_4 / p_2 p_3$ is the odds ratio. We denote this distribution, called Fisher's non-central hypergeometric distribution, by $\text{Hyp}(n_1, n_2, r, \theta)$. In the special case where $\theta = 1$, this is the (central) hypergeometric distribution $\text{Hyp}(n_1, n_2, r)$ with probability function

$$\binom{n_1}{j}\binom{n_2}{r - j} / \binom{n_1 + n_2}{r}.$$

There is a corresponding result for a trinomial: if

$$(X_1, X_2, X_3) \in \text{Mult}_3(n, p_1, p_2, 1 - p_1 - p_2),$$

then the conditional distribution of X_1 given $X_1 + X_2 = t$ follows a $\text{Bin}(t, p_1/(p_1 + p_2))$ distribution. In fact, the conditional probability that $X_1 = j$ is

$$\frac{\binom{n}{j \, (t - j) \, (n - t)} p_1^j p_2^{t-j} (1 - p_1 - p_2)^{n-t}}{\binom{n}{t}(p_1 + p_2)^t (1 - p_1 - p_2)^{n-t}} = \binom{t}{j}\frac{p_1^t p_2^{t-j}}{(p_1 + p_2)^t}.$$

When the results of the above-mentioned study became public, another study was performed, with a slightly different design. In this study the investigators started out with 175 Hodgkin's lymphoma patients, for each of whom they looked for a sibling of the same sex and with an age difference of at most 5 years. Not every patient had such a sibling, but the investigators ended up with 85 sibling pairs where one had Hodgkin's lymphoma and the other did not. Let H denote the event that the Hodgkin's lymphoma patient has had a tonsillectomy,

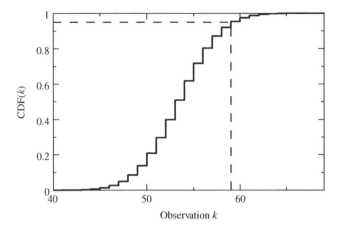

Figure 4.2   The CDF of the Hyp(101,107,110) distribution.

and let S denote the event that the sibling has. The study outcome can be summarized in the following four pairs:

$$(H, S) = 26, \qquad (H, S^c) = 15, \qquad (H^c, S) = 7, \qquad (H^c, S^c) = 37.$$

We can reorganize these data into a table analogous to the one analyzed above. For example, for the cases C (Hodgkin's lymphoma patients) we find $26 + 15 = 41$ with tonsillectomy and $7 + 37 = 44$ without. The reorganized table is as follows:

	E	E^c	Total
C	41	44	85
C^c	33	52	85
Total	74	96	170

If we perform the same analysis as above we get a p-value of 0.14. This is not much support for the conclusion made from the previous study. In fact, the only conclusion we can draw is that this study does not provide sufficient evidence to conclude that E and C are associated. There are at least three reasons for this result:

1. The events E and C are in fact independent.

2. The events are dependent, but the association is so weak that we need more data to demonstrate it.

3. We have not used our data in an optimal way.

In this case it is mainly the last reason, because we cannot reproduce the paired data from the reorganized table. Therefore the table we have analyzed has wasted some of the information, and there must be a better way to analyze these data. The key step is to write down the appropriate table, which is

	S	S^c	Total
H	26	15	41
H^c	7	37	46
Total	33	52	85

and then to pose the appropriate question: is the probability of a tonsillectomy the same for Hodgkin's lymphoma patients as it is for their siblings? In probability notation this means that $P(S) = P(H)$. But $P(S) = p_{11} + p_{21}$ and $P(H) = p_{11} + p_{12}$, so this question is really equivalent to the question whether the off-diagonal probabilities are equal. If they are, the conditional distribution of x_{12}, given that $x_{12} + x_{21} = 15 + 7 = 22$, follows a Bin(22, 0.5) distribution (see Box 4.4), from which we can compute that the probability of observing 7 or fewer among the 22 is 0.067. This is still not statistically significant at the conventional significance level 0.05, but we see that the effect of ignoring information really did weaken our test, and, contrary to the first analysis, we find some support for the result of the first study also in this study. The test we used here is called McNemar's test.

For future reference let us note the following. The odds ratio computed from the data of the first study is $(67 \cdot 64)/(43 \cdot 34) = 2.93$, and the odds ratio from the corresponding analysis of the second study is 1.47. In the second analysis of the second study, we constructed a table for which the probability entries are as follows:

Controls (C^c)

		E	E^c
Cases	E	$P(E|C)P(E|C^c)$	$P(E|C)P(E^c|C^c)$
(C)	E^c	$P(E^c|C)P(E|C^c)$	$P(E^c|C)P(E^c|C^c)$

We therefore see that the ratio of the two elements off the diagonal is actually the odds ratio, so this is estimated by $15/7 = 2.14$ in this analysis. We will see later why it is that the first estimate must be smaller than the second, and why the second is the estimate we are looking for.

4.5.2   General discussion about statistical tests

In this section we will do the opposite to what we did in the previous section: we will give a high-level discussion on statistical tests. This is essentially a discussion about tails in particular CDFs. More precisely, a statistical test is based on a test statistic, for which there are two relevant CDFs. The first is the distribution $F(x)$ of the test statistic under the null hypothesis. The second is the true distribution $G(x)$ of the test statistic. In statistics we know $F(x)$, which can often be well approximated by one of the Gaussian distributions defined in Box 4.5 and further discussed later in this chapter. Knowledge about $F(x)$ is necessary for us to be able to compute p-values etc., as was illustrated in the previous section.

The true distribution $G(x)$, on the other hand, is *always unknown*, though it is the ultimate purpose of the experiment to learn something about it. Knowledge about $G(x)$ would have been useful in the design of the experiment, because it would have made it possible for us to

---

Box 4.5 The univariate Gaussian distribution

The most important distributions in statistics are the Gaussian distributions. The standard Gaussian, or normal, distribution is denoted by $N(0, 1)$ and has the cumulative distribution function

$$\Phi(x) = \int_{-\infty}^{x} \frac{1}{\sqrt{2\pi}} e^{-y^2/2} dy.$$

If a stochastic variable X can be written $X = m + \sigma Z$, where Z has a $N(0, 1)$ distribution, we have $X \in N(m, \sigma^2)$, with CDF given by

$$F(x) = \Phi\left(\frac{x - m}{\sigma}\right).$$

We can compute the parameters m and σ from the CDF as

$$m = \int_{-\infty}^{\infty} x \, dF(x), \quad \sigma^2 = \int_{-\infty}^{\infty} (x - m)^2 dF(x).$$

Here we use the concept of a Stieltjes integral (see Box 4.8). These parameters are defined for a general distribution (provided the integrals converge) and are called the mean and the variance of $F(x)$.

One important reason why Gaussian distributions are so important is that they often are reasonable approximations of distributions that are symmetric around some value m, if we adjust σ appropriately. They also appear as approximations to the distribution of many test statistics, because of a famous theorem called the central limit theorem (CLT), about which we will have more to say in Section 4.7.

---

accurately compute the risk β of a Type II error of a particular experiment. The number $1 - \beta$, which we called the specificity for the diagnostic test, is called the power of the statistical test. But since we do not know $G(x)$ we need to approach that problem in a slightly different way, which will be described in more detail in the next chapter. (If we actually knew $G(x)$ there would not be much statistics to be done.)

As already mentioned, a statistical test can be one-sided or two-sided. For a one-sided test we compute the p-value from only one tail of $F(x)$, whereas for a two-sided test we use both tails. Essentially the difference is the same as the difference between the two questions 'is A better than B?' (one-sided) and 'is there a difference between A and B?' (two-sided). If X is a test statistic (which is a stochastic variable), let x be its observed value, computed from the data we have collected in the experiment. For a one-sided test, for which large values of x are indicative of a false null hypothesis, the p-value is computed from $p = 1 - F(x-) = P(X \geq x)$. The notation $F(x-)$ denotes the left-hand limit of $F(x)$ and is only needed if $F(x)$ is not continuous at the point x (by definition $F(x)$ is continuous from the right). If instead small values of x are indicative of a false null hypothesis we compute the p-value from $p = F(x)$. For the two-sided test we usually have a distribution that is symmetric around $x = 0$. A Gaussian distribution with mean zero is one important example. The two-sided p-value is in such a case computed as the probability that $\{X \leq -x\}$ plus the probability that $\{X \geq x\}$, in other words as $p = F(-x) + 1 - F(x-) = 2(1 - F(x-))$ (see Box 4.6). This can often be simplified by

---

Box 4.6 Distributions symmetric around zero

One of the key properties of the standard normal distribution is that it is symmetric around $x = 0$, which means that its probability density has the same value at the two points $\pm x$ for all x. Expressed in terms of the CDF this means that

$$F(-x) = 1 - F(x-),\qquad(4.1)$$

which is equivalent to the probabilistic statement that it is as probable to get an observation to the left of $-x$ as it is to get one to the right of x: $P(X \le -x) = P(X \ge x)$.

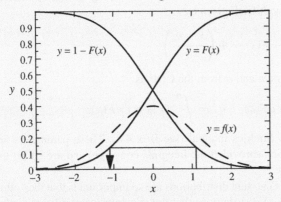

The figure above explains why equation (4.1) expresses symmetry around zero for a continuous distribution (in fact, for a Gaussian distribution). The dashed function is the density function and we see that $F(x)$ is the same as the reflection in the y-axis of $1 - F(x)$. The rectangle at the bottom illustrates that the symmetry relation in equation (4.1) at the point $x = 1.1$.

---

using the square X^2 as the test statistic instead, because the two-sided test for X becomes a one-sided test for X^2. For the normal distribution this turns the problem into one involving a chi-square distribution (see Appendix 5.A.2).

Just as there may be more than one way to detect a specific chemical in a fluid, so the same data can often be approached by different statistical tests. First of all, we can define different test statistics. If we choose to do a t-test, our test statistic is based on a mean difference, whereas the corresponding non-parametric test, the Wilcoxon test, is based on a rank sum. But also when we have decided on the test statistic there are sometimes a few options in the details of the computation of the p-value. A further aspect of statistical testing is concerned with choosing between an unconditional and a conditional tests. An example is Fisher's exact test for a 2×2 table, discussed above. It starts with two binomial distributions (the two rows) and we consider the result conditional on the outcome of the first column total. This way we reduce two parameters, the two binomial proportions, to one, the odds ratio. The general situation is that we are interested in a parameter (say, the odds ratio) and there are other so-called nuisance parameters in the distributions (the first binomial proportion, for example). It is then sometimes possible to derive a conditional distribution which defines a distribution (under the null hypothesis) without the nuisance parameters. The down-side of this is that such a test is a conditional test, conditional on the value of this other test statistic. Even though the other test statistic has nothing to do with our primary problem, it is still the case that any p-value will be computed under the assumption that this other test statistic takes on a particular value. Statisticians are divided as to how much this allows us to generalize (most medics probably do not care). For Fisher's exact test, can you generalize to all 2×2 tables that can be derived from this experiment, or only to those that have the same margins?

4.6   Making statements about a binomial parameter

4.6.1   The frequentist approach

We now wish to address what appears at first to be a very simple problem. Suppose that we have studied $n = 101$ individuals and found that $x = 67$ of these have a particular property; what we can say about the binomial parameter p which is the fraction of individuals in the population with this property? That we have the point estimate $p^* = 67/101 = 0.66$ of p is obvious, but what can we say about how accurate this estimate is? A discussion about this simple problem will allow us to compare the frequentist and Bayesian approaches to statistics.

Here is how the frequentist approaches the problem. He bases his analysis exclusively on what he knows. If p is the true parameter value, the number of individuals in the sample with the property has the $Bin(n, p)$ distribution. The CDF for this distribution is given by

$$F(x, p) = \sum_{k \leq x} \binom{n}{k} p^k (1 - p)^{n-k},$$

which, like the hypergeometric distribution in the previous section, is a step function. We can then define the function (where x is the observed outcome)

$$C(p) = 1 - F(x-, p),$$

which is the probability $P(X \geq x | p)$ that we get an observation at least as large as the observed value x, when the true parameter value is p. This function $C(p)$ is an increasing function, starting at zero when $p = 0$ and ending at one when $p = 1$. It is called the *confidence function* for the binomial parameter p based on the outcome of this experiment. From the frequentist's point of view this function contains all the information about p that is provided by the experiment. For the data above, the confidence function $C(p)$ is illustrated as the solid curve in Figure 4.3. The graph also contains a dashed curve, which shows what the confidence function would look like if we had five times as many observations but with the same observed proportion (i.e., $335/505$). We have added this curve to illustrate how a larger sample size increases the knowledge about p, by making the confidence curve steeper.

What can the confidence function tell us about the binomial parameter p? First, we note that the point estimate 0.66 corresponds to the confidence value 0.5 for both confidence curves (because we assumed that we had the same point estimate in the two cases). Next we consider the solid confidence function. For this we have that $C(0.63) = 0.24$, which means

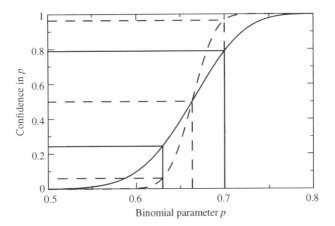

Figure 4.3   Two confidence functions for a binomial parameter, where the dashed one has five times as many observations as the solid one.

that this is the one-sided p-value if we test the hypothesis that $p \leq 0.63$.[2] Similarly, we have $C(0.70) = 0.79$, so the one-sided p-value for the test of the hypothesis $p = 0.70$, or larger, is 0.21. In this way we can compute the p-values, for all kinds of one-sided hypotheses regarding the parameter p.

The numerical observations above also mean that the interval $(0.63, 0.70)$ is a confidence interval with confidence level 0.55 ($= 1 - (0.24 + 0.21)$). A confidence interval has confidence degree $1 - \alpha$, if the method we use to compute it guarantees that it contains the true value p in the fraction $1 - \alpha$ of experiments. We will discuss such intervals in more detail in the next chapter. The corresponding p-values for confidence functions based on the larger sample size can be read off from the dashed curve as $C(0.63) = 0.060$ and $1 - C(0.70) = 0.036$, so in this case the same interval has confidence level 0.904.

Actually, we have cheated a little here. There is a minor complication due to the discrete nature of the binomial CDF $F(x, p)$ as a function of x for fixed p. The complication is that

$$P(X \leq x) = F(x, p) \quad \text{but that} \quad P(X \geq x) = 1 - F(x - 1, p),$$

which represents an asymmetry between the two tails, and justifies why we should redefine $C(p)$ as the compromise

$$C(p) = 1 - \frac{1}{2}(F(x - 1, p) + F(x, p)). \tag{4.2}$$

This modification is called the mid-P adjustment, and is actually the confidence function we have used above. It agrees to a very high precision with what we obtain if we apply a large-sample approximation, namely

$$C(p) = \Phi\left(\frac{np - x}{\sqrt{np(1 - p)}}\right). \tag{4.3}$$

---

[2]More precisely, the null hypothesis is $p = 0.63$ and the alternate hypothesis is $p > 0.63$.

Box 4.7 Confidence intervals for a binomial parameter

The confidence intervals we derive from the discussion in Section 4.6.1 are not the ones most often used. The method discussed, for a symmetric confidence interval of confidence degree $1 - \alpha$, is equivalent to finding the two solutions to the equation

$$\frac{(p^* - p)^2}{p(1 - p)/n} = z^2_{\alpha/2}, \tag{4.4}$$

where z_α denotes the $(1 - \alpha)$th percentile for the standard Gaussian distribution. The solution to this is given by

$$\frac{p^* + \frac{z^2_{\alpha/2}}{2n}}{1 + \frac{z^2_{\alpha/2}}{n}} \pm \frac{z_{\alpha/2}}{\sqrt{n} + \frac{z^2_{\alpha/2}}{\sqrt{n}}} \sqrt{p^*(1 - p^*) + \frac{z^2_{\alpha/2}}{4n}}.$$

These intervals were introduced in the late 1920s by Wilson, and differ from the standard confidence intervals, which are obtained by using the estimate $p^*(1 - p^*)/n$ for the variance and solving equation (4.4):

$$p^* \pm z_{\alpha/2} \sqrt{\frac{p^*(1 - p^*)}{n}}.$$

An important difference between the Wilson intervals and the conventional ones is that the former are always restricted to $[0, 1]$. In fact, it can be seen from the quadratic equation that its limits are such that the interval for the odds $p/(1 - p)$ is symmetrical around the estimate on a multiplicative scale.

These are, however, only two of a number of different suggestions for how to compute confidence intervals for a binomial proportion. The exact version, called the Clopper–Pearson intervals, is derived directly from the binomial distribution (without the mid-P adjustment). It is conservative in terms of actual coverage level, a fact derived from its discreteness. That reason, and the complexity of computing the exact interval (prior to the computer age), has led to the emergence of a number of approximative intervals, of which the standard one and the Wilson interval are only two.

This approximation is called de Moivre's theorem and is justified in Section 4.7. (It is common to use a mid-x adjustment instead of a mid-P adjustment to correct for the discreteness of the binomial distribution, which means that the Gaussian approximation is computed not at the point x but at the point $x + 1/2$. This is called a continuity correction or half-correction. We will later see other arguments for why the mid-P adjustment might be the right thing to do.) We have used a geometric approach to the description of how to derive knowledge about a binomial proportion, but the algebra involved is discussed in Box 4.7 (using the large-sample approximation).

At this point it is worth commenting on how to produce confidence intervals for a function of a parameter. Suppose we want to make statements about the odds $p/(1 - p)$ instead of about p itself. This is simple: all we need to do is to compute the confidence function as before and plot it, not versus p, but versus $\theta = p/(1 - p)$. Confidence intervals for the odds are derived in the same way as described above, which is equivalent to applying the odds function $f(p) = p/(1 - p)$ to the limits of the confidence intervals in Figure 4.3.

There are some interesting observations to make about this. To estimate θ from the observation x, the natural estimate is probably $\theta^* = x/(n - x)$, the empirical odds. But the corresponding test statistic is not unbiased; its expected value is not θ. In fact, its mean value does not even exist, because there is a positive probability of $x/(n - x)$ becoming infinite, which occurs when $x = n$. However, for large n this is unlikely, unless p is very close to one, and there is a useful approximation to the confidence function, namely

$$C(\theta) = \Phi \left(\frac{\ln(\theta) - \ln(\theta^*)}{\sqrt{1/x + 1/(n - x)}} \right). \tag{4.5}$$

In order to justify this, recall from basic calculus that $\ln(1 + x) \approx x$ for small x, from which we deduce that

$$\ln \left(\frac{x}{n - x} \right) \approx \ln \theta + \frac{x - np}{np(1 - p)}.$$

De Moivre's theorem then implies that the right-hand side has (asymptotically) a Gaussian distribution with mean zero and a variance given by

$$\frac{1}{np(1 - p)} = \frac{1}{np} + \frac{1}{n(1 - p)}.$$

Finally, we replace np with the observation x and $n(1 - p)$ with the observation $n - x$.

4.6.2   The Bayesian approach

We saw in Section 1.10 that the concept of probability is not restricted to probabilities which can be interpreted as frequencies, and that there is a whole school of statisticians who think of probabilities more in terms of an subjective assessment – the Bayesians. So how would a Bayesian statistician approach the problem of obtaining knowledge about a binomial parameter?

Before we discuss this, let us stress one particular aspect of the foregoing frequentist analysis. The data from the experiment define a confidence function, from which p-values and confidence intervals are derived. This type of confidence is something that is very reminiscent of a probability; it is always between 0 and 1, for a start. But its interpretation as a probability is more involved. For example, for something to be a 90% confidence interval means that if we repeat the experiment very many times, we expect the true parameter to be somewhere within these intervals in 90% of runs. It is not a probability statement about the parameter but about the computational method for the interval. However, to make probability statements about the parameter is precisely what Bayesian statisticians do. They do this following a well-defined process that captures the inductive aspect of science: we learn more as we do more experiments.

Another point before we start the discussion is the need for a convenient notation that conveys the important mathematical aspects of the discussion, which is about sums. To avoid the conventional distinction between discrete and continuous stochastic variables we will use the integral notation outlined in Box 4.8, which was also mentioned in Section 2.5. Now to the Bayesian analysis.

Box 4.8 Notation: Differentials and the Stieltjes integral

Statistics is often divided into a discrete and continuous version, with the former expressed in probability functions and sums and the latter having probability densities and integrals (and therefore being more involved mathematically). In order to avoid this distinction we will use a uniform notation based on the mathematical concept of the Stieltjes integral.

If $F(x)$ is a right-continuous function (in our case almost always a CDF), we denote by $dF(x)$ either the jump $\Delta F(x) = F(x) - F(x-)$ at that point, if there is one, or the differential at that point (we assume that a function that does not have a jump at a point is differentiable there). The latter means that $dF(x) = f(x)dx$, where $f(x)$ is the probability density at that point. At a jump point, the probability function is $\Delta F(x)$. On occasion we even let $dF(x)$ denote the density $f(x)$ at continuity points x.

Sums and conventional integrals are now replaced by the Stieltjes integral

$$\int_a^b g(x)dF(x).$$

This represents a sum – the notation $\int$ for an integral is actually only a slanted S for the first letter in sum – and the notation means that we sum the $g(x)$ using weights $dF(x)$. It is defined, by first defining it for a step function $g(x)$ corresponding to the partition $a = x_0 < x_1 < \ldots < x_n = b$ with values g_i in the intervals $(x_{i-1}, x_i]$, as

$$\int_a^b g(x)dF(x) = \sum_{i=1}^n g_i(F(x_i) - F(x_{i-1})).$$

Repeating the standard definition of the Riemann integral (sandwich a general function between two step functions with the same partition), we obtain, by making the partition finer and finer, a limit that defines the integral. For a continuous distribution with $dF(x) = f(x)dx$, this is the conventional integral $\int g(x)f(x)dx$, and if the distribution is discrete, it is the sum $\sum_k g(x_k)\Delta F(x_k)$.

Before he starts the experiment, the Bayesian statistician must have a prior opinion about the value of the parameter p. In order to describe his prior belief he defines a CDF $Q(p)$, called the *a priori* distribution. If he is certain of the value of the parameter, he should assign all probability to one point. The opposite extreme is complete ignorance, in which case he might assign equal probability to everything between 0 and 1. We should not confuse $Q(p)$ with the heterogeneity distribution which was discussed in Section 2.5; they are not related. The heterogeneity means that there are different probabilities for different individuals, whereas the Bayesian assumes a common p for all, but is uncertain about its value and wants that uncertainty to be propagated through the calculations. Given the prior, the probability for obtaining a particular outcome is computed by the formula

$$P(X = x) = \int_0^1 P(X = x|p)dQ(p),$$

where $P(X = x|p)$ is the binomial probability function. The probability $P(X = x)$ is called the *predictive probability* for the different outcomes of the experiment.

Having decided on his prior opinion, the Bayesian statistician performs the experiment. We assume he gets the same data as in the previous section ($n = 101$, $x = 67$ for a binomial distribution). Using this information, he updates his belief about what p is. The updated distribution $F(p|x)$ is called the *a posteriori* distribution of p (based on the observation x) and is obtained from Bayes' theorem as

$$dF(p|x) = \frac{P(X = x|p)dQ(p)}{P(X = x)} = C(x)p^x(1 - p)^{n-x}dQ(p).$$

Here $C(x) = 1/P(X = x)$ is independent of p and makes $F(p|x)$ a CDF for p.

In case of complete certainty, when $Q(p)$ has probability one at a single point, the *a posteriori* distribution will also have probability one at that point; there is nothing in the outcome of the experiment that will change your view.

There is a particular prior of special interest, called the *uninformed prior* (flat or uniform prior would probably be a better description). This is the prior distribution for which every value of p in $(0, 1)$ is equally probable, and therefore means that $dQ(p) = dp$. Not surprisingly, the uninformed prior implies that the predictive probability for all possible values is the same:

$$P(X = x) = \int_0^1 \binom{n}{x} p^x(1 - p)^{n-x}dp = \frac{1}{n + 1}, \quad x = 0, \ldots, n.$$

Furthermore, the assumption implies that the *a posteriori* distribution becomes

$$dF(p|x) = (n + 1)p^x(1 - p)^{n-x}dp.$$

Such a distribution, for which the probability density $f(p)$ is proportional to $p^{a-1}(1 - p)^{b-1}$, is called the beta distribution with parameters a and b; we denote its CDF by $\beta(p; a, b)$. We therefore have that

$$F(p|x) = \beta(p; x + 1, n - x + 1).$$

The expected value for p based on this *a posteriori* distribution can be calculated as

$$\int_0^1 p\, dF(p|x) = \frac{x + 1}{n + 2}, \tag{4.6}$$

a result known as 'Laplace's rule of succession' (see Box 4.9). As an application of this, if you have no prior idea of what percentage of swans are white, you may assign the uninformative prior to this. After having observed n swans, all white, your prediction is that the fraction $n/(n + 1)$ of all swans are white. (The frequentist would obtain this estimate had he seen n white and one colored swans.) This is your inductive conclusion, and essentially means that since you have only seen white swans, you believe most (or almost all, depending on n) swans are white.

However, the point about Bayesian statistics is not really to use the uninformed prior. The point is to synthesize prior knowledge into a more structured, but still subjective, prior distribution $Q(p)$. Assume that we believe that the true probability is somewhere in the vicinity of 0.4. To make it precise we assume that our subjective prior information can be summarized

---

Box 4.9 Laplace rule of succession and the sunrise problem

If we have n independent observations of a Bernoulli experiment in which the event has occurred k times, what is the probability that it will occur in the next, independent, experiment? Conventional reasoning would estimate the unknown probability as k/n, which therefore is the predicted probability that it will occur in the next test.

However, if we have had success on all occasions, so that $k = n$, is it obvious that this is the proper estimate? The prediction would be that we are certain the event will occur next time. But if we toss a possibly biased coin 10 times getting a head each time, is the best predictive probability estimate really that we are certain to get a head next time?

Laplace addressed this as the sunrise problem: 'what is the probability that the sun will rise tomorrow?' He argued as follows. Prior to knowing of any sunrises, one is completely ignorant of the probability p of a sunrise. Laplace takes this ignorance to mean that at that stage of our knowledge about p, it can be described by the uniform distribution on the interval $(0, 1)$. He then derived equation (4.6) and concluded that after the sun has risen on n consecutive days, the updated probability of a sunrise is $p = (n + 1)/(n + 2)$. The larger the number of days that have begun with a sunrise, the higher the plausibility of a sunrise tomorrow. His numerical estimate of this was based on the assumption that Earth was created on October 23, 4004 BC, as the Bible had been thought to imply.

However, we must not think that Laplace seriously believed in this. Laplace was the author of the masterpiece *Traité de Méchanique Céleste* in five volumes, which described the solar system in deterministic mathematical equations. Concerning the sunrise problem he reflected: 'But this number is far greater for him who, seeing in the totality of phenomena the principle regulating the days and seasons, realizes that nothing at present moment can arrest the course of it.' Or, in other words, the plausibility of a sunrise depends on how much you know, and this varies from person to person. This is at the heart of Bayesian statistics: the only good probability is a conditional probability taking into account what one knows.

---

by the $\beta(p; 4, 6)$ distribution (which has mean 0.4). In such a case the *a posteriori* distribution is proportional to

$$p^x(1 - x)^{n-x}d\beta(p; 4, 6) = p^{x+4-1}(1 - p)^{n-x+6-1}dp,$$

which is the $\beta(p; x + 4, n - x + 6)$ distribution. (Bayesian statistics is often numerically hard and often requires computer simulations, but for this particular case the beta distribution makes it simple.)

All this is illustrated in Figure 4.4, using the data above. In this graph we have also reproduced, as the dashed curve, the confidence function in Figure 4.3. Indicated is its median value, which is the old estimate 0.66, together with the 90% confidence interval $(0.58, 0.74)$. In addition there are two CDFs, both drawn as solid curves. The rightmost of these, which is very close to the graph of the confidence function, shows the *a posteriori* distribution for p when we use the uninformative prior. We have also indicated its median value, together with what would correspond to a 90% confidence interval. However, in this setting, this corresponds to a true probability statement – the probability that the value of p lies in this

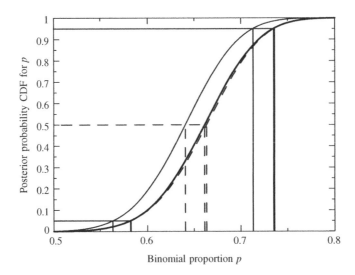

Figure 4.4   Bayesian analysis of a single binomial proportion. The solid curves are Bayesian *a posteriori* distributions, the dashed curve is the frequentist's confidence function. The right solid curve corresponds to the uninformed prior, the left one to an informed prior centered on $p = 0.4$.

interval is 90%. The interval is therefore not called a confidence interval, but a credibility interval (both are conveniently abbreviated CI). Numerically we obtain the estimate 0.66 with 90% credibility interval (0.58, 0.74) – the same to two decimal places as found by the frequentist.

The left solid curve in Figure 4.4 describes the *a posteriori* distribution when we use the informed prior distribution above. Recall that the *a priori* distribution was centered on $p = 0.4$, which is far to the left in this graph, so we see how the data have moved our subjective assessment considerably toward larger values of p, and into a distribution for which we have the median value 0.64 with 90% credibility interval (0.56, 0.71). This illustrates how the Bayesian view of the world works: you update your collected knowledge when you get new information.

Note the close agreement between the *a posteriori* distribution of p when using the uninformed prior and the confidence function. This is sometimes interpreted as implying that if we have no prior opinion on p, we are almost in the frequentist situation. This is, however, debatable. What is true is that in many situations you can find a prior that is such that the corresponding *a posteriori* distribution closely resembles the confidence function. What is debatable is whether it reflects ignorance: if we are ignorant about the value of p we are also ignorant about the value of p^2, but if p is uniformly distributed on (0, 1), then the CDF for p^2 would be $F(x) = \sqrt{x}$, which is not uninformative. However, things gets more bizarre, because the CDF for the $\beta(p; x + 1/2, n - x + 1/2)$ distribution actually improves the approximation to the confidence function. This case corresponds to taking as the *a priori* distribution the $\beta(1/2, 1/2)$ distribution, which (in this case) is called Jeffrey's prior. Its probability density function is $1/\sqrt{p(1 - p)}$, which means that it puts very large weight on very small and large values of p, and much less on intermediate values. So the best approximation to the confidence function is not obtained by ignorance, but by putting weights on the boundaries. In passing, we may note that this observation gives us what is actually a rather robust way to obtain traditional confidence intervals for p, by approximating the confidence function with this beta distribution. (We may also note that if Laplace had used this prior for his law of succession, he would have found the odds to be $2n + 1$ to 1 instead of $n + 1$ to 1 as in the discussion above.)

4.7   The bell-shaped error distribution

The Italian priest and astronomer Guiseppe Piazzi discovered early in 1801 what he thought was a new planet. He gave it the name Ceres, but after a while he lost sight of it when it went behind the sun. There were so few observations made that astronomers were not able to work out its orbit and therefore worried that they would not be able to locate it again, once it emerged from behind the sun. Karl Friedrich Gauss, arguably the greatest mathematician of all time, decided to give this problem his attention. After some lengthy calculations, he predicted where Ceres would reappear. When it duly did so, Gauss's fame spread far and wide. At the time, Gauss did not communicate how he had arrived at his prediction, but it later turned out that the method he used was what we today call the method of least squares. Gauss had invented and used this method years earlier before he applied it to this particular astronomical problem. He subsequently produced two proofs for the method. It is his first version that is of interest to us here, because it justifies the choice of the Gaussian distribution as a distribution for errors.

At the time it was customary in the natural sciences to use the arithmetic mean of repeated observations, taken under essentially the same circumstances, as the estimate of the true value of the phenomenon in questions. However, this method seemed not justified by the probability distributions used at the time. Gauss filled this gap between statistical practice and statistical theory by changing both the probability density and the method of estimation. With respect to the former, the question was this: what does the error distribution look like, if the best way to estimate the location is by taking the average? Using an argument, a modern short version of which is given in Box 4.10, Gauss deduced that the distribution of the errors should have a density which takes the form

$$\varphi(x) = \sqrt{\frac{k}{2\pi}} e^{-kx^2/2}$$

for some constant $k > 0$. Gauss pointed out that 'the constant k can be considered as the measure of precision of the observations'. He also made the comment that the distribution cannot truly represent a law of error, since it also assigns positive probabilities to errors outside the range of possible errors; negative errors are often not possible and there may be practical constraints on how large errors can be. Gauss considers this feature unavoidable but of no importance because of the rapid vanishing of $\varphi(x)$ for large values of $|x|$.

This was, however, not the first time the Gaussian distribution had appeared. In 1733 Abraham de Moivre derived it as an approximation to the probability that a binomial variable is observed in a specified interval. His investigation was prompted by a need to compute probabilities of winning in various games of chance, and what he demonstrated

---

Box 4.10 Gauss's likelihood argument for the Gaussian law

Gauss's first derivation of the least squares method included a derivation of the Gaussian distribution which went something like this. Suppose we want to estimate a parameter μ from n independent observations $x_1, \ldots, x_n$, and denote by $\varphi(x)$ the density of the distribution for the errors $x_i - \mu$. Note that this distribution is assumed not to depend on μ. Calculate the probability for these observations, when μ is given, by the product (now called the likelihood for μ)

$$\varphi(x_1 - \mu) \ldots \varphi(x_n - \mu).$$

The most probable value for μ is taken to be the value that maximizes this expression. Gauss's requirement was that this most probable value should be the arithmetic mean $\bar{x}$. This means that when

$$\sum_{i=1}^{n} \frac{d}{d\mu} \ln \varphi(x_i - \mu) = 0,$$

we must have that

$$\sum_{i=1}^{n} (x_i - \mu) = 0.$$

Gauss deduced from this that the components of these two sums must be proportional, that is, that

$$\frac{d}{d\mu} \ln \varphi(x - \mu) = k(x - \mu)$$

must hold true for all $x = x_i$, and therefore all x. This is a differential equation for which the general solution is

$$\varphi(x) = A e^{kx^2/2}.$$

For this to be a probability density function we must have $k < 0$ and $A = \sqrt{-k/2\pi}$.

---

was that if X has a $\mathrm{Bin}(n, p)$ distribution, its CDF is well approximated by the expression in equation (4.3). In fact, de Moivre only obtained this result for the special case $p = 0.5$; the full generality was obtained by Pierre Simon de Laplace, and published in his monumental work *Théorie Analytique des Probabilités* in 1812, in an investigation we describe next.

Gauss's justification for the Gaussian distribution was based on finding a symmetric error distribution for which the arithmetic mean was the natural estimator of the unknown center. Laplace turned this question around, and asked what is the distribution of the arithmetic mean of independent observations from a common distribution. His analysis indicated that for any error distribution with mean m, it is true that the arithmetic mean $\bar{X}$, at least approximately, has a Gaussian distribution, provided the sample is large enough. More precisely, we have the central limit theorem (CLT): If $X_1, \ldots, X_n$ are independent, identically distributed variables

---

Box 4.11 An alternative derivation of the Gaussian distribution

---

Suppose that we shoot at a two-dimensional target, aiming at a particular point. All deviances are assumed to represent a random deviation. Let us introduce a coordinate system with the bull's-eye at the origin and coordinates x and y. The errors in the different dimensions are assumed independent and the probability of a particular error depends only on its distance from the origin. This means that we assume that

$$\varphi(x)\varphi(y) = g(\sqrt{x^2 + y^2}),$$

and if we take $y = 0$, we see that $\varphi(x)\varphi(0) = g(x)$ for $x > 0$. It follows that $\varphi(x)$ should satisfy the functional equation

$$\varphi(x)\varphi(y) = \varphi(0)\varphi(\sqrt{x^2 + y^2}),$$

an equation which has the solution $\varphi(x) = Ae^{mx^2}$. Again this leads to the Gaussian distribution.

A consequence of this analysis, and some induction, is the following observation. If the variables $X_1, \ldots, X_n$ are all independent, they have a common Gaussian distribution with mean zero precisely when their joint distribution has a distribution that is rotationally symmetric (the density depends only on the distance from the origin).

---

with mean m and (finite) variance σ^2, then $\bar{X} \in AsN(m, \sigma^2/n)$. In other words,

$$P(\bar{X} \le x) \approx \Phi\left(\frac{\sqrt{n}(x - m)}{\sigma}\right).$$

The notation As in front of the normal distribution means that the statement is asymptotically true, that the larger we choose n, the more the true distribution of $\bar{X}$ resembles the referenced Gaussian distribution.

The first application of the CLT is de Moivre's observation above, because if X has a $Bin(n, p)$ distribution, we can write it as a sum of independent simple 0–1 variables $X = X_1 + \ldots + X_n$, where each X_i represents the outcome in individual experiments (it is 1 with probability p, and 0 with probability $1 - p$). Such a distribution is called a Bernoulli distribution. Each X_i therefore has mean p and variance $p(1 - p)$, so the CLT implies de Moivre's result. From this starting point the CLT has been generalized in various directions that are important for its application in statistics, some of which we will encounter later. A brief outline of parts of the CLT history is given in the appendix to this chapter. The main message is that the CLT constitutes the main reason why the normal distribution so often appears as the distribution for a test statistic.

It is important to note that the CLT only states what is asymptotically true. It does not state how large n needs to be to achieve a certain precision in this approximation. If the original distribution is reasonably symmetric around its mean, n does not need to be very large, whereas if it is very skew, we may need to take n very large. For example, for the binomial distribution with p close to 0.5 only a small n is needed, whereas when p is close to zero or one, we need a very large n. In fact, in these extreme situations the binomial distribution is usually compared to the Poisson distribution instead.

An important and noteworthy consequence of the CLT is that it explains why it is that when you add two independent Gaussian distributions, the new distribution is also Gaussian. This is an extremely important property of the Gaussian distribution, and holds also if the distributions are not independent. To be more precise, assume that $X \in N(m_1, \sigma_1^2)$ and $Y \in N(m_2, \sigma_2^2)$ are two independent stochastic variables; then

$$aX + bY \in N(am_1 + bm_2, a^2\sigma_1^2 + b^2\sigma_2^2)$$

for any real numbers a and b. This is easily proved by brute force, using some probability theory and completing some squares, but that it has to be valid can also be deduced from the CLT: if we have a series of independent variables X_i such that their sum is asymptotically distributed as $N(m_1, \sigma_1^2)$ and another series of variables Y_i for which the sum is asymptotically distributed as $N(m_2, \sigma_2^2)$, the sum $\sum_i (aX_i + bY_i)$ must also be asymptotically Gaussian, according to the CLT. Since we can do this to any precision, the result must hold.

4.8   Comments and further reading

The discussion about medical diagnostics was motivated as an analogue to error control in statistics. But it is also the way medical diagnostics is mostly presented in medical textbooks and rests on the assumption that it is the accuracy parameters of sensitivity and specificity that constitute the medical knowledge about the test, when in fact it is the predictive value that is the ultimate goal. Although this view gave us a reason for introducing Bayes' theorem, it has been challenged (Guggenmoos-Holzmann and van Houwelingen, 2000). The natural statistical approach might instead be to determine the predictive values (positive and negative) directly from the data set used to estimate the accuracy parameters. From the predictive values we can derive the specificity and sensitivity if we wish, using Bayes' theorem, provided we know the proportion of positive tests. (Which shows that if the predictive values are fixed, the accuracy parameters will depend on the fraction of positive tests and may therefore vary between subpopulations.) Our discussion is therefore not to be viewed as textbook material about gathering diagnostic information, but as a way into an understanding of the alpha and beta of hypothesis testing in statistics.

The prosecutor's fallacy discussed on page 93 has been analyzed in some detail by Dawid and Mortera (1996). By assuming a finite population of $N + 1$ individuals with one perpetrator, they take $P(G) = 1/(N + 1)$ and carry out the computations in a variety of important situations, reflecting different search strategies for the police. Because of the need for a known population size, this problem is called the *island problem*.

The studies investigating the relationship between tonsillectomy and Hodgkin's lymphoma have previously been discussed (Miller et al., 1980) much along the same lines as above. The original publications are Vianna et al. (1971) for the fist study, and Johnsson and Johnson (1972) for the second. Concerning one-sided and two-sided p-values, we have tacitly avoided a discussion about which to choose (Senn, 2007, Chapter 12).

At first sight it is surprising how difficult it is to get accurate confidence intervals for a single binomial parameter and how many methods are available (Newcombe, 1998). The underlying reason is the discreteness of the binomial distribution, which means that whichever method we suggest, its true coverage probability is not always the nominal one. For more on this, see Brown et al. (2001) and/or Agresti (2003).

Bayesian statistics actually pre-dates the frequentist approach (Feinberg, 2006) in the practice of statistics. Thomas Bayes may have been the first to formulate the inverse probability formula, but he had no influence on its future applications (just as James Lind did not really change the treatment of scurvy). The most influential person when it comes to Bayes' theorem, and early probability theory in general, is without doubt Laplace. In his time it became customary to make statements about a parameter based on the probability of the outcome given the parameter, which meant they used the uniform prior for the parameter and Bayes' theorem. This was called the inverse probability method and was introduced by Laplace in 1774. However, it was not without its critics, including Laplace himself who seems to have moved away from it when he introduced the CLT. It was realized early that its use included a confusing double meaning of probability, which is sometimes taken to be objective and sometimes taken to be subjective. It was also noted at the time that the application of the rule of succession leads to a 'futile and illusory conclusion': if you toss a coin twice and get heads both times, few bet two to one that one gets heads at the next toss (Hald, 2007). Which, of course, is a reflection of the fact that the uniform prior is often irrelevant.

The discussion on Gauss's justification of the Gaussian distribution is based on the description in (Eisenhart, 1982), where the original references can be found. To claim that Gauss invented the least squares method is controversial; some attribute it to the French mathematician Legendre. The importance of the Gaussian distribution in statistics is as an approximation to the distributions for various test statistics, for reasons often traceable to the CLT. It is also true in biostatistics that the data themselves, or a transformation thereof, often have a distribution similar to the Gaussian distribution. Sometimes the argument is that what we see is the net result of many small, random, entities that add up to the outcome variable measured, which is another appeal to the CLT. It does not imply that the Gaussian is appropriate for all kind of data. In many areas of human affairs, such as distributions of wages or time to completion of tasks, it may well be that fundamentally different distributions (Taleb, 2007) are more appropriate, such as heavy-tailed power law distributions of the form $F(x) = 1 - x^{-D}$ for some $D > 0$. These are distributions with very different properties than those of the bell-shaped Gaussian distribution. The key is that they are scale-invariant, which is related to the theory of fractals in modern mathematics.

References

Agresti, A. (2003) Dealing with discreteness: making 'exact' confidence intervals for proportions, difference of proportions and odds ratios more exact. *Statistical Methods in Medical Research*, **12**, 3–21.

Brown, L.D., Cai, T.T. and DasGupta, A. (2001) Interval estimation for a binomial proportion. *Statistical Science*, **16**(2), 101–133.

Dawid, A.P. and Mortera, J. (1996) Coherent analysis of forensic identification evidence. *Journal of the Royal Statistical Society, Series B*, **58**(2), 425–443.

Eisenhart, C. (1982) *Encyclopedia of Statistical Sciences* vol. 4 John Wiley& Sons, Inc. chapter Laws of Error II: Gaussian Distribution, pp. 547–560.

Feinberg, S.E. (2006) When did Bayesian inference become 'Bayesian'?. *Bayesian Analysis*, **1**(1), 1–40.

Guggenmoos-Holzmann, I. and van Houwelingen, H.C. (2000) The (in)validity of sensitivity and specificity. *Statistics in Medicine*, **19**(13), 1783–1792.

Hald, A. (1998) *A History of Mathematical Statistics from 1750 to 1930* Wiley Series in Probability and Statistics. New York: John Wiley & Sons, Inc.

Hald, A. (2007) *A History of Parametric Statistical Inference from Bernoulli to Fisher, 1713–1935* Sources and Studies in the History of Mathematics and Physical Sciences. New York: Springer.

Johnsson, S.K. and Johnson, R.E. (1972) Tonsillectomy history in Hodgkin's disease. *New England Journal of Medicine*, **287**, 1122–1125.

Le Cam, L. (1986) The Central Limit Theorem around 1935. *Statistical Science*, **1**(1), 78–96.

Miller, R.G., Efron, B., Brown, B.W. and Moses, L.E. (1980) *Biostatistics Casebook* Wiley Series in Probability and Mathematical Statistics: Applied Probability and Statistics. New York: John Wiley & Sons, Inc.

Newcombe, R.G. (1998) Two-sided confidence intervals for the single proportion: comparison of seven methods. *Statistics in Medicine*, **17**(8), 857–872.

Senn, S. (2007) *Statistical Issues in Drug Development*. Chichester: John Wiley & Sons, Ltd.

Taleb, N.N. (2007) *The Black Swan. The Impact of the Highly Improbable*. London: Penguin.

Vianna, N.J., Greenwald, P. and Davies, J. (1971) Tonsillectomy and Hodgkin's disease: the lymphoid tissue barrier. *Lancet*, **1**, 431–432.

4.A   Appendix: The evolution of the central limit theorem

The central limit theorem (CLT) is the generic name for a number of different mathematical statements to the effect that the asymptotic distribution of properly normalized cumulative sums of variables is Gaussian. The actual term 'central limit theorem' was introduced by George Pólya in 1920, with 'central' originally referring to its central role in probability theory. It can also be interpreted to refer to the fact that the statements are about the centers of distributions, as opposed to tail behavior. From a statistical perspective its importance is that it explains why many (most) statistical tests involve the Gaussian distribution and, perhaps, why biological data very often have an approximative normal (or lognormal) distribution. In this section we will outline the historical development of some aspects of the CLT that are important in statistics.

We mentioned Section 4.7 that Laplace turned Gauss's arguments for the bell-shaped distribution around into a general theorem about the asymptotic distributional behavior of the arithmetic mean of independent observations from the same distribution. Laplace looked at specific examples (mainly in astronomy, where he wanted to understand the distribution of inclination angles for comets), but his methods can be generalized. In his time, the early nineteenth century, there was no probability theory as we know it today, but only applications of probability concepts to specific real-life problems. The tool Laplace used was the *characteristic function* (Fourier transform) $\psi(t) = \int_{-\infty}^{\infty} e^{itx} dF(x)$ of the distribution, but he only considered examples with discrete distributions. Laplace's finding received little attention in his own time and it was not until the Russian mathematician Aleksandr Lyapunov published an exposition on the 'theorems of Laplace' in 1900–1901 that the arguments used by Laplace were turned into a rigorously proved mathematical theorem. Subsequently attempts were made to see how its assumptions could be relaxed, leading to versions that in many cases have practical applications in statistics.

The CLT is a statement about cumulative sums, properly normalized. The Gaussian distribution is not the only possible limit of such sums, there is a whole family of so-called stable laws that can arise. However, if the limit is not a Gaussian law then we must be dealing with distributions with rather heavy tails. If the variance is finite, which is mostly the case in biology, the Gaussian distribution is what is seen asymptotically and we consider only this case. One formulation of the CLT considers a triangular array of random variables, by which we mean that for each n there is a sequence $X_{n1}, \ldots, X_{nk(n)}$ of $k(n)$ independent random variables, all with zero means and with σ_{nj}^2 denoting the variance of X_{nj}. The CLT statement is that

$$S_n = \sum_{j=1}^{k(n)} X_{nj} \in AsN(0, \sigma^2) \quad \text{when} \quad \sum_{j=1}^{k(n)} \sigma_{nj}^2 \to \sigma^2 \quad \text{as } n \to \infty.$$

Laplace's result is the special case when we take $k(n) = n$ and $X_{nj} = (X_j - m)/\sqrt{n}$, for which we have that $S_n = \sqrt{n}(\bar{X} - m)$, so this formulation contains the basic assumption of independent and identically distributed (i.i.d.) variables. This is what subsequent work tried to relax.

Lyaponov explored the method of his (or Laplace's) proof to formulate the CLT also without requiring that the variables were i.i.d., but the more important development in that direction are the conditions formulated by the Finish mathematician Jarl W. Lindeberg in

1922. He provided the first elementary proof (not using the characteristic function) of the theorem, assuming that, for every given $\epsilon > 0$,

$$\sum_{j=1}^{k(n)} \int_{|x| \geq \epsilon \sigma} x^2 dF_{nj}(x) \to 0 \quad \text{when } n \to \infty.$$

In the case of i.i.d variables the Lindeberg criterion reads

$$\sum_{1}^{n} \int_{|x| \geq \epsilon} x^2 dF(\mu + \sqrt{n}x) = \frac{1}{n} \sum_{1}^{n} \int_{|x-\mu| \geq \epsilon \sqrt{n}} (x - \mu)^2 dF(x)/\sqrt{n} \leq \frac{\sigma^2}{\sqrt{n}},$$

which goes to zero as $n \to \infty$. Later, in 1935, it was shown by William Feller and P. Lévy that when each of the random variables in this sum is small, this criterion is not only sufficient but also necessary for a CLT to hold. The size condition is that $\max_j \sigma_{nj} \to 0$ when $n \to \infty$. (Actually their proof was not complete until Harald Cramér proved a final missing link the year after, namely that if the sum of two independent stochastic variables is Gaussian, the terms are so also.)

There is an immediate extension of these univariate CLTs to multivariate counterparts. To make this extension one uses an observation by Cramér and Wold, combined with the fact that the p-vector X is distributed as $N_p(m, \Sigma)$ precisely when all linear combinations aX of the components have the distribution $N(am, a\Sigma a^t)$.

Another development path for the CLT relaxed the assumption of independence of the sequence of variables X_{nj}, $j = 1, \ldots, k(n)$, for fixed n. There are different ways to do this. One important approach was formulated in 1935 by P. Lévy in terms of what we now call *martingales*. In the CLT, each X_{nj} is assumed to have mean zero, and one way to relax this condition is to replace the mean by a mean computed conditionally on previous variables. To be more specific, assume that the index j represents time, and that we recursively observe the different X_{nj} with increasing j. Introduce the known history $\mathcal{F}_{nj}$ at time j in sequence n, which is the information collected so far. As we go along, we collect more and more information, so we have that $\mathcal{F}_{n(j-1)} \subset \mathcal{F}_{nj}$. We now replace the condition that the mean should be zero with the condition that the conditional distribution given the past should be zero:

$$E(X_{nj}|\mathcal{F}_{n(j-1)}) = 0 \quad \text{for all } j. \tag{4.7}$$

If the variable is independent of all previous history this is the mean value, which means that sequences of independent variables represent one important example. If we assume that equation (4.7) holds, the cumulative sums $\{S_{nk} = \sum_{j=1}^{k} X_{nj}\}$ satisfy the martingale criterion

$$E(S_{nk}|\mathcal{F}_{nj}) = S_{nj} \quad \text{for all } j < k. \tag{4.8}$$

This would be the case for the fortune of a gambler in a fair game, and a sequence $\{S_{nk}\}$ for fixed n is called a martingale in discrete time. This term had been used for some time in gambling theory for the particular strategy of doubling the stake after each loss, and was adopted for statistics by J. L. Doob in 1953. To formulate a CLT for such martingales, we introduce the notation $E_{j-1}(X_{nj}) = E(X_{nj}|\mathcal{F}_{n(j-1)})$ and $\sigma_{nj}^2 = E_{j-1}(X_{nj}^2)$. The latter is a stochastic variable, not the variance parameter of X_{nj}; to get the variance we take the expectation of σ_{nj}^2. With this notation we have the following version of the

CLT: if it is true for all $\epsilon > 0$ that $\sum_{j=1}^{k(n)} E_{j-1}(X_{nj}^2 I(|X_{nj}|) > \epsilon) \to 0$ in probability, then $S_n \in AsN(0, \sigma^2)$, provided that $\sum_{j=1}^{k(n)} \sigma_{nj}^2 \to \sigma^2$ in probability.

The condition is essentially Lindeberg's condition, and one proof for this CLT is an adaption of his method. The theorem is more general, since we can allow $k(n)$ to be a stochastic variable, as long as it is what is called a stopping time. This means you can make conditions on how many terms to include, based on a rule defined from the history of the process.

Martingales in continuous time are defined in a completely analogous way. Assume (for simplicity) that the time interval is $[0, 1]$ and consider a stochastic process $\{x(t); t \in [0, 1]\}$, which means that $x(t)$ is a stochastic variable for each t. The criterion for the process to be a martingale is the same as before, namely that

$$E(x(t)|\mathcal{F}_s) = x(s), \quad s < t.$$

Here $\mathcal{F}_t$ is again the information obtained at time t, the history, so the requirement is that, conditional on what we know at time s, we should expect no change at a later time t. If we have a martingale sequence $S_n = \{S_{nk}, k = 0, \ldots, n\}$ we can define a stochastic process $\{x(t); t \in [0, 1]\}$, for which the paths are all continuous, by defining $x(k/n) = S_{nk}$ with linear interpolation in-between. This gives us a technique to obtain statements about stochastic processes in continuous time from similar statements in discrete time, though the details are more complicated than this. In particular, we can obtain a very important CLT for martingales in continuous time. For this we first need the concept of a predictable process, which is a process such that

$$E(x(t)|\mathcal{F}_{t-}) = x(t).$$

In words: if we know all history before now, we expect no sudden change immediately, which gives the process some local predictability. Moreover, associated with a martingale in continuous time there is an increasing and predictable process $\{\langle x \rangle(t); t \in [0, 1]\}$, called the compensator, which is such that the process $\{x(t)^2 - \langle x \rangle(t); t \in [0, 1]\}$ is also a martingale. This process takes for martingales the role played by the variance for stochastic variables.

An extremely important example of a martingale is the Wiener process $\{w(t); t \in [0, 1]\}$ for which the compensator is $\langle w \rangle(t) = t$. This process plays very much the same role for stochastic processes as the standard Gaussian does for univariate variables. A fuller discussion is given in Appendix 6.A.2.

The version of a CLT for martingales in continuous time we are interested in is similar to the one mentioned above for martingales in discrete time, and will be applied to counting processes in Appendix 11.A. The theorem says that for a sequence of martingales $\{x_n(t); t \in [0, 1]\}$, for which the compensator is such that $\langle x_n \rangle(t) \to \tau(t)$, in probability, and if it is also true for all $\epsilon > 0$ that the condition $\sum_{s \le t} (\Delta x_n(s))^2 I(|\Delta x_n(s)| > \epsilon) \to 0$ holds in probability, then

$$\{x_n(t); t \in [0, 1]\} \to \{w(\tau(t)); t \in [0, 1]\} \quad \text{in distribution.}$$

We recognize again the second criterion above as the Lindeberg criterion. Not surprisingly, there are some additional technical details to sort out.

If we know that a sequence of stochastic processes $x_n = \{x_n(t); t \in [0, 1]\}$ converges in distribution to the limit process $x = \{x(t); t \in [0, 1]\}$, it is also true, for a continuous function $f(x)$ of such processes, that $f(x_n)$ has the same asymptotic distribution as $f(x)$. This is often called the invariance principle. The classical CLT is obtained by taking $f(x)(t) = x(1)$, and another important choice is $f(x) = \max_{0 \leq t \leq 1} x(t)$, which will be used when we derive the Kolmogorov test in statistics (see Appendix 6.A.3).

For further reading on the history of the CLT, see the review by Le Cam (1986) and the extensive description in Hald (1998), which is a general account of the history of statistics before 1930.

5

Learning about parameters, and some notes on planning

5.1 Introduction

In this chapter we continue the discussion in the previous chapter on how to obtain knowledge about parameters defining distributions. It will be a more abstract discussion about how statistical tests are designed and how p-values and confidence intervals for such parameters are derived. We will follow the ideas introduced in the previous chapter and use the confidence function, which is essentially a graphical alternative to the standard approach with a heavier use of mathematical formulas (though we need some of these also). We will illustrate this approach with a few examples, both with a single parameter and when there are nuisance parameters present.

Among the single-parameter examples we will study is the odds ratio in a single 2×2 table, which will be extended to the stratified situation where we adjust for confounders. This leads us to the celebrated Mantel–Haenszel methodology, so important in epidemiology.

The last part of the chapter will be devoted to planning aspects and will introduce the power curve, which is the analysis of the Type II error of a proposed experiment. This discussion is what allows us to size our experiment properly, but does in itself contain a few mysteries. To understand the basic considerations involved is, however, essential for anyone who is about to design a clinical study.

Overall, this chapter is mathematically somewhat more involved than the previous ones, not necessarily in terms of complicated mathematics but in its use of mathematical formulas. This is more or less necessary, and the reason why we gave a softer introduction to the concepts in the previous chapter. However, what is discussed in this chapter is fundamental to the understanding of statistics, and the mathematically less inclined reader is advised to try to extract the important ideas out of it, despite the mathematical formulas.

Understanding Biostatistics, First Edition. Anders Källén.
© 2011 John Wiley & Sons, Ltd. Published 2011 by John Wiley & Sons, Ltd.

5.2   Test statistics described by parameters

The principal task of statistics is to reduce data to something that is interpretable. We see our data, which are observations of an outcome variable, as a representative sample from a hypothetical infinite population in which the corresponding data are described by a distribution function. This distribution is specified up to a few unknown parameters. For example, the ubiquitous Gaussian distribution depends on the parameter vector $\theta = (m, \sigma^2)$, where m is the mean and σ^2 the variance. For the purpose of this discussion we will denote a general distribution function by $F(x, \theta)$, where θ may be a parameter vector, but will still be called a parameter. The aim of the statistical analysis is to estimate this parameter using an estimator, which itself has a distribution because its value varies with the sample we draw from the population.

Since this will be a discussion about tests, our primary interest is not in the CDFs that describe the population data, but in the CDFs of test statistics (which are summaries of the sample data). In such a case the parameter θ may be multidimensional and contain one part that is of interest (such as the odds ratio) but also some *nuisance parameters*. For example, if we want to test for a mean difference (our parameter of interest) in Gaussian data, the original model also contains a nuisance parameter, the variance. Such nuisance parameters cannot be ignored; a way to handle them must be found (this is how the t distribution emerges: we use a particular estimate for the variance, which replaces the original Gaussian distribution by a t distribution). To simplify the present discussion we suppress any nuisance parameters from the notation, so that the CDF $F(x, \theta)$ of the test statistic contains only one parameter. The null hypothesis of the test we are interested is written $\theta = \theta_0$, where θ_0 is a specified value of the parameter, usually zero or one. This means that the distribution $F(x)$ under the null hypothesis is the same as $F(x, \theta_0)$.

When it comes to the parameter θ, statistics addresses two different problems. The first is what we actually know about θ, and the second is what is the best estimate of θ. These two problems must not be confused: a small sample may produce an accurate estimate, but our confidence in it may still be rather low. The two problems can be described as follows.

Hypothesis testing. We want a test of the hypothesis that $\theta = \theta_0$. The confidence in the rejection of this hypothesis is inferred from the p-values which we compute from the distribution $F(x, \theta_0)$. For our discussion we mostly assume that this test is one-sided and that the p-value is computed from the value x^* of the test statistic X obtained in the experiment as the probability $P(X \geq x^* | \theta_0)$, that is,

$$p = 1 - F(x^*-, \theta_0).$$

The different p-values based on this observation, as a function of θ_0, are what constitute the confidence function.

Parameter estimation. This is about finding the 'best' value for θ, based on the information the data give us. One estimation method is the *likelihood method*, which addresses the problem by asking, given the data we have, what is the most likely value of θ. Mathematically this means that we want to find the θ that maximizes the probability $L(\theta)$ for what we have observed. We do not use $L(\theta)$ as a true probability, and it is therefore referred to as the *likelihood function* of θ. The estimation method is called the *maximum likelihood method*. Though probably the most important estimation method,

---

Box 5.1   On Bayes, Fisher and the maximum likelihood concept

---

The maximum likelihood theory was created single-handedly by R. A. Fisher between 1912 and 1922, culminating in his fundamental paper 'On the mathematical foundations of theoretical statistics'. Prior to this time most statistical reasoning had been based on Bayes' theorem which gives the posterior CDF $P(\theta|x)$ for θ from the CDF $F(x, \theta)$ of the test statistics and a prior distribution $Q(\theta)$ for the parameter as

$$dP(\theta|x) \propto dF(x, \theta)dQ(\theta).$$

Here x is the observed data, and the proportionality constant makes the integral one. In the nineteenth century the terminology was that $dF(x, \theta)$ was a direct probability (for the outcome x, given θ) and $dP(\theta|x)$ an 'inverse probability' (for θ, given the observation x). Laplace, at least, used the terminology that θ was the cause and x the effect, and looked for the most probable cause based on the observed effect. Most statistics was done under the assumption that $dP(\theta|x) \propto dF(x, \theta)$, which means that a uniform prior $(dQ(\theta) = d\theta)$ was used. The major argument against such a choice of prior in the physical sciences is that nature does not randomly select parameter values, but even if you accept the argument about the prior (which Fisher did not) the integrals involved in probability statements about θ are still complicated to compute in many cases, and Fisher therefore maximized the direct probability instead. With this he eliminated all reference to a subjective prior distribution, an idea Fisher hated. This led to a confusion of terminology which was not cleared up until 1921 when Fisher introduced the term 'likelihood' for the density/probability function $dF(x, \theta)$ when considered as a function of θ, given data. A year later he also introduced the term 'maximum likelihood estimate'. This constitutes the birth of the frequentist school of statistics.

On many occasions frequentists need to replace the log-likelihood with what is called a *penalized log-likelihood* in order to get sensible parameter estimates. Such a likelihood is obtained by subtracting from the log-likelihood a penalty function $\phi(\theta)$. Put in the context above, this corresponds to having a prior density of the form $dQ(\theta) = e^{-\phi(\theta)}d\theta$.

---

it is not the only one. For a short discussion on how this concept is connected with the Bayesian approach to statistics, see Box 5.1.

Knowledge about a parameter θ is derived from the distribution $F(x, \theta)$ of the test statistic with the observed value of the test statistic inserted, and not directly from the likelihood function. However, in many cases there is a close connection between these, because we use the likelihood method to derive the test statistic, and then the approximation of the distribution for this is derived from the likelihood function.

To pursue the question of what we know about the parameter, if we want to test if $\theta = \theta_0$ at a particular significance level α, we first compute the (one-sided) p-value and then compare it to our chosen α. Alternatively, we can first determine the critical value x_{crit} defined by the relation

$$1 - F(x_{\text{crit}}-, \theta_0) = \alpha,$$

and then reject the hypothesis when the observed value x^* of the test statistic is such that $x^* \geq x_{\text{crit}}$. The set of outcomes $\{x \geq x_{\text{crit}}\}$ defines a region in the space of test outcomes, which is referred to as the *critical region* for the test. How to modify this for one-sided tests in the other tail and for two-sided tests is left to the reader. The implicit assumption is that there is one particular value, θ_{true}, which is such that the true distribution of the test statistic is $F(x, \theta_{\text{true}})$. (This distribution was denoted $G(x)$ in Section 4.5.2.) Once we have obtained the data from the experiment we can estimate θ_{true} and also describe what we actually know about it.

5.3   How we describe our knowledge about a parameter from an experiment

Suppose that we have performed an experiment and have obtained the value x^* for the test statistic. The distribution of the test statistic is assumed to depend on a parameter θ such that large values of θ correspond to large values of the test statistic. Traditionally the knowledge this observation provides us with about θ is summarized in a confidence interval with a specified confidence level $1 - \alpha$. As mentioned in Section 4.6.1, when we discussed a single binomial parameter, such an interval has the interpretation that in the long run a fraction $1 - \alpha$ of them will contain the true parameter θ_{true}. Not all, only the fraction $1 - \alpha$. They have an intimate relationship with p-values, and, in fact, it is all better explained by introducing the *confidence function*, from which both pieces of information can be deduced.

We define the (one-sided) confidence function for θ by inserting the observed value x^* in the distribution function for the test statistic

$$C(\theta) = 1 - F(x^*-, \theta).$$

The proper interpretation of this is as the proportion of experiments that will produce an observation that is at least x^* when θ is the true value, which means it is the (one-sided) p-value for that hypothesis. This function summarizes the knowledge we obtain about the true parameter value from the observation x^*.

One important special case is when we have a location model, which means that we have that $F(x, \theta) = F(x - \theta)$ for a CDF $F(x)$ of one variable. If we introduce the notation $\theta^* = x^*$, we can in this case[1] write

$$C(\theta) = 1 - F(\theta^* - \theta) = F(\theta - \theta^*),$$

so the confidence function is obtained from the CDF under the null hypothesis (which is $F(x)$) by a horizontal shift of magnitude θ^*. Such a function may look like the solid curve in Figure 5.1. This plot shows how we derive p-values and confidence intervals from $C(\theta)$:

- The value $C(\theta_0)$ gives the p-value for the null hypothesis $\theta \leq \theta_0$.[2] (In Figure 5.1 this takes the value 0.12.)

- $1 - C(\theta_0)$ defines the p-value for the hypothesis $\theta \geq \theta_0$.

---

[1] As long as the distribution is continuous, otherwise we replace θ^* in the middle by its left-hand limit.

[2] We use the notation $\theta \leq \theta_0$ to denote the situation where the null hypothesis is $\theta = \theta_0$ and the alternative is $\theta > \theta_0$, with a similar convention for the opposite one-sided hypothesis.

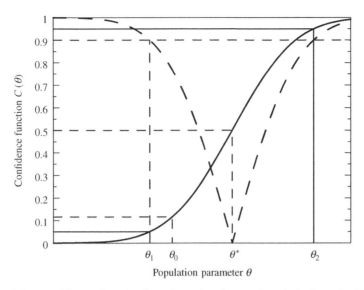

Figure 5.1   The confidence function based on the observed statistical result θ^*. The solid curve shows the one-sided confidence function, whereas the dashed curve corresponds to the two-sided one. See text for details.

- Taking the inverse image of an interval of length d on the y-axis, we obtain a *confidence interval* for the parameter θ with *confidence level d*. In the graph there is a symmetric 90% confidence interval (θ_1, θ_2) around θ^*, obtained by finding the points on the θ-axis with confidence function values equal to 0.05 and 0.95 (i.e., $C(\theta_1) = 0.05$ and $C(\theta_2) = 0.95$, respectively).

We noted in Section 4.5.2 that for symmetric two-sided confidence statements (which confidence intervals typically are) we can use the distribution function of the square of the test statistic. Since many one-sided tests are based on the Gaussian distribution, this means using the χ^2 distribution (see Appendix 5.A.2). We will use the notation $\chi_n(x)$ for the CDF of the $\chi^2(n)$ distribution. This is illustrated in Figure 5.1, where the dashed curve shows the confidence function for two-sided alternatives. Inspecting it, we find that (1) the estimate θ^* corresponds to the minimum, and (2) the 90% confidence interval for θ is obtained by reading off where the dashed curve intersects the line $y = 0.9$ (the horizontal dashed line). The one-sided confidence function is the one with most information, because when we square, we lose track of the sign, but the two-sided version is required in many problems.

A first illustration of this was given in Section 4.6.1, to which we want to add two further examples, both of which are concerned with one-parameter problems. In the next section we will address the more complicated problem with two parameters when we are only interested in some particular combination of these, which will leave us with one nuisance parameter.

Example 5.1   This is a continuation of the Fisher's exact test discussed in Section 4.5.1 and uses the data of the first study there. Fisher's test can be expressed as a test of the null hypothesis that the odds ratio is one, and we now want to exploit this test further in

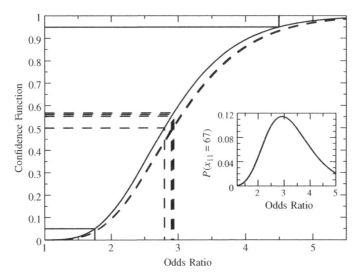

Figure 5.2   The exact confidence function for the odds ratio for the first Hodgkin's lymphoma table is shown as the solid curve, whereas some approximations based on natural estimates of the odds ratio are shown as the almost overlapping dashed curves – see text. The inset graph shows the likelihood function that determines the maximum likelihood estimate.

order to obtain knowledge about the true odds ratio from available data. What we do is essentially to repeat Fisher's test for arbitrary values θ of the odds ratio. For each θ we compute the probability that the observation is 67 or larger, which gives us the confidence function $C(\theta)$ that is plotted as the solid curve in Figure 5.2. Also illustrated is the symmetric 95% confidence interval for the odds ratio, which is $(1.75, 4.49)$, and is indicated by the solid lines in Figure 5.2.

There is more to be said about this example, also illustrated in Figure 5.2. The computation of the confidence interval for the odds ratio was done without reference to a point estimate. How we obtain a point estimate of the odds ratio is a separate question. There are quite a few options available.

- One estimate would be the point that gives 50% confidence, that is, the solution of the equation $C(\theta) = 0.5$. In our case this point estimate is 2.79.

- Intuitively the most obvious choice is probably the empirical odds ratio which was discussed at the end of Section 4.5.1, which we know is 2.93.

- A further choice is to use the θ for which the observed count equals the expected count. This is the θ which solves the equation $x_{11}^* = E_\theta(x_{11})$ and gives us the estimate 2.92.

- A final choice is to use the maximum likelihood estimate, which in this case is 2.90.

All these estimates are shown in Figure 5.2, where we see that all but the first of these estimates correspond to more than 50% confidence on the solid curve. We also see that the variability for the other three estimates is very small. Each of these suggested point estimates for the odds ratio has its own rationale and to each of them there is associated a confidence function, also indicated in Figure 5.2 as the dashed curves. These confidence curves agree well with each other, but not with the original, solid, one. Why this is will be explained after we have discussed how the alternative confidence functions are obtained.

As can be seen from the discussion in Appendix 5.A.1, and will be further detailed in Example 13.3, under the assumption of a (non-central) hypergeometric distribution for x_{11} we have that if $\hat{\theta}$ is the empirical odds ratio, then

$$\ln(\hat{\theta}) \in AsN\left(\ln(\theta), \frac{1}{\mu} + \frac{1}{n_{1+} - \mu} + \frac{1}{n_{2+} - \mu} + \frac{1}{n_{2+} - r + \mu}\right),$$

where $\mu = E_\theta(x_{11})$ and r is the column sum we condition on. Inserting the observation x_{11}^* for μ provides us with an approximative confidence function for the odds ratio, based on the empirical odds ratio. With our data we have that the estimated log-odds ratio is 1.076, and its variance is estimated as $1/67 + 1/34 + 1/43 + 1/64 = 0.083$. This gives a 90% confidence interval of $(0.602, 1.551)$ for the logarithm of the odds ratio, which upon exponentiation gives the corresponding interval for the odds ratio as $(1.82, 4.71)$.

The next choice for an estimate of the odds ratio in the list above is the solution θ of the equation $x_{11}^* = E_\theta(x_{11})$. For this estimator we can obtain an approximative confidence function (by appealing to large-sample theory) as

$$C(\theta) = \Phi\left(\frac{E_\theta(x_{11}) - x_{11}^*}{\sqrt{V_\theta(x_{11})}}\right).$$

How to compute the mean and variance is discussed in Appendix 5.A.1.

Finally, we have the maximum likelihood approach, in which we compute, for different values of the odds ratio θ, the probability of obtaining *exactly* the observed value 67. This gives us a function of θ which is illustrated in the inset graph in Figure 5.2, a function which has its maximum at the point 2.90, which therefore is the maximum likelihood estimate. (The general method of obtaining the confidence function for a maximum likelihood estimator is the subject of Section 13.3.)

It remains to explain why the dashed curves differ from the solid one in Figure 5.2. This explanation has less to do with the fact that the three alternatives build on approximations, appealing as they do to large-sample theory. The root cause of the difference is the discrete nature of the hypergeometric distribution. In this case we have not used the mid-P adjustment in equation (4.2), which we did for the binomial parameter in the previous chapter. With the mid-P adjustment the solid curve moves to the dashed ones, and the confidence interval with it.

The next example describes another one-parameter problem.

Example 5.2 We wish to study a certain adverse event, which is likely to occur with long-term treatment using a particular drug. In order to estimate the risk for this adverse event, we planned to follow 20 subjects for 5 years. The actual observation times for the patients varied between 4.4 and 6.9 years and the event occurred during this observation period in 12 out of the 20 subjects. This particular adverse event can appear anywhere during treatment, so we assume there is a constant hazard λ for it. If subject i is observed for time T_i, the probability that he experiences the event while being observed is $p_i(\lambda) = 1 - e^{-\lambda T_i}$. Denote by x_i the outcome for subject i: it is one if the event has occurred, otherwise zero.

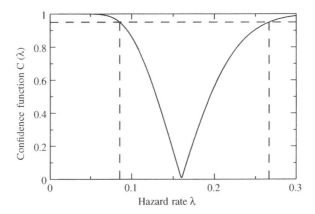

Figure 5.3   Two-sided confidence function for a hazard rate.

The total number of events, $x = x_1 + \ldots + x_n$, is not an observation of a binomial distribution, even though each x_i is an observation of a Bin$(1, p_i(\lambda))$ distribution. We can, however, still appeal to the CLT to find that the total number of events is asymptotically Gaussian in distribution:

$$x = \sum_{i=1}^{n} x_i \in AsN \left(\sum_i p_i(\lambda), \sum_i p_i(\lambda)(1 - p_i(\lambda)) \right).$$

From this we derive the two-sided (approximate) confidence function

$$C(\lambda) = \chi_1 \left(\frac{(x - \sum_i p_i(\lambda))^2}{\sum_i p_i(\lambda)(1 - p_i(\lambda))} \right).$$

We estimate λ as the value λ^* for which $C(\lambda) = 0$ (i.e., solve $\sum_i p_i(\lambda) = x$) and obtain confidence limits for λ as shown in Figure 5.3. We see that the hazard is estimated as 0.16 with 95% confidence interval $(0.09, 0.27)$. We can also compute the predicted probability that an individual experiences an adverse event within 5 years as $1 - e^{-5 \cdot 0.16} = 0.55$, with corresponding 95% confidence interval $(0.35, 0.74)$.

We have used the confidence function in a way that makes it only a graphical tool for obtaining confidence intervals and p-values. We have tacitly assumed that we are only dealing with situations in which such things can be computed. However, looking at the one-sided confidence curve in Figure 5.1 it is tempting to see it as a CDF for the parameter θ and, subsequently, make probabilistic statements about θ. There are two problems with this:

1. There is in general no guarantee that the one-sided confidence function is monotonically covering the entire interval $(0, 1)$, which is a necessary condition for it to be a CDF.

2. There is a considerable conceptual leap involved in assigning a probability to θ. What the confidence function does is to assign probabilities to where we find θ, which is fundamentally different.

We will encounter a famous example of the first problem later, in Example 7.7. When it comes to the second problem, we may note that Fisher himself argued for such a probabilistic viewpoint. He did this in 1930, three years before Kolmogorov formulated his axioms for probabilities and four years before Neyman introduced the concept of the confidence interval. Fisher also assigned a special name to this probability, *fiducial*. He set about introducing it by finding a substitute for the subjective priors used in Bayesian theory. Fisher had a distaste for subjectivity and wanted probabilistic statements derived from objective data only. The concept led to a lot of controversy and more or less died with Fisher himself. As already indicated, there are situations where very specific assumptions on the prior in the Bayesian approach are equivalent to this fiducial probability, the prime example being the *t*-test (and shift models in general).

The controversy around fiducial probabilities is better left to the theoretical statisticians; we only use the confidence function as a convenient tool to summarize the knowledge obtained from an experiment. It is, however, interesting to note that the difference between using the confidence function and using confidence intervals is essentially a continuation of the debate about the nature of the *p*-value outlined in Box 1.3; the confidence function is much closer to the view advocated by Fisher, whereas a single confidence interval, with its prespecified confidence level, is more of a Neyman–Pearson tool for decision making.

5.4 Statistical analysis of two proportions

5.4.1 Some ways to compare two proportions

For a more complicated situation where we want to derive confidence about a parameter, we next discuss ways to compare two proportions. We assume we have observations x_i, $i = 1, 2$, of two independent binomial distributions $\text{Bin}(n_i, p_i)$, respectively, and we want to compare p_1 and p_2. For the purposes of the discussion we assume that the first group consists of individuals exposed to something, and the second group is a control group. We want to summarize the group comparison in a single parameter, which we denote θ in general. The most important choices for θ are:

1. the difference $\Delta = p_1 - p_2$, which measures the additive contribution of the exposure;

2. the relative ratio $RR = p_1/p_2$, closely related to the relative contribution of the exposure $(p_1 - p_2)/p_2 = RR - 1$;

3. the odds ratio $OR = p_1(1 - p_2)/[p_2(1 - p_1)]$, which is the ratio of the odds for the two groups.

(We may note in passing that the odds ratio can be written as a relative risk, because (by independence)

$$OR = \frac{P(X_1 = 1 \text{ and } X_2 = 0)}{P(X_1 = 0 \text{ and } X_2 = 1)} = \frac{P(X_1 > X_2)}{P(X_1 < X_2)}.$$

It is therefore the relative probability of obtaining a larger value on the first variable, to that of a smaller. This is a formulation that can be extended to more general data.) The three measures discussed above are geometrically described in Figure 5.4, and are the ones we will focus on. There are, however, other useful combinations of p_1 and p_2.

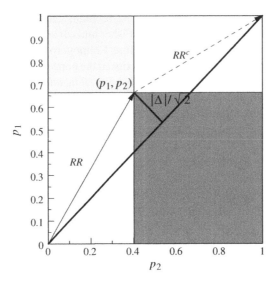

Figure 5.4 Geometrical illustration of parameters comparing two binomial probabilities. RR is the slope of the solid arrow, $|\Delta|/\sqrt{2}$ the distance to the identity line, and OR the ratio of the dark gray to the light gray rectangle (we can also see it as the ratio of the slopes RR and $RR^c = (1 - p_1)/(1 - p_2)$).

Example 5.3 If we measure the frequency of a particular outcome in an exposed and an unexposed group, respectively, we get the *attributable fraction among those exposed* as

$$\frac{p_1 - p_2}{p_1} = \frac{RR - 1}{RR}.$$

This measures the proportion of cases that would be eliminated if we eliminated the exposure, should those in the exposed group have the same risk for being a case as those in the unexposed group (see Box 5.2). Related to the attributable fraction is the *excess risk q*, which is the probability that an exposed individual becomes a case, had he not been one when unexposed, which means that it is derived from the equation $p_1 = p_2 + q(1 - p_2)$ or $q = (p_1 - p_2)/(1 - p_2)$.

Another alternative transformation of p_1 and p_2 is the *number needed to harm*, which refers to how many individuals we need to expose in order to get one extra case. It is defined by the relation $N \cdot \Delta = 1$ and is therefore given by the inverse of Δ. It is often used by clinicians as a way to quantify risks. However, it is a measure with drawbacks. Suppose we have two different exposures with probabilities p_1 and p_2 and let p_0 denote the risk for an unexposed individual. Then the extra risk of the first exposure over the second exposure is the difference of the risk differences to the unexposed, but the relation for the numbers needed to harm is more complicated:

$$p_1 - p_2 = (p_1 - p_0) - (p_2 - p_0) \quad \Leftrightarrow \quad \frac{1}{N_{12}} = \frac{1}{N_{10}} - \frac{1}{N_{20}},$$

Box 5.2 The attributable risk in the population

The *attributable risk*, also known as the *etiologic factor* (or fraction) is a widely used measure for assessing the public health consequences of an association between an exposure (E) and a disease (C). It was first introduced by Morton B. Levin in 1953 as a way to quantify the impact of smoking on lung cancer occurrence, and is defined as

$$AR = \frac{P(C) - P(C|E^c)}{P(C)}.$$

It measures the proportion of disease cases that can be related to the exposure, and can therefore be used to assess the potential impact of a prevention program. It takes into account both the strength of the association between exposure and disease, and the prevalence of the exposure. It should be recognized that it assumes that the probability of the disease in an exposed subject when the exposure is removed is the same as that for an unexposed subject, an assumption we can seldom prove.

If we replace $P(C)$ with $P(C|E)$ we get the attributable risk $AR(E)$ among those exposed. Its relation to the attributable risk is that the latter is obtained by multiplying by the proportion of those with the disease that has been exposed:

$$AR = AR(E)P(E|C).$$

This is derived using Bayes' theorem. For rare diseases we can estimate AR from a case–control study, if we approximate the relative risk with the odds ratio.

---

where N_{ab} denotes the numbers needed to harm comparing groups a and b. This property makes it doubtful that this parameter really is easily interpretable. When the event is positive we talk about the *number needed to treat* (NNT).

In order to build the statistical machinery for the analysis of any of the parameters discussed above, we start from the large-sample $N(p, p(1 - p)/n)$ approximation for the distribution of the estimator $\hat{p} = x/n$ of p. This information can be used in two different ways. First, we may note that

$$\frac{n(\hat{p} - p)^2}{p(1 - p)} = \frac{(x - np)^2}{np(1 - p)} \in \chi^2(1),$$

and since our two groups are independent, the sum of these expressions for the two groups will have a $\chi^2(2)$ distribution:

$$\frac{(x_1 - n_1 p_1)^2}{n_1 p_1(1 - p_1)} + \frac{(x_2 - n_2 p_2)^2}{n_2 p_2(1 - p_2)} \in \chi^2(2).$$

How to use this observation in order to derive confidence statements about different functions of p_1 and p_2 will be discussed in Section 7.7. That discussion will allow us to simultaneously obtain confidence information for any number of derived parameters we want. It is, however, more involved mathematically than the approach we will discuss here.

Instead we will build on the observation that the difference of two Gaussian stochastic variables also has a Gaussian distribution and therefore that

$$\hat{p}_1 - \hat{p}_2 \in N\left(p_1 - p_2, \frac{p_1(1-p_1)}{n_1} + \frac{p_2(1-p_2)}{n_2}\right), \tag{5.1}$$

at least approximately for large n_1 and n_2. In order to make inference about the risk difference $\Delta = p_1 - p_2$, we rewrite equation (5.1) in Δ and one further parameter, for which we choose p_2:

$$\hat{\Delta} = \hat{p}_1 - \hat{p}_2 \in N\left(\Delta, p_2(1-p_2)\left(\frac{1}{n_1} + \frac{1}{n_2}\right) + \frac{\Delta(1-2p_2-\Delta)}{n_1}\right).$$

The problem when we use this to derive a confidence function for Δ is that we need to do something about the nuisance parameter p_2. One suggestion is to estimate it from data. This is an approximation, but it is only the variance that depends on p_2, and only mildly so, so we do not really need a high-precision estimate of it. We will illustrate this in the next section.

5.4.2 Analysis of the group difference

To see how we can do the statistical analysis of the difference Δ we use the data of the first Hodgkin's lymphoma example in Section 4.5.1, with $p_1^* = 67/101 = 0.66$ and $p_2^* = 43/107 = 0.40$. The difference is therefore estimated to be $\Delta^* = 0.26$. Based on the discussion in the previous section, we can derive knowledge about the true difference Δ from the confidence function

$$C(\Delta) = \Phi\left(\frac{\Delta - \Delta^*}{\sigma(\Delta, p_2^*)}\right),$$

where we have inserted the estimate p_2^* for p_2 in the variance expression

$$\sigma^2(\Delta, p) = p(1-p)\left(\frac{1}{n_1} + \frac{1}{n_2}\right) + \frac{\Delta(1-2p-\Delta)}{n_1}. \tag{5.2}$$

The function $C(\Delta)$ is graphed as the solid curve in Figure 5.5, from which we can deduce, for example, that the symmetric 90% confidence interval around Δ^* is (0.15, 0.37). What is not shown is the one-sided p-value for the null hypothesis of $\Delta \leq 0$. This is given by $C(0) = \Phi(-\Delta^*/\sigma(0, p_2^*)) = 0.00008$ and is too small to show up in the graph. We note instead that if we test the hypothesis $\Delta \leq 0.15$ we get a p-value of 0.05; this is part of the information found in the (symmetric) 90% confidence interval above.

This is actually not the standard p-value computed for the null hypothesis $\Delta \leq 0$. What is usually done is to compute the standard deviation using $\bar{p} = (67 + 43)/(101 + 107) = 0.53$ instead of p_2^*. This is because when $\Delta = 0$, this should be a better estimate of the parameter p_2 than p_2^*. Similarly, it is common practice to replace Δ in the expression for the variance with its estimate Δ^* (cf. Box 4.7). The confidence function obtained with these substitutions is shown as the dashed curve in Figure 5.5. We see that there is a difference, but it is very small and of no consequence in this particular case.

We may also note the following. A short calculation shows that

$$\frac{\Delta^*}{\sigma(0, \bar{p})} = \frac{x_1 - E_1(x_1)}{\sqrt{V_1(x_1)}},$$

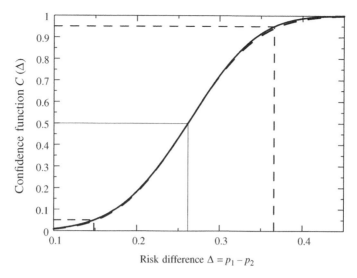

Figure 5.5 Confidence functions for the two-group binomial test. The two functions differ in how the variance is estimated. For the solid curve we vary Δ and use the frequency in the second group to estimate p_2, whereas the dashed curve uses the probability estimates in the variance estimate.

where the mean and variance on the right are computed under the assumption of a hypergeometric distribution for x_1 (as part of a 2×2 table with fixed margins). Actually this is not quite true, there is a factor $\sqrt{n/(n-1)}$ in front of the right-hand side, but we can eliminate it if we bias-correct the estimator $\sigma^2(0, \bar{p})$ (see page 158). This means that, when we use the large-sample approximations, the binomial test and Fisher's exact test give identical p-values for the hypothesis of no difference. The difference between the tests is in their small-sample properties, which is when the different assumptions play a role.

We next turn our attention to the relative risk $\theta = p_1/p_2$. We can obtain a confidence function for this ratio in a way very similar to what we did for the difference by using a trick due to Fieller (see Box 7.7), which uses the observation that

$$\hat{p}_1 - \theta \hat{p}_2 \in AsN\left(0, \frac{p_2\theta(1 - p_2\theta)}{n_1} + \theta^2 \frac{p_2(1 - p_2)}{n_2}\right).$$

As was the case when we analyzed Δ above, it is only the variance that depends on p_2, and it does so only mildly. Carrying through the analysis as we did for Δ above, we can deduce that the estimated relative risk is 1.65 with 90% confidence interval (1.32, 2.08).

In order to get a reasonable estimate of the odds ratio θ we proceed along a different route, building on the observation made at the end of Section 4.6.1 about the large-sample approximation to the logarithm of the odds $x/(n-x)$. Since the logarithm of the odds ratio is the difference of two independent log-odds, we see that, for the empirical odds ratio $\hat{\theta} = x_1(n_2 - x_2)/x_2(n_1 - x_1)$,

$$\ln \hat{\theta} \in AsN\left(\ln(\theta), \frac{1}{n_1 p_1} + \frac{1}{n_1(1 - p_1)} + \frac{1}{n_2 p_2} + \frac{1}{n_2(1 - p_2)}\right).$$

If we insert estimates into this variance expression, this becomes the same analysis as we did in the previous section, though at that time we assumed a hypergeometric distribution, and we found there that the estimate and 90% confidence interval for the odds ratio are 2.93 and (1.82, 4.71), respectively.

We conclude this section with a comment about how a Bayesian statistician might approach this problem. Theoretically he would do the obvious follow-up on what he did for a single proportion, possibly using different priors for the two proportions. Once the priors are decided, the data will give him a two-dimensional distribution for the two probabilities, and with some mathematical tricks he can use standard probability theory to make any kind of probability statements he wants for any particular function of these parameters (with no adjustments made for multiplicity issues).

To be more specific, let us use the same data as above and assume that we believe, *a priori*, that there is no difference between the two groups and that the two proportions have independent and identical *a priori* distributions. If the common prior has a $\beta(x; a, b)$ distribution, we get an *a posteriori* distribution for the pair (p_1, p_2) as the product of two beta distributions: $\beta(x; x_1 + a, n_1 - x_1 + b)$ and $\beta(x; x_2 + a, n_2 - x_2 + b)$. From this we can compute the distribution of any function of (p_1, p_2) we want to, which is illustrated in Figure 5.6 for the difference $\Delta = p_1 - p_2$. There are two curves. The black one to the left uses independent informed priors $\beta(x; 4, 6)$ for the two parameters, whereas the gray one to the right uses uninformed priors. The median for the distribution of Δ with uninformed priors is 0.25 with 90% credibility interval (0.14, 0.36), which is similar to what the confidence function produced in terms of point estimate and 90% confidence interval. The corresponding estimate with informed priors is smaller, 0.23 with 90% credibility interval (0.13, 0.34). The reason why it is smaller is that our prior belief was that there was no difference between p_1 and p_2, so that Δ was symmetrically distributed around zero, and old beliefs are hard to change. But the evidence was somewhat overwhelming, and we therefore ended up relatively close to what the uninformed priors provide.

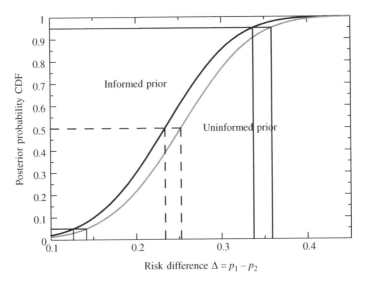

Figure 5.6 Bayesian analysis of the difference of two binomial proportions.

5.5   Adjusting for confounders in the analysis

If we want to investigate an exposure–response relationship defining a 2×2 table and we have known and measurable confounders, one way to adjust for them in the analysis is to divide the population into smaller groups for which we have similar values for these confounders. This gives us a number N of 2×2 tables within each of which we can apply Fisher's exact test, as outlined in Section 4.5.1. The assumption is that within each table there is homogeneity to such an extent that a common odds ratio within that group is a reasonable description. The problem is to summarize this information into a common odds ratio applicable to all groups or, if that is not possible, a simple description of the variability of the odds ratios between groups. As already mentioned, these groups are called strata, and the process of dividing the population up into strata is called stratification.

For example, if we want to see whether there is an effect of a particular exposure on the occurrence of Down's syndrome, we know from previous examples that we should stratify with respect to the age of the mother. This means that we define a sequence of age classes, and do the comparison between exposed and non-exposed pregnancies within each age class. In order to define the age spans of the different strata, it may be helpful to look at the risk function for the event as a function of age, as shown in Figure 2.2(a). We can also define strata by combining a set of different confounders, such as age, smoking habit and social class.

A particular example is when we want to do a meta-analysis and try to combine the information from a number of different studies, possibly of different designs, each providing a 2×2 table from which we want a collated summary. The strata in this case are the individual studies.

Example 5.4   In Section 4.5.1 we discussed two studies investigating tonsillectomy as a risk factor for Hodgkin's lymphoma. The second of these had much more data than was used in the analysis, because the investigators first ascertained the tonsillectomy history of 174 patients with Hodgkin's lymphoma and all of their 472 siblings. They then matched the data in a way that actually ignored more than 70% of these data. We want to use all 646 patients in this study and combine them with those of the first study in a meta-analysis. The data are given in the following table:

	Patients		Controls		
	+	−	+	−	OR
Study 1	67	34	43	64	2.93
Study 2	90	84	165	307	1.99
Sum	157	118	208	371	2.37 (1.83, 3.00)

The last row contains the marginal table, in which we have ignored which study the data come from. It also contains the corresponding empirical odds ratio and its 90% confidence interval. In this table we ignore the familial relations in the second study. To do a traditional meta-analysis we need a way to combine the information about a common odds ratio from those of the individual tables. Some ideas about how to obtain knowledge about the odds ratio within a single table were addressed in Section 5.3, and our objective is to extend one of these ideas to a series of tables.

The method to generalize in Section 4.5.1 is the one that equated the observation of x_{11} with its mean. There are two ways to extend this to two tables. Both methods are based on the idea of looking at the squared difference between what we observe and what a model predicts, normalized to what we expect this squared difference to be. The difference between the methods is whether you sum the observations first, or sum the normalized squares:

1. If we first sum the observations from the two tables, and then normalize with the variance for the sum, we get the function

$$Q_{MH}(\theta) = \frac{(x_{11} + y_{11} - E_\theta(x_{11}) - E_\theta(y_{11}))^2}{V_\theta(x_{11}) + V_\theta(y_{11})}.$$

   We will call this the Mantel–Haenszel quadratic form (it is a quadratic form in the observations, not the parameter, but we study it as a function of the parameter).

2. If we first normalize each squared difference with its variance and then compute the sum, we get the function

$$Q_{BD}(\theta) = \frac{(x_{11} - E_\theta(x_{11}))^2}{V_\theta(x_{11})} + \frac{(y_{11} - E_\theta(y_{11}))^2}{V_\theta(y_{11})},$$

   which we call the Breslow–Day quadratic form.

With large numbers, $x_{11} + y_{11}$ is approximately Gaussian in distribution, and we should have that $Q_{MH}(\theta)$ is distributed approximately as $\chi^2(1)$. We can use this to define a pooled odds ratio for the two tables by finding the θ for which the quadratic form is zero, and we can also derive a two-sided confidence function for θ, which is shown as the solid function in Figure 5.7 for our data. From this we get the estimate of the pooled odds ratio as 2.22, and the symmetric 90% confidence interval is given by (1.73, 2.85).

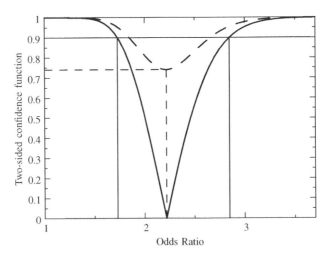

Figure 5.7   The two curves in this graph are the $\chi^2(1)$ maps of the quadratic forms $Q_{MH}(\theta)$ and $Q_{BD}(\theta)$ discussed in the text. The solid curve is the one for the Mantel–Haenszel form, and defines a two-sided confidence function for θ. The dashed one is for the Breslow–Day form, and is used to test for homogeneity as indicated by the dashed lines.

The Breslow–Day quadratic form $Q_{BD}(\theta)$ is used differently, namely to assess how reasonable it is that a particular pooled odds ratio θ is a good description of the data. If a particular value θ applies to both tables, this quadratic form should also be zero. The amount it differs from zero is therefore a measure of the heterogeneity of the two tables. To translate this to a probability we need to compare its value to the $\chi^2(1)$ distribution. The corresponding function is shown in Figure 5.7 as the dashed curve and by taking one minus the confidence in the estimate for the pooled odds ratio, we get a p-value for the hypothesis of homogeneity of the odds ratios in the two studies. In this case it turns out to be 0.26.

However, this example does not really describe how most epidemiologists work with pooled odds ratios. In order to understand their preferred method we first generalize the present discussion to a general series of 2×2 tables. In doing this we change the notation so that we have N different strata, and in

	E	E^c
C	a_k	b_k
C^c	c_k	d_k

each a 2×2 table with data as shown in the table shown on the right. We can then define the following two quadratic forms:

1. The Mantel–Haenszel quadratic form is defined as the function

$$Q_{MH}(\theta) = \frac{(\sum_{k=1}^{N} a_k - E_\theta(a_k))^2}{\sum_{k=1}^{N} V_\theta(a_k)}.$$

   A pooled estimate θ^* is obtained as the point where this is zero, and a confidence function for θ is obtained as $C(\theta) = \chi_1(Q_{MH}(\theta))$.

2. The Breslow–Day quadratic form is defined as the function

$$Q_{BD}(\theta) = \sum_{k=1}^{N} \frac{(a_k - E_\theta(a_k))^2}{V_\theta(a_k)}.$$

   It is used to test for homogeneity among strata of the individual odds ratios by comparing $Q_{BD}(\theta^*)$ to a $\chi^2(N-1)$ distribution.

Example 5.5 Before we continue, let us see what the quadratic form $Q_{MH}(\theta)$ means in the case where we have a one-to-one matched case–control study. In such a study, the investigator selects for each case a control with a specified set of characteristics that are the same as those of the case, and then determines, for both case and control, whether there has been an exposure or not. Each such pair produces a small 2×2 table, in which the total number in each row is one. In this case, what is the quadratic form $Q_{MH}(\theta)$?

Each table has row sums $n_1 = n_2 = 1$, and for the sum of the elements in the first column, r, we can have only three possible values, $r = 0, 1, 2$. However, when $r = 0$ neither case nor control was exposed, and the expected value of a_k must be zero, and when $r = 2$ both case and control were exposed, so the expected value of a_k is necessarily one. In both cases we have $a_k - E_\theta(a_k) = 0$, so the terms involved in $Q_{MH}(\theta)$ are only non-zero if $r = 1$; only those pairs contribute to the analysis for which either the case or the control is exposed, but not both. Let the number of such tables be n.

For a table with $r = 1$ we have (see Appendix 5.A.1) that $E_\theta(a_k) = p = \theta/(1+\theta)$ and that $V_\theta(a_k) = \theta/(1+\theta)^2 = p(1-p)$. It follows that if $z = \sum_k a_k$ is the total number of cases that were exposed, and if we express Q_{MH} as a function of p, $Q_{MH}(p) = (z - np)^2/np(1 - p)$. This is the same test as the large-sample approximation for a $Bin(n, p)$ distribution, and the test for the null hypothesis $\theta = 1$ is therefore McNemar's test mentioned in Section 4.5.1.

Using $Q_{MH}(\theta)$ for the estimation of θ is, however, somewhat complex and it is not clear how the pooled estimate relates to the empirical odds ratio of the individual tables, given by $\theta_k^* = a_k d_k/b_k c_k$. Epidemiologists have therefore largely chosen to use a different pooled estimate, one that is explicitly derived from the individual θ_k^*s. When a_k follows a hypergeometric distribution we have that $E_\theta(a_k d_k - \theta b_k c_k) = 0$ and if we specify stratum weights w_k, we can therefore define a pooled odds ratio from the equation

$$E_\theta \left(\sum_k w_k(a_k d_k - \theta b_k c_k) \right) = 0.$$

The corresponding pooled odds ratio estimate is obtained by inserting observed table data, from which we derive the pooled estimate $\hat{\theta}_{pool} = (\sum_{k=1}^{N} w_k a_k d_k)/(\sum_{k=1}^{N} w_k b_k c_k)$. Each choice of weights gives its own pooled odds ratio. The particular choice $w_k = 1/N_k$, where $N_k = a_k + b_k + c_k + d_k$ is the total number of observations in stratum k, provides the Mantel–Haenszel pooled odds ratio

$$\hat{\theta}_{MH} = \frac{\sum_{k=1}^{N} R_k}{\sum_{k=1}^{N} S_k}, \quad R_k = \frac{a_k d_k}{N_k}, \quad S_k = \frac{b_k c_k}{N_k}.$$

This is a weighted sum of the individual empirical odds ratios, $\hat{\theta}_{MH} = \sum_{k=1}^{N} w_k \theta_k^*$, where $w_k = S_k/\sum_{k=1}^{N} S_k$. We know that the variances of the individual empirical odds ratios are approximately given by

$$\frac{1}{a_k} + \frac{1}{b_k} + \frac{1}{c_k} + \frac{1}{d_k} = \frac{(a_k + d_k)\theta_k + b_k + c_k}{b_k c_k} \approx \frac{1}{S_k},$$

where the last approximation is valid when $\theta_k \approx 1$, which means that the Mantel–Haenszel pooled odds ratio is close to being the most efficient weighting of the individual table odds ratios. It is so if the individual odds ratios are one.

As an example, consider the one-to-one matched case–control study. In each cell we have that $a_k d_k = 1$ if and only if case is exposed and control is not, whereas $b_k c_k = 1$ if the reverse is true. In all other cases these products are zero, so the Mantel–Haenszel odds ratio becomes the ratio $(\sum_{k=1}^{N} b_k)/(\sum_{k=1}^{N} c_k)$ of the off-diagonal elements in the overall summary table (cf. page 98).

In order to obtain a confidence function for the Mantel–Haenszel estimator we need its variance. As we will see in Example 13.3, we should do the analysis on the log scale, with the variance of $\ln \hat{\theta}_{MH}$ well approximated by

$$\frac{1}{2(\sum_k R_k)^2} \sum_{k=1}^{N} N_k^{-2}(a_k d_k + b_k c_k \theta_{MH})(a_k + d_k + (b_k + c_k)\theta_{MH}). \tag{5.3}$$

(There are other approximations for the variance of $\hat{\theta}_{MH}$, on the original scale and on the log scale, but experience and simulation studies indicate that the variance given by equation (5.3) is to be preferred over others.) If we apply these results to the data in Example 5.4, we find the pooled odds ratio to be 2.23 with 90% confidence interval given by (1.73, 2.86), very close (but not identical) to what was obtained with the first method.

We noted above, and it is worth repeating, that the Mantel–Haenszel pooled estimate of the odds ratio is obtained as a weighted average of the individual empirical odds ratios. The weights were so determined that they were essentially optimal from a statistical perspective. They had no other, independent, meaning. From a practical point of view this means that the estimate obtained is really only meaningful if all individual odds ratios (the true ones, not the empirical ones) are equal. In such a case the pooled estimate will estimate this number in an efficient way.

There are alternative ways to estimate a pooled odds ratio from a stratified analysis. A common alternative to the Mantel–Haenszel method is derived from the problem of estimating a pooled odds ratio from the equation $Q_{MH}(\theta) = 0$, which was mentioned in Example 5.4. The equation to solve is $G(\theta) = \sum_k (a_k - E_\theta(a_k)) = 0$, which is a nonlinear equation, so a solution must be obtained by iterative methods. If we write $\theta = e^\beta$ and consider only one iteration (starting with $\beta = 0$), we get the equation $\sum_k (a_k - E_1(a_k) - \beta V_1(a_k)) = 0$, and if we solve this we arrive at Peto's suggestion for a pooled odds ratio,

$$\hat{\theta}_{Peto} = \exp\left(\sum_k (a_k - E_1(a_k))/V_1(a_k)\right).$$

How to obtain confidence intervals, etc., is left to the reader. Even though the Peto estimate is expected to behave well when the true odds ratio is close to one, it is not known for good properties when this is not the case.

Pooling the odds ratios is the only option for case–control studies, but in a cohort study we might prefer to estimate pooled proportions for exposed and non-exposed groups, respectively, and derive a pooled relative risk instead. Alternatively, we might consider rates. The corresponding Mantel–Haenszel methodology is outlined in Box 5.3. Again the weights are obtained for optimal statistical efficiency, and the value of the pooled group estimates depends on the relative contribution of the individual strata. If the stratum-specific probabilities/rates differ, the pooled ones may be a rather meaningless estimate of some population probability/rate.

To be more specific, consider rates and assume that we have stratified on age, so that we have a rate $\lambda = \sum_a w_a \lambda_a$, where λ_a denotes the rate in age stratum a. If we want the λ to represent the overall risk in the population, we should take weights w_a corresponding to the size of the age class a in the population. However, that may not be the age distribution in the study, unless we have representative sampling. The weights of the pooled analysis may therefore provide an estimate for a population with a different age structure than the one we want to relate to. In order to give meaning to pooled measures, we therefore need to define the weights properly, a process called *standardization*. Note that the standardized rate will typically be estimated to a lower precision than the pooled one. Epidemiologists often talk about the *standardized mortality ratio*, which is the rate ratio obtained when both nominator and denominator use the weights of the exposed group. It represents the ratio of the number of observed cases, divided by the number of expected cases.

Box 5.3 Mantel–Haenszel type estimate of a pooled relative risk

---

For a cohort study we can adjust the Mantel–Haenszel method of obtaining a pooled estimate to obtain one for a relative risk. This can be applied equally well to proportions and rates. Suppose that we have observed x_{ij} events in group j in stratum i and assume that one of the following two cases applies:

1. $x_{ij} \in \mathrm{Bin}(T_{ij}, p_{ij})$, where T_{ij} is total number of observations;

2. $x_{ij} \in \mathrm{Po}(\lambda_{ij}T_{ij})$, where T_{ij} is observation time.

Let the relative risk in stratum i be denoted θ_i, so that $\theta_i = p_{i1}/p_{i2}$ in the first case and $\theta_i = \lambda_{i1}/\lambda_{i2}$ in the second case. We wish to define a pooled relative risk θ_{MH} from these these.

The relative risk in stratum i satisfies the equation $E_{\theta_i}(x_{i1}/T_{i1} - \theta_i x_{i2}/T_{i2}) = 0$ so a pooled relative risk could be defined from an equation of the form

$$E_\theta \left(\sum_i w_i(x_{i1}/T_{i1} - \theta x_{i2}/T_{i2}) \right) = 0.$$

If, for each i, we take as weights the harmonic mean of the T_{ij}, that is, $w_i = 2T_{i1}T_{i2}/T_{i+}$, where $T_{i+} = T_{i1} + T_{i2}$, we arrive at the Mantel–Haenszel relative risk estimate

$$\hat\theta_{\mathrm{MH}} = \frac{\sum_i(x_{i1}T_{i2}/T_{i+})}{\sum_i(x_{i2}T_{i1}/T_{i+})}.$$

This corresponds to the pooled probability estimates $p_j^* = \sum_i w_i' p_{ij}^*$ for group j, where $w_i' = w_i/\sum_i w_i$, and similarly for rates. Much as in Example 13.3 the variance of $\ln\hat\theta_{\mathrm{MH}}$ is approximately given by

$$\frac{\sum_i(T_{i1}T_{i2}x_{i+}/T_{i+}^2)}{\theta_{\mathrm{MH}}\left(\sum (T_{i1}T_{i2}x_{i+}/T_{i+}(T_{i1} + T_{i2}\theta_{\mathrm{MH}}))\right)^2}$$

We are appealing to Gaussian approximations here, so we can alternatively use the Fieller approach (see Box 7.7) to obtain a confidence function for θ_{MH}.

5.6 The power curve of an experiment

In the planning phase for an experiment it is important to figure out how large it should be – how many patients to include in a clinical study. If the experiment is too small, so that the statistical analysis has no realistic chance of detecting a difference, it may be a waste of money to make the experiment at all (or unethical, when it comes to clinical trials). On the other hand, if it is too large, providing more evidence than necessary, there is also some concern about money spent unnecessarily (and possibly patients put at risk). To find the right size of an experiment we need to investigate the statistical properties of a proposed size. To do that, we first need to be explicit about the objective of the experiment.

The standard protocol for this is as follows. We want to reject the null hypothesis $\theta = 0$ at a specified significance level α. That means that we consider the experiment a success if we get a p-value which is less than α. That part is the actual statistical test, and controls the Type I error. As noted at the end of Section 5.2, this defines a particular value, x_{crit}, such that if the value of the test statistic is equal to this, the p-value is α. (We continue to consider one-sided tests where a large value of the test statistic is evidence against the null hypothesis.)

We also want to understand the Type II error. This involves the true distribution $G(x) = F(x, \theta_{\text{true}})$ for the test statistic. We want to know what our chances are of a successful outcome of the experiment. However, we do not know θ_{true}, so we investigate the probability

$$\beta(\theta) = 1 - F(x_{\text{crit}}, \theta)$$

for a range of θ where we expect θ_{true} to be. This function is called the *power function* of the experiment and is such that $\beta(0) = \alpha$ by construction. Actually the notation is unconventional here; usually $\beta(\theta)$ is used to denote the Type II error, so the power function should be $\beta^c(\theta)$.

If we vary x_{crit} we get a family of different power curves. Each such curve corresponds to a particular choice of significance level, which can be read off from its value at $\theta = 0$. Some further insight into the nature of power curves is obtained in the following example. (This is the slightly confusing case when we estimate a location parameter, so that the experiment outcome x works on the same scale as the parameter θ, and we therefore label values on the x-axis in two different ways.)

Example 5.6 Assume that the parameter θ is a location parameter (such as a mean) and that the statistical model is a shift model, which means that $F(x, \theta)$ takes on the form $F(x - \theta)$ for some CDF $F(x)$, assumed to be symmetric around $x = 0$. This special case has some geometrical properties which are illustrated in Figure 5.8. For a symmetric and continuous distribution we have that $F(-x) = 1 - F(x)$, which implies not only that $F(0) = 0.5$, but also that

$$\beta(\theta) = 1 - F(x_{\text{crit}} - \theta) = F(\theta - x_{\text{crit}}),$$

so the power function is obtained from a horizontal shift of magnitude x_{crit} of the null distribution $F(x)$. If we introduce the alternative notation θ_c for the critical value x_{crit} we can write $\beta(\theta) = F(\theta - \theta_c)$, and we have that $\beta(\theta_c) = 0.5$. In other words, the value of θ which gives 50% power defines the critical value x_{crit} for the test statistic.

The dashed curve on the left in Figure 5.8 is a part of the graph of $F(\theta)$. The solid curve is this function shifted by the amount θ_c. This is an increasing function which starts with the value α at $\theta = 0$, passes through the value 0.5 at $\theta = \theta_c(= x_{\text{crit}})$, and reaches the level $1 - \alpha$ for $\theta = 2\theta_c$, after which it increases asymptotically to one. To be successful, the experiment must produce an observation of the test statistic which is larger than $x_{\text{crit}} = \theta_c$. When θ is less than θ_c, this probability is less than 50%, and when it is greater than θ_c, the probability is greater than 50%.

So far we have discussed the power curve for a particular experiment. Most often we want to use power considerations to find the appropriate size of the experiment. We therefore introduce the study size n into the notation, so that the CDF for the test statistic becomes $F_n(x, \theta)$. To determine n we must hold something fixed, and this can be done in different ways. We can, for example, decide that we want the test on significance level α and specify

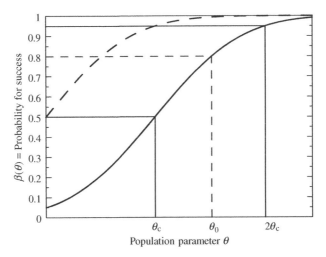

Figure 5.8 The geometric derivation of a power curve when the null distribution is symmetric around zero.

the critical value x_{crit} to be x_c. This gives us the equation $1 - F_n(x_c, 0) = \alpha$, which fixes n (recall that $x_c = \theta_c$). It means that only observations x of the test statistic such that $x > x_c$ will be 'statistically significant'. For a shift model this can also be written $\beta(\theta_c) = 0.5$, so that the size of the experiment is such that the power for the critical value θ_c is 50% (so if θ_c is the true value, only 50% of experiments will be successful). However, the conventional way to determine the sample size uses a different type of argument. We first decide on a value θ_0 which is such that we 'do not want to miss' it. The precise meaning of this is that we specify the probability of getting a successful outcome of the experiment, if θ_0 is the true value. In Figure 5.8 this has been set to 80% and is shown by the dashed lines. It means that we select n so that we obtain the power function that starts at α and satisfies $\beta(\theta_0) = 0.8$. The choice of θ_0 defines how steep the power curve will be, since it is anchored at level α at zero. The smaller the θ_0, the steeper the curve, which means a larger study. Usually this computation involves a Gaussian approximation to the CDF of the test statistic, and the analytic computations needed to derive n are outlined in Box 5.4.

We should also note the following. For the shift model we have that the p-value (for the null hypothesis $\theta = 0$) based on the observation x of our test statistic is $p = 1 - F(x) = 1 - \beta(x + \theta_c)$, which shows that the p-value will be α if $x = \theta_c$, but also that it is considerably smaller than α if $x = \theta_0$ (the effect size we did not want to miss).

It is important to understand that the power $\beta(\theta)$ gives the probability that the study will produce a p-value less than α if θ is the true value. Since we do not know this true value, this is a very crucial assumption, on which the result depends. To get an estimate of the probability that the study as such turns out to be successful requires information on how likely it is for different θ to be the true parameter value. If we have such information, which amounts to an *a priori* distribution for θ, we can use Bayesian ideas to compute an overall probability for a successful study outcome. To emphasize the Bayesian nature of this probability, it is often called assurance instead.

There is a geometrical relationship between the power function $\beta(\theta)$ and the confidence function $C(\theta)$, in particular in the case where we have a shift model (for the location model

---

Box 5.4 Sample size computations

In order to determine the sample size n of an experiment, we usually start with a test statistic $\hat{\theta}$ which estimates a parameter θ about which we want to make claims, together with a model which describes how the distribution of $\hat{\theta}$ depends on n. This is usually of the form

$$\hat{\theta} \in N(\theta, \sigma^2(\theta)/n).$$

For a one-sided test $\theta \leq 0$ this means that the power function is

$$\beta(\theta) = \Phi\left(\frac{\sqrt{n}(\theta - \theta_c)}{\sigma(\theta)}\right),$$

which we solve for n under a criterion of the form $\beta(\theta_0) = q$. Here q is usually 0.8 or 0.95, corresponding to power 80% or 95%, respectively.

To solve for n in this equation, write $q = 1 - \beta$ and note that

$$1 - \beta = \Phi\left(\sqrt{n}\frac{\theta_0 - \theta_c}{\sigma(\theta_0)}\right) = \Phi\left(\frac{\sqrt{n}\theta_0}{\sigma(\theta_0)} - z_\alpha R\right), \quad R = \frac{\sigma(\theta_c)}{\sigma(\theta_0)}.$$

Here $z_\alpha = \Phi^{-1}(1 - \alpha)$ and we have used the fact that $\sqrt{n}\theta_c/\sigma(\theta_c) = z_\alpha$. Solving for n, we get

$$n = (z_\alpha R + z_\beta)^2 \left(\frac{\sigma(\theta_0)}{\theta_0}\right)^2.$$

When σ does not depend on θ, so that $R = 1$, we see that this is the product of two terms, the first entirely concerned with confidence (choice of significance level and power) the second the inverse of the square of θ_0/σ. To use this in practice, we can first compute an n for the commonly used two-sided test at significance level $2\alpha = 0.05$ and power 80%, using the formula

$$n = \frac{7.85\sigma^2}{\theta_0^2}.$$

We then multiply by the factor in the following table to obtain n for other levels of error control:

			Power (%)			
2α	50	60	70	80	90	95
0.01	0.85	1.02	1.22	1.49	1.90	2.27
0.05	0.49	0.62	0.79	1.00	1.34	1.66
0.10	0.34	0.46	0.60	0.79	1.09	1.38
0.20	0.21	0.30	0.42	0.57	0.84	1.09

To illustrate the use of this, assume we want to compare the mean for two groups of sizes rn and $(1 - r)n$, respectively, where $0 < r < 1$. The standard deviation σ above is then given by $\sigma_0/\sqrt{r(1 - r)}$, where σ_0 is the standard deviation of the observations in one group. With equal numbers in the two groups we have $\sigma = 2\sigma_0$. Note that we get best power if the groups are equal in size.

and a one-sided test, the power curve that has 50% power at θ^* is identical to the confidence curve corresponding to the observation $x^* = \theta^*$); it is important to note that the power function is only of interest before the experiment is performed, whereas the confidence function is defined by the outcome of the experiment.

For the rest of this section we will apply this general discussion to the particular problem of planning a study which will compare two independent binomial proportions. The key part of this planning is to decide on how large we need the study to be. Let us assume that we want to draw samples of equal size from two groups, and we propose to use a sample of size $n = 100$ for each group. We want to understand what we can expect from such a study in terms of what the true difference in proportions needs, to be for the test to 'provide a statistically significant result' (our success criteria).

First, we need to decide on a significance level for the test; we use the conventional level $\alpha = 0.05$ for a one-sided test, $\Delta \leq 0$. For fixed p, the critical value Δ_c is obtained as the solution to the equation $1 - \Phi(\Delta/\sigma(0, p_2)) = \alpha$, where $\sigma^2(\Delta, p)$ is defined in equation (5.2) with both n_i equal to $n = 100$. If the observed difference Δ^* is greater than Δ_c, we will reject the null hypothesis. This Δ_c will depend on p_2, so we write it as $\Delta_c(p_2)$. For fixed p_2 this gives us the power function

$$\beta(\Delta) = \Phi\left(\frac{\Delta - \Delta_c(p_2)}{\sigma(\Delta, p_2)}\right).$$

Figure 5.9 gives a graphical illustration of the situation for two different values of p_2. In the upper left part we have plotted, with thinner lines, $\Phi(\Delta/\sigma(0, p_2))$ as a function of Δ. They both start at the level 0.5. On the Δ-axis we have marked where these curves intersect the horizontal line at level $1 - \alpha = 0.95$; these intersections define the two $\Delta_c(p_2)$s. We have also plotted, with thick lines, the power functions $\beta(\Delta)$ for the two choices of p_2; note that they both have 50% power at the respective points $\Delta_c(p_2)$. They are, however, not simple translations of the thinner curves because the scale factor $\sigma(\Delta, p_2)$ depends on Δ.

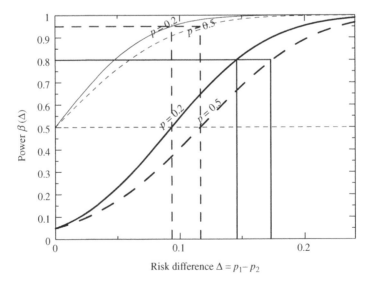

Figure 5.9 Power curves for a two-group binomial test with 100 patients in each group.

We see that the proposed size of the trial would have 80% chance of success (meaning a statistically significant result) if the true value of Δ is 0.15, provided the proportion in the control group is $p_2 = 0.2$, whereas the corresponding number is 0.17 when $p_2 = 0.5$. The latter number is larger, because the variance $\sigma(0, p_2)$ varies with p_2 in such a way that it has its maximum when $p_2 = 0.5$, and a larger variance will require a larger sample size for the same effect and power.

5.7   Some confusing aspects of power calculations

As mentioned above, power considerations are not fully understood by everyone, in particular not by those who do not take the trouble to try. Some of this confusion stems from the extremely common problem of mistaking an estimate of a parameter for knowledge about it. For this discussion we will assume that the model is a shift model, so that the test statistic is an estimate of θ, and that this measures an effect. In the discussion on power in Section 5.6 we introduced two effect sizes of particular interest:

the critical level θ_c, which is such that if the estimate is greater than this, we have sufficient evidence to claim an effect;

the effect we do not want to miss θ_0, which is a number, chosen by us, such that if this is the true effect, there should be a high probability of obtaining sufficient evidence that there is an effect.

We also had the true, but unknown, effect size, denoted θ_{true}. In addition to these, medical science often introduces one further number:

the minimal clinically important difference which we denote θ_{MID} and call MID for short.

The (medical) meaning of MID is that unless $\theta_{\text{true}} > \theta_{\text{MID}}$, the effect is not sufficiently large to merit clinical interest. (Actually it is slightly more complicated: the parameter θ is usually a mean difference, and in order to transform this into a test statistic we need to divide the observed mean by a standard deviation. In order to carry calculations through we therefore will need to assume a value for this standard deviation, which allows us to translate means to means over standard deviations. However, this distinction is of no importance for the present discussion.)

In terms of planning it may be wise to take $\theta_0 = \theta_{\text{MID}}$; we do not want to miss the minimal clinically important effect. Once the experiment is done, we may find an estimate θ^* of θ, which is such that $\theta_c < \theta^* < \theta_{\text{MID}}$. To some people this is confusing, because the study was a success ($\theta_c < \theta^*$) but the effect was not clinically important ($\theta^* < \theta_{\text{MID}}$), despite the fact that we planned the study to find clinically important effects.

The underlying reason is the confusion of estimates with knowledge. The estimate in itself does not help us understand what we actually know about θ; in an extreme situation the estimate may be obtained from a single observation. The knowledge about θ is contained in the confidence functions, and the confidence intervals defined by it. If the appropriate confidence interval actually misses the MID, we have evidence that $\theta < \theta_{\text{MID}}$, but most likely it will contain the MID and we cannot say, with confidence, that the true effect is not as good as that defined by the clinically important difference.

Box 5.5   The (ir)relevance of the MID

The concept of a minimal important difference in clinical research has much in common with the significance level in statistics. Sometimes this author wonders if the MID concept was introduced because the Type I error concept was too difficult for those making decisions on biostatistical data. This box, which is my personal opinion, explains this viewpoint.

The MID should, like the significance level, be context-dependent. Here it is not about what level of evidence one wants, but about a cost–benefit assessment. Suppose there is a drug that prolongs life by one month. Is that clinically relevant? If it is free of cost and totally devoid of side-effects, who would say no? If, on the other hand, it costs half of what you earn and gives you constant diarrhoea, chances are you would not consider it worth the cost. In this sense what is clinically important must depend on other factors, including various types of costs. It may be meaningful for some outcomes to define an MID for a specific purpose, such as the degree of improvement at which the community is prepared to pay for a new drug. But that again must depend on the prize.

The real usefulness of the MID concept is for planning studies, as input to how to decide on the sample size, and not for deciding the clinical importance of the drug. But that leads to a catch-22 situation, since the estimate of the treatment effect should not be compared to the MID for decision making. Which may be difficult to avoid, since it was used at the planning stage. But the study is done for a purpose, and in that particular context a MID may be defined. It is the universality of the concept that should be cast in doubt.

If showing that $\theta > \theta_{\text{MID}}$ is what we need to do, we must test the null hypothesis $\theta = \theta_{\text{MID}}$ instead, which is equivalent to carrying out the analysis above in the new parameter $\theta - \theta_{\text{MID}}$ instead. But the result of this samples size calculation too often produces numbers far too large for studies to be affordable.

Unfortunately, estimates are regularly used as evidence by various types of reviewers of clinical trials, including health authorities when they decide whether they can approve drugs. Scientific arguments will often not convince the ignorant, and we may need to consider actually powering our study so that its success criteria are that we obtain both a statistical significant result, and a point estimate larger than MID. This does not really provide any new problems in itself from a mathematical point of view; we define the power function as

$$\beta(\theta) = F(\theta - \theta_{\text{MID}})$$

instead. This is the power function for a test of the null hypothesis $\theta = 0$ at the smaller significance level $\alpha = F(-\theta_{\text{MID}})$, so the power decreases compared to the case without this extra requirement. From a sample sizing perspective, we need larger studies to maintain power.

Since the requirement that the estimate should come out above the MID is essentially equivalent to a reduction in the significance level for the test, we might ask if we could not retain the significance level and instead decrease the size of the study and pick the size so that $\theta_c = \theta_{\text{MID}}$. If we do that, and the true value of θ is equal to θ_{MID}, we only have a 50% chance of success. In other words, the drug needs to do substantially better than MID for this trial to have a reasonable power to justify the investment in it. From that observation alone it should be clear that this way of reasoning is not fruitful, and the reason should also be obvious: *do not take an estimate alone as reliable evidence*.

5.8 Comments and further reading

The way we have introduced confidence intervals and p-values, using the confidence function, is not standard in textbooks on statistics. Nor is it a new idea, since it was suggested in 1958 by David Cox (Bender et al., 2005). It must be emphasized that it is only a graphical description of the mathematics that is usually presented.

When we compared two binomial proportions, we did not expand further on the inaccuracy that is a consequence of the fact that the underlying distributions are discrete distributions. It is present, as it was for a single proportion, and is discussed by Agresti (2003). As was the case for a single binomial parameter, there are a number of different methods to compute the confidence intervals (Newcombe, 1998). See Eide and Heuch (2001) for a discussion about the epidemiologically important concept of the *attributable fraction*.

The Mantel–Haenszel pooled odds ratio for a series of 2×2 tables was first proposed by Mantel and Haenszel (1959), together with the test based on the quadratic form $Q_{MH}(1)$. The same test had been suggested five years earlier by Cochran, except that he used a binomial model instead of a hypergeometric model. The test is therefore often called the Cochran–Mantel–Haenszel chi-squared test. Since then various extensions have led to a whole family of 'Mantel–Haenszel methods', some indicated in Box 5.3. As a test of association it is easily extended to different situations with more than two levels of one or both of the two factors; see Kuritz et al. (1988) which also gives a review of different ways to estimate the variance of the pooled Mantel–Haenszel odds ratio. As we will discuss in a later chapter, the Mantel–Haenszel technique is for all practical purposes equivalent to the popular logistic regression approach, as long as we analyze the same model (i.e., use the stratum variable as a factor in the case of logistic regression), but the latter method allows us to design more complex models. We will encounter another application of the Mantel–Haenszel methodology when we discuss survival analysis in Chapter 12.

References

Agresti, A. (2003) Dealing with discreteness: making 'exact' confidence intervals for proportions, difference of proportions and odds ratios more exact. *Statistical Methods in Medical Research*, **12**, 3–21.

Bender, R., Berg, G. and Zeeb, H. (2005) Tutorial: Using confidence curves in medical research. *Biometrical Journal*, **47**(2), 237–247.

Eide, G.E. and Heuch, I. (2001) Attributable fractions: fundamental concepts and their visualization. *Statistical Methods in Medical Research*, **10**(3), 159–193.

Kuritz, S.J., Landis, J.R. and Koch, G.G. (1988) A general overview of Mantel-Haenszel methods: applications and recent developments. *Annual Review of Public Health*, **9**, 355–367.

Mantel, N. and Haenszel, W. (1959) Statistical aspects of the analysis of data from retrospective studies of disease. *Journal of the National Cancer Institute*, **22**, 719–748.

Newcombe, R.G. (1998) Interval estimation for the difference between independent proportions: comparison of eleven methods. *Statistics in Medicine*, **17**(8), 873–890.

Phillips, A. and Holland, P.W. (1987) Estimators of the variance of the Mantel-Haenszel log-odds-ratio estimate. *Biometrics*, **43**, 425–431.

5.A Appendix: Some technical comments

5.A.1   The non-central hypergeometric distribution and 2 × 2 tables

Fisher's hypergeometric distribution is often introduced through an urn model: you have n_1 red balls and n_2 white balls and randomly draw r of these without replacement, in which case the number X of red balls drawn follows a hypergeometric distribution. Fisher's non-central distribution occurs when picking red balls and picking white balls are not exchangeable; one is θ times more likely to pick a red ball than a white one. (From a sampling perspective it is important that it just happened to be r balls drawn; r is observed after the experiment. If we instead draw balls one by one until r are drawn, we get a slightly different non-central distribution, often referred to as Wallenius' non-central hypergeometric distribution. In the central case when $\theta = 1$ these are the same.) Fisher's non-central hypergeometric distribution $\text{Hyp}(n_1, n_2, r, \theta)$ is defined using the hypergeometric function (which is not a CDF)

$$F(\theta) = \sum_{j=0}^{\min(n_1, r)} \binom{n_1}{j} \binom{n_2}{r-j} \theta^j.$$

In fact, the probability for the outcome j is obtained by taking the corresponding term in this sum and divide it by $F(\theta)$. To compute the mean and variance of this distribution there is a trick which uses the function $\kappa(\beta) = \ln F(e^\beta)$. In fact, if $\beta = \ln \theta$, then

$$E_\theta(X) = F(e^\beta)^{-1} \sum_j \binom{n_1}{j} \binom{n_2}{r-j} j\, e^{\beta j} = \kappa'(\beta)$$

and, similarly, $V_\theta(X) = \kappa''(\beta)$. These formulas are no accident, they hold because the non-central hypergeometric distribution belongs to the exponential family, discussed in Section 9.7. For the central hypergeometric distribution we can compute the mean and variance explicitly as

$$E_1(x) = \frac{n_1 r}{n_1 + n_2}, \quad V_1(x) = \frac{n_1 n_2 r (n_1 + n_2 - r)}{(n_1 + n_2)^2 (n_1 + n_2 - 1)}.$$

Fisher's hypergeometric distribution appears, as was explained in Box 4.4, in a 2 × 2 table with the three constraints $x_{11} + x_{12} = n_1$, $x_{21} + x_{22} = n_2$ and $x_{11} + x_{21} = r$. In such a case the stochastic variable X_{11} has the $\text{Hyp}(n_1, n_2, r, \theta)$ distribution, where θ is the odds ratio in the table. Using the function $F(\theta)$ and properties of the binomial coefficients, it can be shown that

$$E_\theta(X_{11}^{(a)} X_{12}^{(b)} X_{21}^{(c)} X_{22}^{(d)}) = \theta^e E_\theta(X_{11}^{(a-e)} X_{12}^{(b+e)} X_{21}^{(c+e)} X_{22}^{(d-e)}), \tag{5.4}$$

where $x^{(k)} = x(x-1)\ldots(x-k+1)$. The first consequence of this is that

$$E_\theta(X_{11} X_{22} - \theta X_{21} X_{12}) = 0, \quad \text{or} \quad \theta = \frac{E_\theta(X_{11} X_{22})}{E_\theta(X_{21} X_{12})}.$$

If we replace the expectations here with the observations, this means that we estimate θ with the empirical odds ratio. Another way to express this is to say that we estimate θ using the estimating equation

$$U(\theta) = x_{11}x_{22} - \theta x_{21}x_{12} = 0,$$

and knowledge about θ can then be obtained from the study of the confidence function $C(\theta) = \Phi(U(\theta)/\sqrt{V_\theta(U(\theta))})$. To do this we need to be able to compute the variance of $U(\theta)$, which can be done using equation (5.4) and some algebra (Phillips and Holland, 1987), as

$$V_\theta(U(\theta)) = \frac{1}{2n^2} E_\theta((X_{11}X_{22} + \theta X_{21}X_{12})(X_{11} + X_{22} + \theta(X_{21} + X_{12}))),$$

where n is the table total. This is estimated by

$$\hat{V}_\theta(U(\theta)) = \frac{1}{2n^2}(x_{11}x_{22} + \theta x_{21}x_{12})(x_{11} + x_{22} + \theta(x_{21} + x_{12})).$$

5.A.2 The gamma and χ^2 distributions

If X is a stochastic variable with CDF $F(x)$, the CDF for X^2 is given by

$$G(x) = P(X^2 \le x) = F(\sqrt{x}) - F(-\sqrt{x}), \quad x > 0,$$

which equals $2F(\sqrt{x}) - 1$ if the distribution $F(x)$ is symmetric around zero, as for the Gaussian distribution. For the standard Gaussian distribution this means that the density is given by

$$G'(x) = \frac{1}{\sqrt{x}}\varphi(\sqrt{x}) = \frac{1}{\sqrt{2\pi x}}e^{-x/2},$$

and the distribution belongs to the general family of gamma distributions, in which $\Gamma(p, a)$ is defined by the probability density function

$$\frac{a^p}{\Gamma(p)}x^{p-1}e^{-ax}.$$

This family has the property that the sum of two independent such variables with a common a is also of the same type:

$$\Gamma(p_1, a) + \Gamma(p_2, a) = \Gamma(p_1 + p_2, a).$$

The most important cases are when $p = n/2$ and $a = 1/2$, which defines the $\chi^2(n)$ distribution. This means that if $X \in N(0, 1)$, then $X^2 \in \chi^2(1)$. The addition formula then shows that the sum of n squares of independent standardized Gaussian variables has a $\chi^2(n)$ distribution.

The observation that the distribution of a sum of squares of Gaussian variables is a χ^2 distribution holds true in more generality. A multidimensional Gaussian distribution in p variables is one for which the probability density is proportional to $e^{-Q(x)/2}$, where $Q(x)$ is a quadratic form in $x = (x_1, \ldots, x_p)$ such that $Q(x) \ge 0$ for all x. If X has such a distribution, $Q(X)$ has a χ^2 distribution. To see why, we note that, by integration over slices $\{y; Q(y) = r\}$,

$$P(Q(X) \le x) = C \int_{Q(y) \le x} e^{-Q(y)/2}dy = C' \int_0^x r^{p-1}e^{-r/2}dr.$$

The factor r^{p-1} comes from the $(p - 1)$-dimensional 'areas' of the slices. The formula is actually only true if the quadratic form is positive definite (i.e., never zero except at the origin); if not p should be the dimension of the largest linear space in which it is (this allows for singular Gaussian distribution). The result is that $Q(X) \in \chi^2(p)$, where p is the dimension of the maximal space on which $Q(x)$ is positive definite. Notation-wise we have that $X \in N_p(m, \Sigma)$ if the quadratic form can be written in matrix language as

$$Q(x) = (x - m)^t \Sigma^{-1} (x - m)$$

for some matrix $p \times p$ matrix Σ, which may be non-singular (in which case the matrix inverse is used) or singular (in which case a generalized inverse is used).

6

Empirical distribution functions

6.1 Introduction

So far we have discussed rather general aspects of the role and purpose of statistics in clinical research. We have done so against the background of 2×2 tables which, although simple, contain most of the statistical issues that are not directly connected with specific types of data. In this chapter we will begin the rest of this book, which is about different statistical methods. The first step in this is to understand how to describe non-dichotomized data in an appropriate way.

The underlying philosophy is that the values of what we measure as the outcome variable follow a distribution in the population we study. If we take a representative sample from this population we can use these data to compute an empirical distribution, which is an estimate of this (true) distribution. In this chapter we will discuss the relation between the unobserved true distribution and the observed one, and how we can define and estimate parameters in the population distribution. The way we do this gives us a general method to obtain sample counterparts of population parameters. However, statistics is not primarily about estimated parameters, but rather about describing the knowledge we have acquired about the population parameters. For this we also need to understand how we obtain knowledge about the true distribution from the empirical one.

In addition to understanding how to obtain knowledge about a distribution function the mathematical way (deriving formulas), we will also discuss some related areas. One is how to obtain confidence the non-mathematical way, with a technique called bootstrapping which replaces mathematics (theory) with *in silico* experiments (computer simulations). The other is a short discussion about meta-analysis and heterogeneity. The story below will explain the connection.

6.2 How to describe the distribution of a sample

So far we have analyzed proportions. The data consist of a set $\{x_1, \ldots, x_n\}$ of data points, such that each x_i is either 0 or 1, and the estimate p^* of the proportion p is simply the average of

Understanding Biostatistics, First Edition. Anders Källén.
© 2011 John Wiley & Sons, Ltd. Published 2011 by John Wiley & Sons, Ltd.

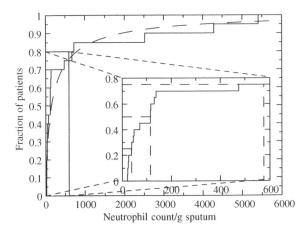

Figure 6.1   Illustration of data as an empirical distribution function. In the main graph, the solid step function is the e-CDF and the dashed curve the population CDF. The inset graph shows a blow-up of the initial part of the e-CDF, illustrating how some percentiles are estimated.

all these x_i. We now wish to look at the more general situation in which the outcome variable can take on a multitude of different values, including the case where (in theory) it can take on all possible values, at least in some interval. Well-known examples are blood pressure and objective lung function measurements such as FEV_1.

Consider the following 20 numbers: 37, 71, 24, 38, 44, 4286, 728, 26, 22, 138, 475, 47, 5408, 115, 131, 2520, 676, 21, 117, 115. They represent the neutrophil cell counts found in sputum samples from 20 COPD patients. The medical meaning of this is not important, only the assumption that these observations constitute a representative sample from some population. We are going to use these data for illustrative purposes extensively in the upcoming discussion. (The numbers are actually simulated, which explains why we can draw certain dashed curves in some graphs below.)

One way to illustrate these 20 numbers graphically is the staircase function shown in Figure 6.1. The data values are on the x-axis, at each of which there is a vertical step of 1/20, leading up to a monotonically increasing function starting at zero and finishing at one. This function is called the *empirical cumulative distribution function* (e-CDF) of the data. We usually denote it by $F_n(x)$, with n being the number of observations it is based on.

In the graph there is also a dashed curve. This represents another function, namely the population CDF $F(x)$, the distribution in the population of such data. This is what the e-CDF $F_n(x)$ tries to estimate. The difference between the e-CDF and the CDF is that if we repeat the experiment there will be a new e-CDF each time, although the CDF stays the same. This means that $F(x)$ is a characteristic of the outcome variable in the population (when measured in the particular way of the specific protocol), whereas $F_n(x)$ is an estimate of $F(x)$ from a particular sample from that population. In general, statistics is divided into two main parts:

descriptive statistics which describes $F_n(x)$, and

inferential statistics which uses $F_n(x)$ to understand $F(x)$.

We have already noted that there are continuous and discrete stochastic variables. For the former the CDF is a continuous function, whereas for the latter it is a step function. Some data are hybrids of discrete and continuous data. For example, in order to assess a person's disability we may use a Visual Analogue Scale (VAS) which means that the disability is described by a mark on a line going from 0 to 1, where 0 means no disability and 1 complete disability. The corresponding CDF may have jumps at both the end points, but be a continuous and increasing function in between.

The value of the e-CDF at the point x is the proportion of elements in the sample $\{x_1, \ldots, x_n\}$ that are at most x in magnitude. It therefore has the analytical expression

$$F_n(x) = \frac{1}{n} \sum_{i=1}^{n} I(x_i \leq x), \tag{6.1}$$

where $I(C)$ denotes the indicator function which is 1 if the condition C is true and 0 if it is false. If precisely k of these x_i are less than or equal to a particular value x, then $F_n(x) = k/n$. If so, and x is one of the observations, we call k the rank of x and denote it by $R_n(x)$ (which therefore is equal to $nF_n(x)$). (If there are ties, a modification is needed.) There is a fundamental theorem in probability theory, called the Glivenko–Cantelli theorem, which says that the e-CDF $F_n(x)$ converges to $F(x)$ (in a uniform way in x), as the sample size n increases.

The e-CDF is defined as a (right-continuous) step function. For many purposes it would have been more convenient, at least for a continuous CDF, to define it as a piecewise linear function such that the point $(x_k, F(x_k))$ is connected to $(x_{k+1}, F(x_{k+1}))$ by a straight line, instead of via the point $(x_{k+1}, F(x_k))$ (the staircase). We will, however, mostly stick with the convention, except for a few occasions when we point out some benefits of the linearly interpolated version, when this clarifies a particular statistical method.

There is an alternative formula for the e-CDF. Its derivation is based on the observation that for any monotone function we have that $F(x) = F(x-) + \Delta F(x)$, where $\Delta F(x)$ is the jump at the point x and $F(x-)$ is the left-hand limit of $F(x)$. If we rewrite this for the complementary function $F^c(x) = 1 - F(x)$, we get

$$F^c(x) = F^c(x-) - \Delta F(x) = F^c(x-) \left(1 - \frac{\Delta F(x)}{F^c(x-)} \right).$$

This is an observation which only is of interest at jump points, and since the e-CDF consists exclusively of jump points, it is particularly useful for that function. In fact, if we apply the observation repeatedly to the e-CDF, we get the alternative formula for $F_n(x)$, referred to above. The way it is computed is as follows. Order the different values in the sample into a strictly increasing sequence $x_1 < x_2 < \ldots$ and let d_j denote the number of observations with value x_j. Let $r_j = nF_n^c(x_j-)$ be the number of observations that are at least x_j in size. The formula then reads

$$F_n^c(x) = \prod_{j:x_j \leq x} \left(1 - \frac{d_j}{r_j} \right),$$

where $\prod_C a_j$ means that we should multiply all the a_j that fulfill the criterion defined in C. This way of writing the e-CDF is called the *Kaplan–Meier* form of the e-CDF, or the Kaplan–Meier estimate of the CDF. It has the important property that it can be generalized to some situations where there is incomplete knowledge in the data. If we study the time until some particular event occurs, this event may not occur for some patients during the observation period, in which case we only know that the true observation is larger than the observation time. In such a case we can use the Kaplan–Meier form of the e-CDF to estimate $F(x)$, which is useful in the analysis of survival data.

The discussion above applies equally well to continuous and discrete CDFs, but there are some aspects of the latter that should be emphasized. To be able to plot the e-CDF for a discrete distribution, we need the observations to be ordered. For a binomial, for example, we can code no event as $x = 0$, and event as $x = 1$. An observation k from a Bin(n, p) distribution is then the sum of n independent observations of a Bin(1, p) distribution, which is such that the CDF has jump $1 - p$ at $x = 0$, followed by another jump of size p in the point $x = 1$. The e-CDF is the corresponding jump function, for which the jump at $x = 0$ is $1 - k/n$ and at $x = 1$ is k/n. This can be generalized to data that fall into more than two categories, as long as these are ordered so that we actually can draw the graph of the CDF $F(x)$. If so, it is no restriction to code the categories as $k = 1, \ldots, c$. We may, for example, assign a disability index in c categories to patients who have had a stroke, which represents a judgement of how disabled they are, in such a way that a larger score means that the patient is more disabled. Such data are simply described by the probabilities p_j of falling into each category j, and we have that the CDF is a cumulative sum of these p_j, that is, $F(k) = p_1 + \ldots + p_k$. This defines a step function, an example of which is shown in gray in Figure 6.2. In this case the jump points of the CDF and the e-CDF of a particular experiment will coincide, though the actual jump sizes may differ.

Ordered categorical data can often be perceived to be a discrete approximation of some underlying continuous variable. For the disability score above we may assume the existence of a continuous (and probably unmeasurable) disability variable X with a continuous CDF $F(x)$, together with a series of cut-off points $a_0 = -\infty < a_1 < \ldots < a_{c-1} < a_c = +\infty$, such that our ordered categorical variable takes the value k precisely if $a_{k-1} < X \leq a_k$. The probability of falling into category k is then given by $p_k = F(a_k) - F(a_{k-1})$.

We have pointed out that some aspects of statistics would have been conceptually simpler with the e-CDF defined by linear interpolation. This applies to the CDF of a discrete distribution as well, in that it is not clear why the CDF should be right- and not left-continuous

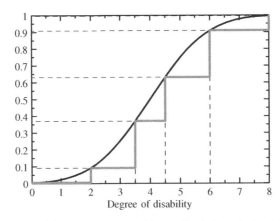

Figure 6.2   Illustration of how we can consider ordinal data derived from an underlying continuous distribution, with cut-off points $2 < 3.5 < 4.5 < 6$.

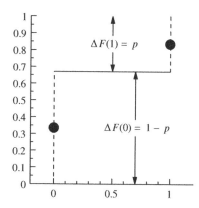

Figure 6.3   The mid-point CDF for a two-point distribution.

at jump points. What corresponds to linear interpolation in such a case would be to take the mid-value of the jump as the definition of the CDF at that point. This provides the same staircase function as the traditional one, except for the values at the actual jumps. If $F(x)$ is the traditional, right-continuous, distribution, then we redefine the CDF to take the average value $(F(x-) + F(x))/2$ at the point x. This is illustrated for a two-point $\text{Bin}(1, p)$ distribution in Figure 6.3. We will later see the usefulness of this modified CDF in connection with the Wilcoxon test for discrete data. We should also note that this is precisely the mid-P adjustment mentioned in equation (4.2), where we constructed the confidence function for a parameter in a discrete distribution.

If we have c ordered categories, labeled $1, \ldots, c$, then the jumps $p_k = \Delta F(k)$ are probabilities, and the CDF is that of the multinomial, $\text{Mult}_c(1, p_1, \ldots, p_c)$, distribution. If we perform n experiments resulting in observed frequencies $x_1, \ldots, x_c$, the e-CDF is the step function with jump x_k/n at the point k. Moreover, there are only finitely many possible outcomes of a multinomial distribution, so there are a finite number of possible e-CDFs, namely $\binom{n+c-1}{n}$. This has implications for how p-values can be computed.

6.3   Describing the sample: descriptive statistics

We will now see how we can describe the e-CDF of a sample $\{x_1, \ldots, x_n\}$ using various summary statistics. Though the most complete way to describe the sample is to list it (not practical unless the sample is small) or plot it, there is often a need to summarize its key aspects in some convenient and easily communicable way.

One way to do this is to choose some value between 0 and 1 and describe where the e-CDF is equal to this particular value. In other words, choose $0 < p < 1$ and solve the equation $F_n(x) = p$ for $x = x_p$. Because $F_n(x)$ is a step function this cannot always be done exactly, and we may need to average two numbers. As an example, take $p = 0.5$. This means that we want the value $x_{0.5}$ for which half of the observations are less than or equal to it. If n is odd, $x_{0.5}$ is obtained by ordering the sample and take the middle number. If n is even, we need to take a number between the two values that are in the middle of the sample, and by convention we take the arithmetic mean of these. We call $x_{0.5}$ the median of the sample. (This is one instance when using linear interpolation to define the e-CDF would help: it would define

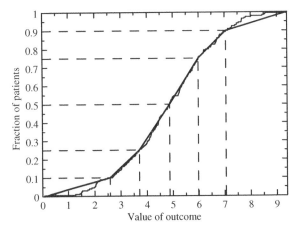

Figure 6.4   Approximating an empirical distribution function, which describes a large data set, by a piecewise linear function defined by percentiles.

a unique sample median, providing a better estimate of the median of the true distribution.) The general x_p is called a percentile, and as for $x_{0.5}$ we have special names for $x_{0.25}$ and $x_{0.75}$: the lower and upper quartiles. By describing a set of percentiles and adding the smallest and largest observations, we get a smaller set of data, which can be used to obtain an approximation of the e-CDF. This is illustrated in Figure 6.4, where the e-CDF of a sample of 200 points from a continuous distribution is approximated by the 10th and 90th percentiles, the quartiles and the median: a stepwise linear function connecting the smallest and the largest observations with break points at these percentiles may be a fairly good approximation to the CDF.

It should be noted that what value a percentile assumes does not depend on all the individual observations in the sample. Take the median as an example. Its value is the value in the middle of the sorted sample, but does not depend on the exact values of those data points that are less than (or larger than) this value, except that they are smaller (or larger).

The most popular way of summarizing a data set is through the (arithmetic) mean

$$\bar{x} = \frac{1}{n} \sum_{i=1}^{n} x_i$$

together with the standard deviation s, which is the square root of the sample variance

$$s^2 = \frac{1}{n-1} \sum_{i=1}^{n} (x_i - \bar{x})^2.$$

The mean is a measure of the location of data and the standard deviation is a measure of the dispersion around the mean. These sample parameters estimate the corresponding population parameters (which were defined in Box 4.5 and will be revisited in the next section).

However, $\bar{x}$ and s are not always good measures for describing the CDF. For the sputum data shown in Figure 6.1 the mean is 752.0 and the standard deviation is 1521.1. Compare this with the three quartiles 37.5, 115.0, 575.5, shown in the small subplot in Figure 6.1. We see that the upper quartile, below which 75% of the data lie, is much smaller than the mean.

Box 6.1 The lognormal distributions

If the logarithm of a stochastic variable X has a Gaussian distribution, $\ln X \in N(m, \sigma^2)$, we say that X has a lognormal distribution. Its CDF is given by $F(x) = \Phi((\ln x - m)/\sigma)$, from which some calculus shows that its mean and variance are given by

$$E(X) = e^{m+\sigma^2/2} \quad \text{and} \quad V(X) = e^{2m+\sigma^2}(e^{\sigma^2} - 1),$$

respectively. The median is e^m, which we can derive from the general observation that for a monotonic transformation $h(x)$ we have that

$$\text{median}(h(X)) = h(\text{median}(X)),$$

here applied to $h(x) = e^x$. The same applies to any other percentile. The coefficient of variation is by definition $CV = \sqrt{V(X)}/E(X)$, and is a dimensionless expression of variability. For the particular case of the lognormal distribution we find that

$$CV = \sqrt{e^{\sigma^2} - 1}.$$

Note that for σ^2 small we have that $CV \approx \sigma = \sqrt{V(\ln X)}$.

The reason for this is obvious from the formulas above in combination with Figure 6.1; the distribution is skewed to the right in the sense that most of the data is concentrated in the left portion of the graph, but with a long tail of large values to the right. In such a case the choice of mean and standard deviation as descriptors of the e-CDF, and ultimately the CDF, may be rather poor and misleading. Implicit in the choice of these sample statistics as descriptors of the e-CDF lies an assumption that the distribution is reasonably symmetric around its mean. When the distribution is symmetric around its mean, the median should be equal to it, so if the difference between these two parameters is large, the distribution is skewed. (This is not to say that you do not want to estimate the mean for a skew distribution, only that more information is needed to understand the CDF.)

When a distribution is skewed to the right, and the data are positive, we may try to symmetrize it by taking the logarithms of the data. This means that we replot the e-CDF on a logarithmic scale, as we have done in Figure 6.5 for the sputum data. The mean in this graph corresponds to the geometric mean 147.2 of the original data. More precisely, this is the arithmetic mean of the logarithmic data, rescaled to the original scale by exponentiation:

$$e^{\overline{\ln x}} = \left(\prod_{i=1}^{n} x_i \right)^{1/n}.$$

This number is much closer to the sample median, which we saw above was 115, and is in this case a meaningful description of the center of the distribution. To describe the variability for a lognormal distribution we can use the coefficient of variation for that distribution, $CV = \sqrt{e^{s^2} - 1}$ (see Box 6.1), where s^2 is the sample variance of $\ln x_i$. Multiply it by 100, and we express the coefficient of variation as a percentage. For our sputum data it is 469%.

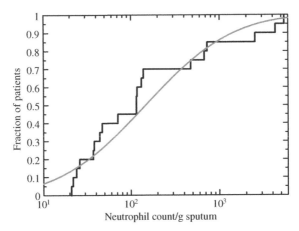

Figure 6.5   Illustrating sputum data as an empirical distribution function with a logarithmic scale. The gray curve is the reconstructed CDF from the observed mean and standard deviation of the logged data.

When we discussed percentiles, we emphasized their ability to reconstruct the e-CDF. How can we reconstruct the e-CDF from the mean and standard deviation? This is done based on the Gaussian distribution in such a way that we reconstruct $F(x)$ as $\Phi((x - \bar{x})/s)$. The dashed curve in Figure 6.5 shows this for the sputum data, albeit on a logarithmic scale. We see that the mean and standard deviation of the logged data (or the geometric mean and coefficient of variation on the original scale) represent a reasonable summary of these data.

6.4   Population distribution parameters

Descriptive statistics have their counterpart for the population CDF $F(x)$. To describe this we use the notation that was introduced in Box 4.8, but we repeat it here because of the fundamental role it will play in our discussion from now onwards. The notation is that we write $dF(x) = f(x)dx$ for a point x where a particular distribution is (absolutely) continuous, and $dF(x) = \Delta F(x)$ for a point x where there is a jump of size $\Delta F(x)$. Based on this, we denote sums and integrals by use of the common notation

$$\int_a^b g(x)dF(x).$$

If $F(x)$ is piecewise differentiable with derivative $f(x)$, this is

$$\int_a^b g(x)dF(x) = \int_a^b g(x)f(x)dx,$$

whereas for a e-CDF $F_n(x)$, computed from the data $\{x_1, \ldots, x_n\}$, it becomes

$$\int_a^b g(x)dF_n(x) = \frac{1}{n}\sum_{i=1}^n g(x_i),$$

since $dF_n(x_i) = 1/n$ in this case. When $F(x)$ is a population CDF the integral represents the average of the values $g(x)$ in the whole population and is therefore a population parameter. It is estimated by replacing $F(x)$ with the e-CDF $F_n(x)$ in the computation, defining the sample parameter.

As a first example, recall that the mean of $F(x)$ is defined as

$$m = \int_{-\infty}^{\infty} x \, dF(x).$$

To estimate it, we replace $F(x)$ with $F_n(x)$ and get the sample mean:

$$\bar{x} = \int_{-\infty}^{\infty} x \, dF_n(x) = \frac{1}{n} \sum_i x_i.$$

The mean is obviously a location measure, being the average of all values. To describe it geometrically, we can rewrite it as

$$m = \int_0^{\infty} (1 - F(x))dx - \int_{-\infty}^0 F(x)dx.$$

This means that we can interpret the mean as the area between the curve $y = F(x)$ and the line $y = 1$ for positive x, minus the area under the curve $y = F(x)$ above the x-axis for negative x. More importantly, the formula gives us a way to visualize the difference in the mean values for two distributions. In such a case, if m_F is the mean for the CDF $F(x)$ and m_G that for $G(x)$, then

$$m_F - m_G = \int_{-\infty}^{\infty} (G(x) - F(x))dx,$$

which is the area enclosed by the CDFs.

Example 6.1   As an example, consider Figure 6.6, which is a reproduction of the graph that illustrated the effect of differential drop-out in Section 3.7. The area between the dashed and solid curves for a group shows the effect that missing data have on the mean value in that group. Inspecting the graph, we see that the area between the two solid curves is smaller than the area between the two dashed curves, showing that, if we analyze completers only, we get a bias toward no effect in the mean.

We can also visually estimate the bias with the LOCF approach. This would mean that we impute the value -1 for missing data, which means that the distribution for LOCF imputed data is zero up to the point $x = -1$, where there is a jump up to the dashed curve. The area between these two CDFs is the area between the dashed curves over the region $x > 1$. The area between the curves to the left of $x = -1$ is therefore the bias we get by the LOCF approach. Comparing areas, we see that in this case LOCF is less biased than analyzing completers only.

To measure the spread, or dispersion, of the distribution, a common measure is the population variance defined by

$$\sigma^2 = \int_{-\infty}^{\infty} (x - m)^2 \, dF(x). \tag{6.2}$$

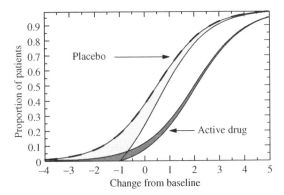

Figure 6.6 Reproduction of Figure 3.1, illustrating how we can assess the effect on the mean by comparing areas.

If we replace $F(x)$ with $F_n(x)$, we see that this should be estimated by

$$\hat{\sigma}_n^2(m) = \frac{1}{n} \sum (x_i - m)^2.$$

However, this only works if we know the mean m. If we do not, we need to estimate it with the sample mean and use $\hat{\sigma}_n^2(\bar{x})$ as the estimate of σ^2. For this to be an unbiased estimator we need to replace the denominator n with $n - 1$, which gives us the sample variance s^2. Intuitively the reason is that though we have n terms, there are only $n - 1$ independent ones, since subtracting $\bar{x}$ means that the sum of the terms $x_i - \bar{x}$ is constrained to be zero. By analogy, with more parameters, say p, to estimate, we would in general have to divide by $n - p$. There are ways to derive the correct denominator directly by eliminating the mean from the problem. One way uses the differences $x_i - x_j$ and is outlined in Box 8.6, another is based on the residuals $x_i - \bar{x}$ and the likelihood theory for Gaussian data and is discussed in Box 13.1. For both of these derivations the situation is complicated by the fact that the data analyzed are dependent.

It is interesting to note that the bias adjustment of the variance made by dividing by $n - 1$ instead of by n is not usually done when we estimate proportions in a binomial distribution. The variance estimator for the frequency $\hat{p}$ is usually taken to be $\hat{p}(1 - \hat{p})/n$, which is biased for the reason above; the unbiased estimator is $\hat{p}(1 - \hat{p})/(n - 1)$.

The mean is a measure of location, and so are the percentiles, which are defined for the population CDF in the same way as the sample percentiles were defined for the e-CDF – as the solution of the equation $F(x) = p$. This defines the population percentiles which we estimate by the sample percentiles, as discussed in the previous section.

6.5 Confidence in the CDF and its parameters

For a given x, the value $F(x)$ of the CDF is a probability and the value of the e-CDF $F_n(x)$ is an observed relative frequency, such that $n F_n(x)$ follows a $\text{Bin}(n, F(x))$ distribution. Since $V(F_n(x)) = p(1 - p)/n$, this means that the set of ps that satisfy

$$\chi_1 \left(\frac{n(F_n(x) - p)^2}{p(1 - p)} \right) \leq 1 - \alpha$$

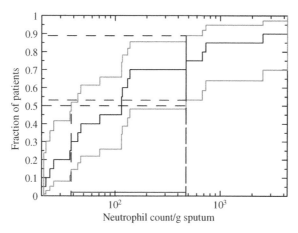

Figure 6.7   Illustration of how to obtain confidence limits for percentiles, using pointwise confidence limits for the CDF.

constitute a confidence interval for $F(x)$ of degree $1 - \alpha$. Figure 6.7 illustrates this for the sputum data. The black step function is the original e-CDF, and the two solid gray step functions above and below it define (the end points of) the 95% confidence limits. One such confidence interval is also illustrated as the thick vertical line above the point $x = 475$. It is important to point out that these intervals are pointwise confidence intervals: to get the correct confidence degree we first need to fix the point x. If we do not prespecify the point x, we have a multiplicity problem. The natural follow-up question is then to ask for limits that by construction contain the complete CDF $F(x)$ with 95% probability. Such a region is made up of simultaneous confidence intervals, and its construction is briefly addressed in Appendix 6.A.2.

In Figure 6.7 we also have a horizontal line which is defined from the gray confidence limit curves at level 0.5. The interval it defines has limits 38 and 475. These are (approximate) 95% confidence limits for the median of $F(x)$. Confidence limits for other percentiles are obtained in the same way. The argument is the following simple observation: finding the confidence interval for a parameter θ such that $F(\theta) = p$ is the same as finding all θ such that

$$\chi_1 \left(\frac{n(F_n(\theta) - p)^2}{p(1 - p)} \right) \le 1 - \alpha$$

(the same as above, except that now p is held fixed).

To obtain confidence intervals for the mean is done in an altogether different way, which is not easily described in geometric terms. The computation is based on the fact that $\bar{x}$ (according to the CLT) asymptotically has a Gaussian distribution with mean m and variance σ^2/n, where σ^2 is the variance of $F(x)$. This suggests the use of the confidence function

$$C(m) = \Phi \left(\frac{\sqrt{n}(m - \bar{x})}{\sigma} \right),$$

which should be at least approximately valid for learning about the mean m. The problem here is that we do not know σ^2, a problem we solve by estimating it by the sample variance s^2.

This means that the (approximate) symmetric confidence interval, of confidence degree $1 - \alpha$, for the mean is computed from the well-known formula

$$\left(\bar{x} - z_{\alpha/2} \frac{s}{\sqrt{n}}, \bar{x} + z_{\alpha/2} \frac{s}{\sqrt{n}} \right).$$

However, whether this approximation is good or not depends on how skewed (i.e., non-symmetrical) the distribution $F(x)$ is (and how heavy the tails are). If n is taken sufficiently large, it can be made as good as we want, but it may require very large values of n. On the other hand, when $F(x)$ is very symmetric around the mean, a relatively small n might suffice. The extreme case here is when $F(x)$ is in fact Gaussian in itself, in which case the confidence function $C(m)$ is exact if σ^2 is known. If σ^2 is not known, but needs to be estimated by s^2, the mathematical analysis has been taken one step further, to obtain an explicit, analytical, description of the true distribution of $\sqrt{n}(\bar{x} - m)/s$. This is the t distribution with $n - 1$ degrees of freedom, which we denote $t(n - 1)$ (a derivation of this is outlined in Appendix 6.A.1). So, if we let $T_n(x)$ denote the CDF of the $t(n)$ distribution, we have in this case the exact confidence function

$$C(m) = T_{n-1} \left(\frac{\sqrt{n}(m - \bar{x})}{s} \right).$$

The corresponding confidence interval is wider than that for the normal approximation; how much wider depends on the size of n. The following example illustrates this.

Example 6.2 To illustrate the confidence functions for the mean we use the data that Student, the inventor of the t distribution, used for illustration. These data are often referred to as the Cushny and Peebles data and consist of the following list of paired observations from 10 subjects: $(0.7, 1.9), (-1.6, 0.8), (-0.2, 1.1), (-1.2, 0.1), (-0.1, -0.1), (3.4, 4.4), (3.7, 5.5), (0.8, 1.6), (0.0, 4.6), (2.0, 3.4)$. In this example we analyze the data obtained by computing the difference for each pair. Figure 6.8 contains two confidence functions for this data set, the conventional t-test and the large-sample approximation which uses the Gaussian CDF $\Phi(x)$. We see that ignoring the extra variability that led to the definition of the t distribution will produce shorter confidence intervals than the correct one, as expected.

Recalling how we obtained confidence intervals for a binomial proportion in Section 4.6.1, it may be natural to ask if we should not use the ratio $\sqrt{n}(m - \bar{x})/\hat{\sigma}_n(m)$ as a test statistic instead. At first glance it may appear that this should have a $t(n)$ distribution, but it does not, because the numerator and denominator are now dependent. Its distribution is given in Appendix 6.A.1, where it is shown that the confidence function for m it defines is equivalent to the $t(n - 1)$-based confidence function above.

As outlined in Box 6.2, the t-based confidence function also has a Bayesian interpretation: if we assume very special prior distributions for m and σ, the function $C(m)$ is the *a posteriori* distribution for m. However, for this we need to use rather peculiar prior distributions for m and σ, at least from a probabilistic point of view.

Box 6.2 Bayesian approach to a Gaussian distribution

Given independent observations $x_1, \ldots, x_n$ from a $N(m, \sigma^2)$ distribution, the sample is an observation of a distribution with probability density

$$\prod_{i=1}^{n} \frac{1}{\sigma} \phi\left(\frac{x_i - m}{\sigma}\right) = \frac{1}{(2\pi\sigma^2)^{n/2}} \exp\left(-\sum_i \frac{(x_i - m)^2}{2\sigma^2}\right) = (2\pi\sigma)^{-n} \exp(-Q/2\sigma^2),$$

where

$$Q = (n-1)s^2 + n(\bar{x} - m)^2 = (n-1)\left(1 + \frac{t^2}{n-1}\right), \quad t = \frac{m - \bar{x}}{s/\sqrt{n}}.$$

If we assume that the prior distributions for m and σ^2 are independent, and that the prior for σ^2 corresponds to the measure $d\sigma^2/\sigma^2$ (which is not a probability measure, but means that the precision variable σ^{-2} has a uniform distribution), we can obtain the marginal distribution for m by first integrating σ^2 out, and then multiply by the prior for m. To integrate, we introduce the new variable $y = 1/\sigma^2$, and get the integral

$$(2\pi)^{-n/2} \int_0^\infty y^{n/2-1} \exp(-Qy/2)dy \propto Q^{-n/2}.$$

This means that the density for m is proportional to $(1 + t^2/(n-1))^{-n/2}$, which defines the $t(n-1)$ distribution. If we therefore take the uninformative prior dm for m, we see that the posterior density for m, given the observed data, is derived from the observation that

$$\frac{m - \bar{x}}{s/\sqrt{n}} \in t(n-1).$$

If we have some prior information on m (but none on σ^2) we simply multiply the $t(n-1)$ density with this *a priori* density and normalize, to obtain the *a posteriori* distribution.

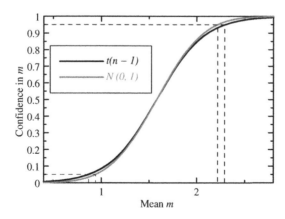

Figure 6.8 Confidence functions describing our information about the mean. The solid black curve corresponds to Student's t-test, the gray curve to the Gaussian approximation.

6.6   Analysis of paired data

We next return to the COPD patients and their sputum data, but now we assume we have two observations on each of the 20 subjects. The second set of data consists of the following numbers (paired with the previous sequence): 18, 19, 2, 2, 5, 515, 30, 10, 3, 55, 127, 2, 260, 16, 8, 301, 443, 1, 26, 24. The original observations (X) were taken after some drug was given, whereas the new data (Y) were obtained under the same conditions, but without this drug. Precisely as with Cushny and Peebles data in the previous section, we wish to analyze the difference $Z = Y - X$. Our intention here is to have a more general discussion about what we can do by exploring the CDF for Z, noting that if X and Y have the same distribution, the distribution for Z would be symmetric around zero (see equation (4.1)). We wish to find ways to explore this symmetry in order to obtain a test for the null hypothesis that there is no effect of the drug.

The immediate consequence of the symmetry is that both the median and the mean of Z are zero. One way to test the null hypothesis is therefore to test if either of these two parameters are zero. For the mean we get the estimate -659 with 95% confidence limits $(-1318, 1)$, and for the median we get the estimate -87 with 95% confidence limits $(-235, -39)$. For the mean there is not sufficient evidence to reject the null hypothesis at the conventional two-sided 5% level, since the interval contains zero, whereas based on the median we have sufficient evidence. However, because of the skewness of the data, the proposed confidence interval for the mean may be quite inaccurate (have the wrong error control). We will see that this is the case in the next section.

The mean and median estimates are quite different, and from our previous discussion we have a fairly good idea of why this is: we should probably log our data before we analyze. The analysis should therefore be on the stochastic variable $Z = \ln Y - \ln X = \ln(Y/X)$ instead. The mean and median of this distribution should then be back-transformed to the original measurement scale by exponentiation. As discussed earlier, when we exponentiate the (arithmetic) mean of the logged data, we get the geometric mean of the original data, whereas when we exponentiate the median of the logged data, we get the median of the original data. For our data the geometric mean estimate is 0.14 with 95% confidence limits $(0.09, 0.21)$, and for the median the estimate is 0.13 with 95% confidence limits $(0.06, 0.22)$. We see that we get more consistent results by analyzing logged data instead of the original data, but the final claim is different: the first is the ratio of the geometric means obtained with and without drug treatment, and the second is the median of the individual ratios.

What if we analyze the ratio Y/X directly? The median is the same, but now the mean is the arithmetic mean of the ratio, estimated as 0.19 with 95% confidence limits $(0.12, 0.27)$. This is not the ratio of the individual means (as is true for the geometric mean) and may therefore not be a natural measure of the location of the data. If we want to discuss a mean ratio we should analyze differences of logged data instead (see Box 6.3).

In Figure 6.9 we have plotted the various e-CDFs discussed above. The largest, outer, graph shows the e-CDF for the difference, which is highly skewed to the left, making measures like the mean more or less meaningless. The middle graph shows the e-CDF for the ratios on the original (linear) scale. This is also slightly skewed, but now to the right. Finally, the innermost graph shows the same e-CDF but now on a logarithmic scale. This is the most symmetric of the e-CDFs, and the scale we should work on for these data (supported by the fact that when we consider cell counts in sputum, it is the relative effect that has clinical meaning).

Box 6.3 When should we log the data in the analysis?

This question is an important one that does not have a simple, definite answer. But there are a few points to be considered.

One situation in which to consider log-transformation is when we have a skewed distribution with a long tail to the right. Log-transformation may or may not symmetrize the distribution in this case; if it does there is still the problem of justifying why we should describe this variable in this way.

To me the main driver behind the choice to log-transform the data is how the experienced clinician, or scientist, sees the variable in question. Does he think of effects as absolute changes, or as relative changes? If the answer is absolute changes, the choice should be to analyze the data on the original scale. If, on the other hand, the answer is relative changes, that means interpreting effects in a relative sense, so that the expression dE/E represents the true effect. In such a case the analysis should be done on $\ln E$. There are no other changes to how we set up the model, but care must be taken with the description of the end result. Usually a mean difference on the logarithmic scale should be back-transformed to a ratio of (geometric) means.

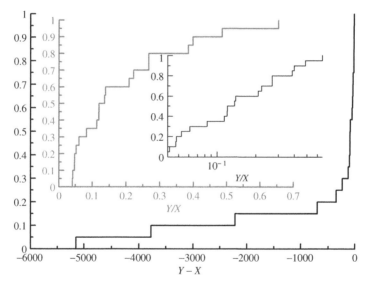

Figure 6.9 The e-CDFs for $Y - X$ and Y/X, the latter on both a linear and a logarithmic scale.

6.7 Bootstrapping

We have discussed how we can describe a population CDF $F(x)$ in terms of various (population) parameters, and how these can be estimated from a random sample. The tricky part is to describe the actual knowledge we obtain about the parameter from our estimate, that is, how close we believe it is to the true value. We have outlined how we can do this in special situations, such as for percentiles and the mean. However, there may be other parameters we wish to estimate, and it may not be trivial to derive confidence statements for these. In such situations a method of considerable conceptual simplicity comes to the rescue, but it is rather computer-intensive (though this is less of a problem now than it used to be).

To specify the problem, from a particular random sample $\{x_1, \ldots, x_n\}$ we wish to compute a particular function $\hat{\theta}(x_1, \ldots, x_n)$, often an estimator of a parameter θ. We can compute the estimate, but how do we get confidence statements about θ? We have seen how such statements can be obtained for percentiles by using the e-CDF $F_n(x)$ together with knowledge about the binomial distribution. The mean, in turn, used a different method which did not work well for the sputum data on the original scale, because the distribution was so skew. How can we get more reliable confidence statements for the mean on the original scale?

To understand the method to be described, we first note that if we knew the true CDF $F(x)$, we could obtain such information with any prespecified precision by employing an *in silico* experiment. We do this by obtaining the distribution of $\hat{\theta}$ (which is a function of a sample) by constructing a very large number of independent samples from the distribution, computing the test statistic for each of these and investigating the distribution of these computed numbers. A single such simulation is how the collection of data from one experiment works, and we can do this on the computer as many times as we wish. The fact that we do not know $F(x)$ poses a problem, but we have data from an experiment, and with it an approximation of $F(x)$ as the linearly interpolated version of the e-CDFs (instead of the step function version; see page 151). We therefore start this discussion by assuming that $F_n(x)$ is a linearly interpolated e-CDF, and that this function is a good description of the true (but unknown) CDF $F(x)$. We then generate new random samples using the following process:

1. Choose n random numbers p_k from the uniform distribution on the unit interval.

2. For each of these, compute the percentiles $x_k = F_n^{-1}(p_k)$.

3. Compute the function $\hat{\theta}(x_1, \ldots, x_n)$.

On the computer we can do this as many times as we wish. This will produce a random sample $\{\theta_1, \ldots, \theta_N\}$ of function values, which can be used to describe the distribution of $\hat{\theta}$. The process therefore defines an approximation to the confidence function for θ, from which we can compute any (approximate) confidence interval or p-value we wish. How well this works obviously depends on how well $F_n(x)$ approximates $F(x)$, but it is expected to improve as we increase the sample.

As already hinted, this is not the method we actually use, because the e-CDF is not defined by linear interpolation, but as a staircase function. If we repeat the procedure described above, we now get our simulated data values only among the original observations. If these are all different, we can get for each random number any of these with equal probability; if there are ties such numbers will turn up proportionally more often. But this means that the procedure above is the same as taking a random sample from the original data set by *sampling with replacement*. This defines the *bootstrap* method and for clarity we repeat how it works.

1. Generate a large number N of samples by sampling with replacement from the data $x_1, \ldots, x_n$.

2. For each sample i, compute the function value as θ_i.

3. Use the (linearly interpolated) e-CDF for the data $\{\theta_1, \ldots, \theta_N\}$ as a confidence function for θ.

The term 'bootstrapping' was introduced by Bradley Efron and is taken from the phrase 'to pull oneself up one's bootstrap', from the famous *Adventures of Baron Munchausen*, where in one story the baron used this method to escape from the bottom of a deep lake.

Example 6.3   To illustrate bootstrapping, consider our original sputum cell count data. Using 2000 samples, the process mentioned above gives us the confidence curves in Figure 6.10 for the median, the geometric mean and the arithmetic mean. Note that the data have been analyzed on the original scale, but we display them on a logarithmic scale for presentational clarity. The graph also illustrates the corresponding symmetric 80% confidence intervals for each of these parameters.

We can use this approach for any parameter we wish, including the value of the CDF $F(x)$ itself at a given point x. We can also provide estimates with confidence intervals for some more unconventional parameters in the same way. We can look at the difference of the mean and the median to demonstrate that the distribution is not symmetric, or we may use the difference between the median and the average of the two quartiles for the same purpose.

If we apply this to the comparison of the two stochastic variables Y and X we discussed in the previous section, we can compare those results with the results obtained using bootstrapping.

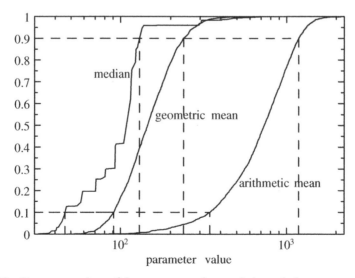

Figure 6.10   Bootstrapped confidence curves for statistics of the sputum count data (obtained on the original scale, but shown on the log scale), demonstrating how we obtain their confidence limits.

Example 6.4 The table below compares what we get using the bootstrap method (Boot), in a run with 2000 samples, with the results we derived in the previous section (NP), for the median and the mean. We make the comparison both for the difference, labeled $Y - X$, and for the difference of the logged data, labeled Y/X. What is shown are estimates and 95% confidence limits. For the bootstrap samples the point estimate is the median of the confidence curve.

| | Analysis of $Y - X$ | | | | Analysis of Y/X | | | |
| | Median | | Mean | | Median | | Mean | |
	NP	Boot	NP	Boot	NP	Boot	NP	Boot
Lower CL	−235	−208	−1318	−1360	0.09	0.09	0.06	0.07
Estimate	−87	−87	−659	−650	0.14	0.14	0.13	0.13
Upper CL	−39	−38	1	−126	0.21	0.20	0.22	0.24

We find consistency in the results for the two methods when it comes to the analysis of the ratio, and also when it comes to the analysis of the median for the difference. However, the upper limit for the mean has been lowered considerably now, which justifies our earlier cautionary remark about the NP confidence interval for the mean with skewed data.

6.8   Meta-analysis and heterogeneity

It may be tempting to view meta-analysis in roughly the same way as we view the bootstrap method: you have a number of different estimates of a parameter in a series of studies, so why not apply the bootstrap idea to obtain the pooled information by looking at the CDF generated from the individual study results? As an example, consider a meta-analysis carried out in the mid 1980s on 22 clinical trials, each of which investigated whether the use of beta-blockers reduces mortality after a heart attack. The 22 odds ratios (treated versus control) for the individual trials are shown in Figure 6.11 as an e-CDF versus logged odds ratio (the staircase curve).

Before we discuss the graph in detail, we perform the conventional Mantel–Haenszel analysis (see Section 5.5) on these data. Such an analysis gives us an estimated pooled odds ratio of 1.20 with 90% confidence interval (1.10, 1.30). The traditional analysis also includes a test for homogeneity, the Breslow–Day test, with p-value 0.063, which is close to being statistically significant at the conventional 5% level. This would be the standard output of a conventional meta-analysis for these data. On the log scale the odds ratio and 90% confidence interval are given by 0.18 and (0.10, 0.26), respectively.

As already mentioned, the e-CDF of the logged odd ratios is shown in Figure 6.11 as the staircase function. This is approximated by the dashed curve, which shows a Gaussian CDF with the mean and standard deviation of the 22 logged odds ratios. Given the rationale for the bootstrap method, we might think that the dashed curve provides us with a confidence function for a common odds ratio on the log scale. However, a quick inspection of the symmetric 90% confidence interval derived from it tells us that it cannot be so. The interval is far too wide, compared to what we found above.

In bootstrapping we reconstruct the confidence function from estimates of a parameter which are given with the same precision, a precision that is defined by the size n of the data sample. The larger the value of n, the steeper the confidence curve will be, because when n is small, some estimates are way off in either direction, giving heavy tails in the

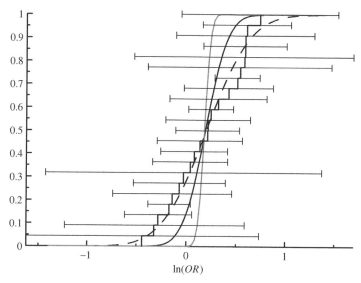

Figure 6.11   A meta-analysis provides both information on a common odds ratio and a description of heterogeneity.

confidence function. In the situation now at hand the individual estimates are computed to different precision, and then the method does not work. This is illustrated in Figure 6.11 by the 90% confidence intervals around the log-odds ratio estimate for each individual study. The appropriate confidence function must be computed by factoring this in, which is what the Mantel–Haenszel methodology does, weighting individual odds ratios according to precision. For reference the confidence curve for the Mantel–Haenszel pooled odds ratio is shown as the gray solid curve in the graph.

To better understand this, we need to be explicit about the assumptions. Let θ_i^* denote the estimate of the odds ratio in study i. Since studies are different in size, the appropriate probability model is that $x_i = \ln(\theta_i^*)$ is an observation of a $N(\ln(\theta), \sigma_i^2)$ distribution, where we have an estimate s_i^2 of σ_i^2. If we substitute s_i^2 for σ_i^2, how would we estimate $\ln(\theta)$? We can use the least squares idea, which means minimizing $Q(\theta) = \sum_i(x_i - \ln(\theta))^2/s_i^2$ (see Section 9.3), which would give $\ln(\theta)$ as a weighted average of the individual log-odds ratios, weighted according to precision: $\ln(\theta) = (\sum_i x_i s_i^{-2})/(\sum_i s_i^{-2}) = W/S$. This is an alternative estimator to the Mantel–Haenszel pooled odds ratio, for which the corresponding confidence curve θ is given by $C(\theta) = \Phi(\sqrt{S}(\ln(\theta) - W/S))$. It is very close to the one derived from the Mantel–Haenszel method in our case (and in general). In summary, the dashed curve in Figure 6.11 would be our choice of confidence function if we only knew the estimates, and not the precision. Since we know the precision, we can improve considerably on this, which is what we have done with either the Mantel–Haenszel method or the method above.

As we saw above, the Breslow–Day test produced a relatively small p-value, so there is some evidence that the true odds ratio actually varies between different studies. We can modify the model to account for this – in other words, to allow for heterogeneity in the studies. For this we assume that for the true odds ratio θ_i of study i we have that

$$\ln(\theta_i) \in N(\ln(\theta), \sigma^2),$$

Box 6.4 Bayes on Gauss

Assume that a characteristic $x \in N(\theta, \sigma^2)$, with σ known, and make the Bayesian assumption that we have the *a priori* distribution $\theta \in N(\mu, \tau^2)$, where μ and τ both are known. What can we then say about θ after we have made the observation x? The density for the *a posteriori* distribution is proportional to $\varphi((x - \theta)/\sigma)\varphi((\theta - \mu)/\tau)$, which means that the distribution is

$$N\left(\frac{x\tau^2 + \mu\sigma^2}{\sigma^2 + \tau^2}, \frac{\sigma^2\tau^2}{\sigma^2 + \tau^2}\right).$$

The best estimate of θ is the mean in this distribution, which we can write

$$\mu + \left(1 - \frac{\sigma^2}{\sigma^2 + \tau^2}\right)(x - \mu).$$

This is an average of the global mean μ and the observation x. It is closer to x the larger the ratio τ/σ is. It is the Bayesian update of θ from the observation x.

The calculation is also relevant for the frequentist, but now θ has a distribution in the population instead. We do not have a single observation, but a sample $x_1, \ldots, x_n$, and we know that the expected value of the average of these observations is μ and that s^2 is an estimate of $\sigma^2 + \tau^2$. The properties of the $\chi^2(n - 1)$ distribution show that the expected value of s^{-2} is $(n - 1)/(n - 3)(\sigma^2 + \tau^2)$, so the frequentist can substitute the Bayesian update above with the corresponding empirical Bayesian update

$$\bar{x} - \left(1 - \frac{n - 3}{n - 1}\frac{\sigma^2}{s^2}\right)(x - \bar{x}).$$

This is actually a better estimate of the value of θ based on an observation x, than the observation itself. Put in the context of point estimation, this result is known as the Stein effect; it is discussed further in Box 7.5.

---

where the variance σ^2 is a measure of the heterogeneity. It then follows that $\ln(\theta_i^*)$ is (asymptotically) an observation of a $N(\ln(\theta), \sigma^2 + s_i^2)$ distribution. If we have an estimate of σ^2, we can, on the one hand, derive a new confidence function for θ and, on the other hand, graphically illustrate the heterogeneity. With respect to the first, the estimate for $\ln(\theta)$ now becomes 0.19, and if we transform this to the odds ratio scale we get 1.21 with 90% confidence interval (1.08, 1.36), which is very similar to the Mantel–Haenszel estimate. The standard deviation σ is estimated at 0.18, and if we draw the CDF for the distribution of the θ_i from this, we get the solid, smooth, curve in Figure 6.11. This curve therefore is a description of how the true odds ratios vary in these types of studies. The range of odds ratios can be described by saying that 90% of the odds ratios from different studies can be found in the interval (0.90, 1.62).

The model we have used also gives us updated estimates of the odds ratios of individual studies, using the information that they all follow a Gaussian distribution. The idea, which is Bayesian except that we replace the parameters $\ln(\theta)$ and σ^2 with estimates from the data, is outlined in Box 6.4. The result is shown in Figure 6.12. These updated estimates are called

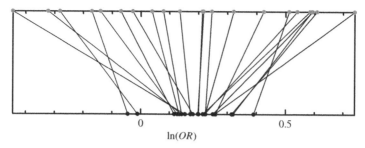

0   0.5

$\ln(OR)$

Figure 6.12   The estimated log-odds (upper row) and how they are shrunk (lower row) if we assume the true log-odds come from a Gaussian distribution.

empirical Bayes estimates for the individual odds ratios, and the fact that we can strip the estimates of some uncertainty this way is an example of a famous paradox pointed out by Charles Stein (see Box 7.5). The paradox is that you can improve the estimate in one study by using the estimates of other studies, which in this case is because we make the assumption that the true log-odds follow a Gaussian distribution. This assumption provides a mechanism for other studies to provide information on a single one.

One loose end remains: how do we get an estimate of σ^2? Simple least squares does not suffice because of the way this parameter appears in the statistical model. The are different ways we can estimate it, the simplest of which is to refer to the general maximum likelihood theory. If we write down the log-likelihood of the problem above, we get a function of the parameters θ and σ^2, which we can maximize. This provides us with point estimates, and what we have done above is use the point estimate of σ^2 from this as a true value in the description.

6.9   Comments and further reading

Most of the material here is very basic statistics, though the format is slightly unconventional. Our main ambition has been to clarify the connection between the true CDF (the Platonic world), and the world we observe, which consists of a limited amount of finite observed data, summarized in the e-CDF. The e-CDF is my preferred substitute for histograms, which is the way distributions usually are displayed. For discrete data histograms are not problematic, describing frequencies as they do, but for continuous data one first has to discretize the data, and there is no unique way to do this. The precise way this is done may have a strong influence on what the distribution looks like in a histogram. The main virtue of the e-CDF is that it works for all kinds of distributions – continuous, discrete and mixed – which also explains why the CDF is preferred over probability functions and probability densities in this book. For an overview of the mathematical properties of the e-CDF, both exact and large-sample theory, see Csáki (1984). Some key large-sample results are summarized in Appendix 6.A.3.

When we introduced the *t*-test, we used the original data that William Gosset, publishing under the pseudonym 'Student', used for illustration of his method. His test, now known as Student's (one-sample) *t*-test, was developed from a need to analyze small data sets accurately. He could not use any of his own data for illustration (his employer, the beer company Guinness, was afraid of disclosing company secrets), so he borrowed some medical data from Cushny and Peebles, describing a crossover study on additional hours of sleep obtained when subjects were using a certain sedative in two different isomer forms. The full story of both these data and the test is described by Senn and Richardson (1994).

Bootstrapping is one, but not the only, computer-intensive method that can help us make reasonable inference in situations where we cannot obtain analytical expressions for the statistics involved. It was pre-dated by a similar method, called the jackknife (named after the Swiss army knife) which from a data set of n observations formed n new data sets of size $n - 1$ by leaving out one observation at a time. As far as bootstrapping is concerned, we have only presented the general idea and ignored the bells and whistles, of which there are many. In particular, we do not discuss how to obtain a more efficient algorithm if we require a confidence interval with good precision in the tails, but not necessarily in the center of the distribution. Both the book by Efron and Tibshirani (1993) and the shorter tutorial by Carpenter and Bithell (2000) discuss such problems.

The discussion in Section 6.8 is about random effects and heterogeneity. This is more or less the normal state of affairs in biology, and we took the opportunity to illustrate how we can approach it model-wise and analysis-wise. More examples will come later, in particular when we discuss modeling in general and dose response in particular, where further important aspects of this will be discussed. The data we used as background material for the discussion here is from Yusuf et al. (1985).

References

Carpenter, J. and Bithell, J. (2000) Bootstrap confidence intervals: when, which, what? A practical guide for medical statisticians. *Statistics in Medicine*, **19**, 1141–1164.

Csáki, E. (1984) *Handbook of Statistics* vol. 4 New York: Elsevier Science chapter Empirical distribution function, pp. 405–430.

Efron, B. and Tibshirani, R.J. (1993) *An Introduction to the Bootstrap* vol. 57 of *Monographs on Statistics and Applied Probability*. New York: Chapman & Hall.

Senn, S. and Richardson, W. (1994) The first t-test. *Statistics in Medicine*, **13**, 785–803.

Yusuf, S., Peto, R., Lewis, J., Collins, R. and Sleight, P. (1985) Beta blockade during and after myocardial infarction: an overview of the randomized trials. *Progress in Cardiovascular Diseases*, **27**, 335–371.

6.A   Appendix: Some technical comments

6.A.1   The extended family of the univariate Gaussian distributions

In this appendix we will outline the basic and classical statistical theory for data with a common Gaussian distribution, and mathematically derive both Student's t distribution and Fisher's F distribution.

The standardized Gaussian distribution $N(0, 1)$ was defined in Box 4.5, where it was also noted that a stochastic variable has the $N(m, \sigma^2)$ distribution if it can be written $m + \sigma Z$, where $Z \in N(0, 1)$. We then have that $Z^2 \in \Gamma(1/2, 1/2)$, where the $\Gamma(p, a)$ distribution was defined in Section 5.A.2. We therefore learn a lot about the Gaussian distribution if we understand the gamma distribution. One simple observation is that if $X \in \Gamma(p, a)$ then $E(X^k) = a^{-k}\Gamma(p + k)/\Gamma(p)$ for all $k > -p$, but the most important relation for the gamma distribution is probably that

$$(X, Y) \in \Gamma(p, a) \otimes \Gamma(q, a) \;\Rightarrow\; \left(X + Y, \frac{X}{Y}\right) \in \Gamma(p + q, a) \otimes \beta'(p, q). \qquad (6.3)$$

Here $\beta'(p, q)$ denotes the β distribution of the second kind and $\otimes$ means that the components are independent. This is derived in Box 6.5, where the density of the $\beta'(p, q)$ distribution is given and its relation to the β distribution of the first kind, introduced in Section 4.6.2, is clarified. Note, in particular, that $X/(X + Y) \in \beta(p, q)$.

To 'take the square-root' of the $\beta(p, q)$ distributions when $p = 1/2$ is of traditional importance, since this is how the t distribution appears. To take the square root of a distribution means to find the distribution which is symmetric around zero, and for which the squared variable has the given distribution. That $F(x)$ is the CDF for the square root of a $\beta'(1/2, q)$ distribution means that

$$2F(x) - 1 = P(X^2 \leq x^2) = B(1/2, q) \int_0^{x^2} t^{-1/2}(1 + t)^{-q-1/2}dt,$$

from which we deduce that the density for X is given by

$$\frac{\Gamma(q + \frac{1}{2})}{\sqrt{\pi}\Gamma(q)}(1 + x^2)^{-q-1/2}.$$

Similarly, we can take the square root of the $\beta(1/2, q)$ distribution, which will give us a distribution with the density function $B(1/2, q)(1 - x^2)^{q-1}$, $|x| \leq 1$.

When this is applied to statistics, at least to the part of statistics that is about Gaussian data, we consider the special cases where $a = 1/2$ and take $p = f/2, q = f'/2$ for integers f and f'. The gamma distributions are then χ^2 distributions and equation (6.3) means that if $(X, Y) \in \chi^2(f) \otimes \chi^2(f')$ then $X + Y$ is distributed as $\chi^2(f + f')$ and is independent of the ratio X/Y. The distribution for the latter is more commonly referenced as

$$\frac{X/f}{Y/f'} \in F(f, f'),$$

where $F(f, f') = (f'/f)\beta'(f/2, f'/2)$ is called the F distribution. Similarly, if X has the distribution of the square root of the $\beta'(1/2, f/2)$ distribution, we consider $\sqrt{f}X$ instead, whose distribution is the $t(f)$ distribution. Its density is obtained by replacing x^2 by x^2/f

---

Box 6.5 Deriving formula (6.3)

It suffices to show the claim when $a = 1$. If $X \in \Gamma(p, 1)$ and $Y \in \Gamma(q, 1)$ are independent, then

$$P\left(X + Y \leq z, \frac{X}{Y} \leq u\right) = \frac{1}{\Gamma(p)\Gamma(q)} \int\int_{x+y \leq z, x/y \leq u} x^{p-1} y^{q-1} e^{-x-y} dx dy.$$

To evaluate the integral, introduce the new variables $s = x + y$, $t = x/y$. The functional determinant is then given by $-s/(1 + t)^2$ and the integrand in the integral in these new coordinates becomes

$$\frac{s^{p+q-2} t^{p-1}}{(1+t)^{p+q-2}} \frac{s}{(1+t)^2} = s^{p+q-1} e^{-s} t^{p-1} (1+t)^{-p-q}, \quad s, t > 0.$$

This is a product of a function in s and one in t, which means that $X + Y$ and X/Y are independent. Furthermore, we see that the function in s is proportional to the density for the $\Gamma(p + q, 1)$ distribution, which shows that this is the distribution for $X + Y$. It now follows that the distribution for X/Y must have density $B(p, q) t^{p-1} (1 + t)^{-p-q}$, where $B(p, q) = \Gamma(p + q)/[\Gamma(p)\Gamma(q)]$. The distribution with this density is called the beta distribution of the second kind, denoted by $\beta'(p, q)$.

The relationship to the beta distribution $\beta(p, q)$ of the first kind is that if $X \in \beta'(p, q)$, then $Y = X/(1 + X) \in \beta(p, q)$. To see this, note that with the substitution $u = t/(1 + t)$ the density becomes $B(p, q) u^{p-1} (1 - u)^{q-1}$ (including the factor $|du/dt| = 1/(1 + t)^2$).

---

in the expression for the density for X, together with the appropriate modification of the coefficient.

Example 6.5   Assume that $X_1, \ldots, X_n$ are independent observations of $N(0, 1)$. Then $n\bar{X}^2 \in \chi^2(1)$ and $S = \sum_i (X_i - \bar{X})^2 \in \chi^2(n - 1)$, and these are independent. A direct consequence of the discussion above is the well-known fact that

$$\frac{\sqrt{n}\bar{X}}{s} = \sqrt{n-1} \sqrt{\frac{n\bar{X}^2}{S}} \in t(n - 1)$$

(the equality means that they have the same distribution), but we also see that

$$\frac{\sqrt{n}\bar{X}}{\sqrt{\frac{1}{n} \sum_i X_i^2}} = \sqrt{n} \sqrt{\frac{n\bar{X}^2}{S + n\bar{X}^2}}$$

has the density function

$$\frac{\Gamma(\frac{n}{2})}{\sqrt{n\pi}\Gamma(\frac{n-1}{2})} \left(1 - \frac{x^2}{n}\right)^{(n-3)/2}, \quad |x| \leq \sqrt{n}.$$

6.A.2   The Wiener process and its bridge

One of the key properties of the Gaussian distribution is that the sum of Gaussian variables itself has a Gaussian distribution. A consequence of this is that if $X \in N(0, 1)$, then we can write $X = S_n = X_{1n} + \ldots + X_{nn}$, where the X_{in} are independent with the same distribution $X_{in} \in N(0, 1/n)$. If we consider all the cumulative sums $\{S_{kn}\}_{k=1}^{n}$ simultaneously, we can plot the different S_{nk} versus k/n and connect these points with straight lines to obtain a polygon starting at the origin and ending up at $(1, X)$. If we view this the opposite way, we have for each t a stochastic variable $x(t)$ such that when $t = k/n$, it is given by S_{kn}, whereas for other values of t it is obtained by linear interpolation of such points. For this stochastic process $x(t)$ we have the following:

- all its paths are continuous and start at zero;

- the increments over disjoint time intervals are all independent (because they are derived from independent X_{kn});

- $x(t) - x(s) \in N(0, |t - s|)$ when $t = k/n$ and $s = j/n$.

There is a stochastic process which has precisely these properties, including the last one for all $s, t \in [0, 1]$. It is called the Wiener process, which we denote by $\{w(t); t \in [0, 1]\}$. It plays much the same role for stochastic processes as the standardized Gaussian plays for stochastic variables.

The Wiener process is named after Norbert Wiener who defined it mathematically in the late 1920s. Intuitively it represents the cumulative sum of infinitely many, infinitely small and independent errors. Its physical realization is, however, much older and is named after Robert Brown who, in 1828, described the apparently random movement of particles suspended in a medium such as a fluid, something we today call *Brownian motion* in his honor. Although it was Wiener who put this process in the context of stochastic processes, it was Albert Einstein who was the first to describe it mathematically. When he described it in the famous Einstein year 1905 (the same year he published the theory of special relativity and discovered the photon) he did not know about Brown's writings. Einstein's mission was different. He felt that if there are molecules (which was only a theory at the time) there should be macroscopic manifestations of their motion, and if these could be observed, this would serve as confirmation of their existence. This was a problem in mathematical physics, and Einstein showed that under certain assumptions, reflecting what should happen on average in a world full of colliding particles, the probability density $\rho(x, t)$ of particles should be governed by the diffusion equation $\partial_t \rho = D\Delta_x \rho$, where Δ_x is the Laplacian (sum of second-order derivatives) and D a diffusion coefficient. From this it follows that if the particle starts at the origin, the density ρ at time t is distributed according to a Gaussian distribution with mean zero and variance Dt. Einstein then derived an expression for the diffusion coefficient and laid out a program for 'proving' the existence of the atom. This program was later carried out by Jean Baptiste Perrin, winning him the Nobel Prize in 1926.

One problem with Brownian motion as described was that the paths, as can easily be guessed from the description, are very irregular. In fact, they are continuous but non-differentiable at every point. This mathematical puzzle was laid to rest by Wiener (and others). (Actually non-differentiable functions had been known to mathematicians for quite some time, since the work of Weierstrass in the late nineteenth century.)

As a stochastic process, the Wiener process has some important properties, including the reflection principle and the strong Markov property. Without being explicit about what these are, one important consequence of them is that

$$P\left(\max_{[0,t]} |w(s)| < x, w(t) \in I\right) = \int_I h(t, y)dy,$$

where

$$h(t, y) = \sum_{-\infty}^{\infty} \frac{(-1)^n}{\sqrt{2\pi t}} e^{-(y-2nx)^2/2t}.$$

A close relative of the Wiener process is the Brownian bridge $\{w^0(t); t \in [0, 1]\}$, which is essentially a process defined for $0 \le t \le 1$, consisting of those paths of the Wiener process which end at zero when $t = 1$. Mathematically it can be obtained from the Wiener process by applying either of the functions $h(x)(t) = x(t) - tx(1)$ or $h(x)(t) = (1 - t)x(t/(1 - t))$ on it, each of which produces another Gaussian process with mean zero, but now the covariance between times s and t is given by $s \wedge t - st$. Among its most important properties is that

$$P\left(\max_{[0,1]} |w^0(t)| \le x\right) = 1 + 2\sum_{1}^{\infty}(-1)^n e^{-2n^2x^2}, \qquad (6.4)$$

which is derived from the formula above by considering a sequence of intervals I decreasing to zero. Together with a large-sample approximation this observation is the Kolmogorov test in statistics, as is described in the next section.

6.A.3   Confidence regions for the CDF and the Kolmogorov–Smirnov test

Let X be a stochastic variable with CDF $F(x)$. We have seen that, for a given x, the CLT implies that

$$\sqrt{n}(F_n(x) - F(x)) \in AsN(0, F(x)(1 - F(x))).$$

In order to extend this to pairs of such values, fix x and y such that $x < y$. Since we must have that $F_n(x) \le F_n(y)$, these two variables cannot be independent. To compute their covariance, note that the bivariate CDF of the pair $U = (I(X \le x), I(X \le y))$ is given by $P(X \le x, X \le y) = F(x \wedge y)$ (where $x \wedge y$ denotes the smaller of x and y), and since the marginal distributions are $F(x)$ and $F(y)$, respectively, it follows that the covariance of U is given by $F(x \wedge y) - F(x)F(y)$, and the covariance of $F_n(x)$ and $F_n(y)$ is therefore this divided by n. When $x = y$ this reduces to the variance formula for $F_n(x)$. It is now a consequence of the multivariate CLT that the bivariate distribution of

$$(\sqrt{n}(F_n(x) - F(x)), \ \sqrt{n}(F_n(y) - F(y)))$$

is asymptotically a Gaussian bivariate distribution with zero mean and a covariance matrix given by

$$\Sigma = \begin{pmatrix} F(x)(1 - F(x)) & F(x \wedge y) - F(x)F(y) \\ F(x \wedge y) - F(x)F(y) & F(y)(1 - F(y)) \end{pmatrix}. \tag{6.5}$$

We can extend this to any vector of values of the e-CDF at different points. But this means that the stochastic process $\{\sqrt{n}(F_n(x) - F(x))\}$ has the same finite-dimensional distributions as the process $\{w^0(F(x))\}$, where $\{w^0(t); t \in [0, 1]\}$ is the Brownian bridge defined in the previous section. Using this, it is not a far-fetched guess that we have that

$$\{\sqrt{n}(F_n(x) - F(x))\} \rightarrow \{w^0(F(x))\} \quad \text{in distribution,}$$

a fact that can be verified in different ways. One way uses the CLT for martingales in continuous time and is described in Appendix 4.A. The observation has many useful consequences, derived from the general fact that we then also have that

$$h(\{\sqrt{n}(F_n(x) - F(x))\}) \rightarrow h(\{w^0(F(x))\}) \quad \text{in distribution,}$$

for any function h which is a continuous function on continuous functions over the interval $[0, 1]$. An example of such a function is $h(x) = \max_{[0,1]} |x(t)|$, so we have that

$$\max_x \sqrt{n}|F_n(x) - F(x)| \rightarrow \max_x |w^0(F(x))| \quad \text{in distribution.}$$

If $F(x)$ is continuous, the right-hand side is $\max_{[0,1]} |w^0(t)|$, the distribution of which was given in equation (6.4). It follows that

$$P\left(\sqrt{n} \max_y |F_n(y) - F(y)| \leq x\right) \approx 1 + 2\sum_{k=1}^{\infty}(-1)^k e^{-2k^2 x^2}, \quad x > 0, \tag{6.6}$$

for large n. The immediate effect of this expression is that it provides us with a statistical test, called the Kolmogorov one-sample test, for the hypothesis that the distribution we have sampled from has CDF $F(x)$. It also provides us with simultaneous confidence intervals for the function $F(x)$, based on the function $F_n(x)$. But these intervals are of fixed width and do therefore not generalize the confidence intervals for $F(x)$ we discussed Section 6.5, since the width of those was proportional to $\sqrt{F(x)(1 - F(x))}$. To obtain a corresponding simultaneous region, just note that

$$\left\{\frac{\sqrt{n}(F_n(x) - F(x))}{\sqrt{F(x)(1 - F(x))}}\right\} \rightarrow \left\{\frac{w^0(F(x))}{\sqrt{F(x)(1 - F(x))}}\right\} \quad \text{in distribution.}$$

For a continuous distribution $F(x)$ we can therefore compute the distribution of $\max_t |w^0(t)|/\sqrt{t(1 - t)}|$ and determine its percentiles in order to get a simultaneous confidence region proportional to the pointwise intervals. We do not pursue this any further.

We can also derive the two-sample Kolmogorov–Smirnov test from this. One way uses the observation that

$$\sqrt{\frac{nm}{n+m}}(F_n(x) - F(x) - (G_m(x) - G(x))) = \sqrt{\frac{m}{n+m}}Y_n(x) - \sqrt{\frac{n}{n+m}}Z_m(x),$$

where $Y_n(x)$ and $Z_m(x)$ are independent processes with Brownian bridge limits. From this we get a Gaussian process that has the same mean and covariance structure as the Brownian bridge when $F(x) = G(x)$. The full Kolmogorov–Smirnov test for the equality of two distributions therefore follows from the corresponding Kolmogorov one-sample test.

7

Correlation and regression in bivariate distributions

7.1 Introduction

In the previous chapter we discussed the CDF of a single stochastic variable and how we estimate it from data – this as a generalization of the problem of estimating a single binomial parameter. The next step is to generalize the analysis of a 2×2 table to general data. As we have seen, there are two closely related but (philosophically) different ways to view a 2×2 table, both of which describe the relation between an exposure and a response. We can either see it as a problem of association, where we consider both the exposure and response as characteristics of members of the population and wish to understand how they are related, or we can see it as an estimation problem, where we wish to assess how much the response changes because of the exposure. We will generalize both these viewpoints to more general situations, starting with association in this chapter. The description in terms of effect sizes is postponed to the next chapter.

The association aspect is a two-dimensional problem, whereas the effect size aspect is about comparing two one-dimensional CDFs. Describing association is about describing the pair (exposure, response), while the effect size is about investigating the response conditional on the exposure status. The former therefore requires the mathematics of functions of two variables, whereas the latter is concerned with functions of one variable only. So association is more technically involved, but we should still start with it, because this discussion will allow us to take into account baseline information when we compare groups in the next chapter.

There are essentially three parts to what we wish to do in this chapter. The first is to discuss how we describe distributions of two variables (i.e., bivariate distributions), with emphasis on the dependence of the two components on each other, expressed as correlation. We also discuss how we can use correlation to our advantage in the design of experimental studies. This part introduces the important bivariate Gaussian distribution, which is uniquely defined by adding a correlation coefficient to the list of parameters describing Gaussian distributions.

Understanding Biostatistics, First Edition. Anders Källén.
© 2011 John Wiley & Sons, Ltd. Published 2011 by John Wiley & Sons, Ltd.

Having introduced this distribution we turn to the all-important concept of regression to the mean, its meaning and consequences. The term refers to a mathematical property of the bivariate Gaussian distribution, which is of great importance in understanding the relationship between two consecutive measurements. Finally, we address the problem of how we use data to obtain knowledge about the two means in a bivariate distribution, which is an extension of the univariate t-test to bivariate data, and we will see how these ideas will allow us to analyze functions of binomial parameters in a way that takes all uncertainty into account.

7.2   Bivariate distributions and correlation

A bivariate stochastic variable (X, Y) is a pair of stochastic variables (possibly two test statistics) and is described by a two-dimensional CDF

$$F(x, y) = P(X \le x, Y \le y).$$

Similar to the univariate CDF, this is a function that starts at zero far in the lower left 'corner' of the xy-plane and increases to one as we move upwards toward the upper right 'corner'. If we denote the CDFs for the components X and Y by $F_1(x)$ and $F_2(y)$, respectively, these can be obtained as the *marginal CDFs*

$$F_1(x) = F(x, \infty), \quad F_2(y) = F(\infty, y).$$

For obvious reasons it may sometimes be convenient to denote these marginal distributions by $F_X(x)$ and $F_Y(y)$, respectively.

There is a relationship between bivariate distributions and 2×2 tables in that if we dichotomize the two variables X and Y according to cut-off points x and y, respectively, the proportions in each cell are as shown in the table to the right. (To dichotomize perfectly valid measurement data in this way is common in medical statistics.) Marginal probabilities can be obtained by choosing cut-off points for the two marginal distributions separately, but in order to get a fit to the complete table we need to determine the association between the variables.

$F_1(x) - F(x, y)$	$F^c(x, y)$
$F(x, y)$	$F_2(y) - F(x, y)$

$$F^c(x, y) = 1 - F_1(x) - F_2(y) + F(x, y)$$

If X and Y are independent stochastic variables we have that $F(x, y) = F_1(x)F_2(y)$, so when this is not the case, the two variables X and Y are dependent in some way. We wish to describe such dependence in a simple way, in a single parameter, in the same way as location is described by the mean and dispersion is described by the variance. The simplest parameter that describes dependence is called the covariance, and is defined as

$$C(X, Y) = \int_{-\infty}^{\infty} \int_{-\infty}^{\infty} (x - m_1)(y - m_2) dF(x, y),$$

where m_1 and m_2 are the means of X and Y, respectively. It is estimated by the sample covariance

$$\frac{1}{n-1} \sum_i (x_i - \bar{x})(y_i - \bar{y}).$$

Box 7.1 Spearman correlation, concordant pairs and Kendall's tau

---

There are alternatives to the Pearson correlation coefficient to describe the association between two stochastic variables X and Y. One is the Pearson correlation coefficient for the pair $(F_1(X), F_2(Y))$ instead of (X, Y). The marginal distributions for this variable are uniformly distributed on $(0, 1)$, so this coefficient, known as the *Spearman correlation*, is obtained as

$$\rho_S = 3 \left(4 \int_0^1 \int_0^1 uv \, dU(u, v) - 1 \right),$$

where $U(u, v)$ is the CDF of $(F_1(X), F_2(Y))$. To estimate it we use e-CDFs, which is equivalent to computing the Pearson sample correlation of the rank-transformed data.

Another way to describe association is obtained if we rewrite the covariance parameter as

$$C(X, Y) = \frac{1}{2} \iiiint (x_1 - x_2)(y_1 - y_2) dF(x_1, y_1) dF(x_2, y_2).$$

Two pairs (x_1, y_1) and (x_2, y_2) are said to be concordant if $(x_1 - x_2)(y_1 - y_2) > 0$ and discordant if the reverse inequality holds true. We can measure correlation by using any function of $(x_1 - x_2)(y_1 - y_2)$. A particular case is the sign function, which leads to the parameter

$$\tau = P((X_1 - X_2)(Y_1 - Y_2) > 0) - P((X_1 - X_2)(Y_1 - Y_2) < 0),$$

which is called *Kendall's tau*. For a continuous variable this is $\tau = 4p_\tau - 1$, where

$$p_\tau = E \left[\iint I(X > x, Y > y) dF(x, y) \right] = \iint F^c(x, y) dF(x, y).$$

The ratio of the two probabilities above, concordant to discordant pairs, defines a quantity that reduces to the odds ratio in a 2×2 table, and is therefore a kind of generalized odds ratio.

The analytical expressions for these correlation coefficients can be found for a bivariate Gaussian distribution:

$$\rho_S = \frac{6}{\pi} \arcsin(\rho/2), \quad \tau = \frac{2}{\pi} \arcsin(\rho).$$

The graph to the right shows these functions as functions of ρ in gray and black, respectively. We have that $\rho_S \approx \rho$ to a very high precision, which means that if estimates of Pearson's and Spearman's correlation coefficients differ substantially, the data are probably not from a bi-

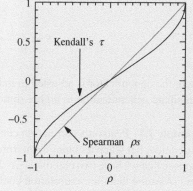

variate Gaussian distribution. The magnitude of Kendall's τ must, however, be interpreted in a different way than how ρ is interpreted.

The covariance is important for much the same reason the mean and variance are important; it will be the key additional descriptor that defines the bivariate Gaussian distribution. The factor that we integrate in the definition is positive when x and y are on the same side of m_1 and m_2, respectively, so the covariance will be positive if $F(x, y)$ puts more weight in those areas than in other areas. A positive covariance therefore implies that there is a net covariation of X and Y, whereas a negative value implies that, on average, whenever one is high, the other is low. It is a mathematical fact that $C(X, Y)^2 \leq V(X)V(Y)$, and therefore it is natural to describe the covariation of two stochastic variables in terms of the unit-free correlation coefficient

$$\rho(X, Y) = \frac{C(X, Y)}{\sqrt{V(X)V(Y)}}.$$

This correlation coefficient ρ is by necessity a number between -1 and 1, and two variables are said to be uncorrelated if $\rho = 0$. It is easy to show that two independent variables are uncorrelated, but it is possible for dependent variables to be uncorrelated (though not for Gaussian variables). As already mentioned, the correlation coefficient is a parameter that is heavily related to a bivariate Gaussian distribution, but it is not the only way to describe correlation. Some alternative measures of correlation are outlined in Box 7.1. To distinguish ρ from other correlation coefficients, it is referred to as the Pearson correlation coefficient.

In passing we may note that if $Z = (X, Y)$ and $m = (m_1, m_2)$ we can define the matrix

$$V(Z) = \int_{-\infty}^{\infty} \int_{-\infty}^{\infty} (z - m)(z - m)^t dF(z), \tag{7.1}$$

which will be a 2×2 matrix ($z - m$ is a column vector in the integral) such that the diagonal elements are the variances of the two components and the off-diagonal elements are the covariance. We call this matrix the variance matrix of the bivariate stochastic variable. To estimate it we use the corresponding discrete entity in which we replace the vector m with the vector of arithmetic means $\bar{z}$, and the CDF with the corresponding e-CDF. The result is the sample variance matrix

$$S = \frac{1}{n - 1} \sum_i (z_i - \bar{z})(z_i - \bar{z})^t,$$

where the denominator is adjusted to make the corresponding variable unbiased. This has an immediate generalization to higher-dimensional stochastic variables.

Example 7.1 Pearson's correlation coefficient is often used in medical statistics to describe association, but when it comes to interpreting it, there is one salient point that is often missed. Suppose that we have two outcomes (X, Y) and we wish to understand the association between them. Assume that their correlation coefficient is ρ and let σ_1^2 and σ_2^2 denote their variances. Each of the variables (X could be a biomarker, and Y a manifestation of a disease) is measured with a measurement error. Assume that the measurement error for X has variance η_1^2 and that for Y has variance η_2^2. Let X' and Y' be the observed data. If measurement error is independent of actual values, we have that $C(X', Y') = C(X, Y) = \rho\sigma_1\sigma_2$, and therefore that the observed correlation coefficient is

$$\rho(X', Y') = \frac{\rho \sigma_1 \sigma_2}{\sqrt{\sigma_1^2 + \eta_1^2}\sqrt{\sigma_2^2 + \eta_2^2}} = \frac{\rho}{\sqrt{(1 + \eta_1^2/\sigma_1^2)(1 + \eta_2^2/\sigma_2^2)}}.$$

If the measurement error is large relative to the variable variability, this means that the correlation we observe gets attenuated. If we do not have precision in our measurements of X and Y, the association may drown in the measurement noise.

We next wish to see how the Pearson correlation coefficient relates to the odds ratio in the simple case of a 2×2 table.

Example 7.2 For a 2×2 table with cell probabilities $p_{ij}, i, j = 1, 2$, the covariance is easily computed as $p_{11} - p_{1+}p_{+1} = p_{11}p_{22} - p_{12}p_{21}$, so that Pearson's correlation coefficient is given by

$$\rho = \frac{p_{11}p_{22} - p_{12}p_{21}}{\sqrt{p_{1+}p_{+1}p_{2+}p_{+2}}}.$$

An estimate of this is called the phi coefficient, and is obtained from cell counts as follows:

$$\phi = \frac{x_{11}x_{22} - x_{12}x_{21}}{\sqrt{x_{1+}x_{+1}x_{2+}x_{+2}}}.$$

To interpret ρ here, introduce δ by the relation $p_{11} = p_{1+}p_{+1} + \delta$. Then we must also have that $p_{21} = p_{2+}p_{+1} - \delta$, $p_{12} = p_{1+}p_{+2} - \delta$, and $p_{22} = p_{2+}p_{+2} + \delta$, from which it follows that $\delta = p_{11}p_{22} - p_{12}p_{21}$. We obtain a relationship between δ and the odds ratio $\theta = p_{11}p_{22}/p_{12}p_{21}$ by inserting the expressions above to get

$$\delta^2 + \delta\left(p_{1+}p_{+1} + p_{2+}p_{+2} + \frac{\theta}{1-\theta}\right) + p_{1+}p_{+1}p_{2+}p_{+2} = 0.$$

Solving this equation, we can express $\rho = \delta/\sqrt{p_{1+}p_{+1}p_{2+}p_{+2}}$ in the odds ratio θ and constants that only depend on margins. This means that we can obtain a confidence interval for ρ conditional on the margins from a corresponding interval for the odds ratio. As example we take the first Hodgkin's lymphoma example, for which the relation is illustrated in Figure 7.1. The 90% confidence interval (1.82, 4.71) for the odds ratio translates into the 90% confidence interval (0.15, 0.37) for ρ.

A related measure of association is Cohen's κ, which is discussed in Box 7.2.

Figure 7.2 describes a geometric interpretation of the Pearson correlation coefficient. It shows how, given two independent and standardized stochastic variables, we can find a linear transformation of these which has correlation ρ to one of them. For future reference, we note that the bivariate distribution $F(y, z) = P(Y \le y, Z \le z)$ of (Y, Z) in Figure 7.2 is computed from

$$\int_{-\infty}^{z} P(Y \le y | Z = \zeta)dF_Z(\zeta) = \int_{-\infty}^{z} P\left(X \le \frac{y - (\sin\theta)\zeta}{\cos\theta} | Z = \zeta\right)dF_Z(\zeta),$$

Box 7.2 Rater agreement and Cohen's κ

Suppose that two pathologists independently examine a set of n slides for the presence or absence of a cell abnormality. One way to express the agreement between the two (as an indicator of the reproducibility or interchangeability of their classifications) is in terms of Cohen's κ (kappa). To define this, arrange the possible outcomes in a 2×2 table so that the rows and columns are the classifications of the two pathologists. If p_{ij} is the probability of a slide falling into cell (i, j), the actual probability of agreement is $p_a = p_{11} + p_{22}$ and the probability of agreement by chance alone is $p_c = p_{1+}p_{+1} + p_{2+}p_{+2}$. Cohen's measure of agreement is now the excess probability of agreement beyond chance,

$$\kappa = \frac{p_a - p_c}{1 - p_c}.$$

To estimate this, we insert the observed cell counts into the expression for κ to get

$$\hat{\kappa} = \frac{2(x_{11}x_{22} - x_{12}x_{21})}{x_{1+}x_{+2} + x_{2+}x_{+1}}.$$

We see that κ is proportional to the Pearson correlation coefficient discussed in Example 7.2 with a proportionality constant that only depends on margins. This means that we can use the discussion in that example to transform a confidence interval for the odds ratio to one for κ. The result will be conditional upon margins, as for Fisher's exact test.

---

which means that

$$F(y, z) = \int_{-\infty}^{z} F_X \left(\frac{y - \rho\zeta}{\sqrt{1 - \rho^2}} \right) dF_Z(\zeta). \tag{7.2}$$

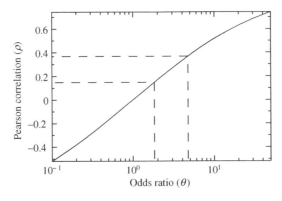

Figure 7.1 The connection between the odds ratio and the Pearson correlation coefficient in a (particular) 2×2 table with given margins.

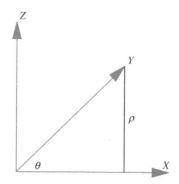

Figure 7.2 From two independent stochastic standardized (mean zero and variance one) variables X and Z we can construct a new standardized variable Y which has correlation ρ with Z by rotation an angle θ such that $\rho = \sin\theta$. This is easily checked from the fact that $Y = (\cos\theta)X + (\sin\theta)Z$. The correlation with X is $\cos\theta$.

For a continuous distribution this means that the bivariate density is given by

$$f(y, z) = \frac{1}{\sqrt{1 - \rho^2}} f_X\left(\frac{y - \rho z}{\sqrt{1 - \rho^2}}\right) f_Z(z)$$

This formula will be used to derive the bivariate Gaussian distribution below.

The description of a general bivariate distribution in terms of parameters is more complicated than in the univariate case. Percentiles, which should be the points defining the same height for the bivariate CDF, corresponds not to points on a line, but to curves in the plane. This makes them not only hard to estimate, but also impossible to describe, except in a graphical form. Therefore, to describe the location of a bivariate distribution the mean (vector) becomes even more important than it was for univariate distributions, and multivariate Gaussian distributions are even more important in multivariate statistics than univariate Gaussian distributions are in univariate statistics.

7.3   On baseline corrections and other covariates

In this section we address the question how we best can utilize information on one variable in the analysis of another when they are dependent. The setting will be the analysis of a clinical trial, for which there is a choice between designing it as a parallel group study and a crossover study. Sometimes there is no such choice, as when a long treatment period is needed, or the effect of the drug is long-lasting, but when there is a choice, we need to understand the pros and cons of the two designs. Some aspects of this are statistical and will now be discussed.

The setting is that we want to compare two treatments, an active drug and a matching placebo, utilizing a particular outcome variable which we do not need to specify. Define the following two stochastic variables:

- X is the value of the outcome variable when the patient is given placebo;

- Y is the value of the outcome variable when the patient is given the active drug.

In theory both of these have a value for each individual, though we may only measure one of them. Assume that (X, Y) has a bivariate distribution with mean (m_1, m_2), variances (σ_1^2, σ_2^2) and (the Pearson) correlation coefficient ρ. If we run a parallel group study, we will observe X for some patients and Y for others, whereas if we run a crossover study we essentially observe both, though not necessarily under identical conditions (which has consequences which will be the subject of Section 8.7).

Our objective is to estimate the treatment difference $\Delta = m_2 - m_1$, which in the case of a parallel group study is the mean difference between the groups. We wish to know which design has the smallest standard error σ (the standard error is the standard deviation of the estimator of Δ). Recall that it is Δ/σ that defines the power of the study, so this determines which design is the most efficient when it comes to demonstrating effects.

If we assume that for the parallel group study each group contains n subjects, we have that the variance of the observed mean difference is

$$V(\bar{Y} - \bar{X}) = \frac{1}{n}(\sigma_1^2 + \sigma_2^2),$$

whereas for the crossover study with n subjects it is

$$V(\bar{Y} - \bar{X}) = \frac{1}{n}(\sigma_1^2 + \sigma_2^2 - 2\rho\sigma_1\sigma_2).$$

If we denote the estimator in the parallel group study by $\hat{\Delta}_{PG}$ and the estimator in the crossover study by $\hat{\Delta}_{XO}$, this means that we have the following relationship between their variances:

$$V(\hat{\Delta}_{XO}) = V(\hat{\Delta}_{PG}) - 2\rho\sigma_1\sigma_2/n.$$

If there is a positive correlation between the variables X and Y, which is to be expected in this context, a two-period crossover study with n subjects is therefore always more efficient than a parallel group study with $2n$ subjects in total. These two study types can be considered equivalent, since they involve the same number of assessments. In the extreme case of $\rho = 0$ they are equivalent from a statistical (though probably not from a cost) perspective, since this condition implies that it does not matter if we reuse subjects in the study.

Conceptually things become clearer if we assume that the variance is the same for X and Y and equal to σ^2. In such a case we have that $V(\hat{\Delta}_{PG}) = 2\sigma^2/n$ and that $V(\hat{\Delta}_{XO}) = 2(1 - \rho)\sigma^2/n$. Split σ^2 into a sum $\sigma^2 = \sigma_b^2 + \sigma_w^2$, with $\sigma_b^2 = \rho\sigma^2$ and $\sigma_w^2 = (1 - \rho)\sigma^2$.

We can think of $X - m_1$ and $Y - m_2$ as constructed from a common stochastic variable Z which has variance σ_b^2, to which we have added a measurement error ξ, which has variance σ_w^2. The values of ξ are different and independent for the two measurements as well as independent of Z. Under this assumption X and Y will have the same variance $\sigma_b^2 + \sigma_w^2$, whereas the difference $Y - X$ will have the same variance as the difference of two independent ξs, which is $2\sigma_w^2$. We call σ_b^2 the *between-subject variance*, and σ_w^2 the *within-subject variance*. In this notation we have that $\rho = \sigma_b^2/(\sigma_b^2 + \sigma_w^2)$ and the variance of $\hat{\Delta}_{PG}$ will be proportional to $\sigma_b^2 + \sigma_w^2$, whereas that of $\hat{\Delta}_{XO}$ will be proportional to σ_w^2. In both cases the proportionality factor is $2/n$.

A more detailed modeling along these lines would suggest that we should write X as above but that for Y, which is obtained on drug treatment, we should have the decomposition $Y = m_2 + Z + \tau + \xi$, with the same Z as before but now adding not only an observational error ξ but also a true treatment effect size τ which varies between subjects (and is independent of the other components) with a variance σ_t^2. This assumption implies that $V(Y) = V(X) + \sigma_t^2$, so we expect the variance in the drug-treated group to be larger than in the placebo group. In statistical modeling it is common practice to assume $\sigma_t^2 = 0$, which corresponds to a fixed treatment effect for all subjects. If we perform a t-test in such a situation, the estimated (common) variance will be a compromise between σ^2 and $\sigma^2 + \sigma_t^2$. To avoid making this assumption, and to allow for heterogeneity in the response to treatment, we need to replace the classical fixed effects models with mixed effects models.

One way to improve the efficiency of a (randomized) parallel group study would be to measure the outcome variable not only at the end of treatment, but also at randomization, providing a baseline measurement. We can then study the change from baseline as a new outcome variable. The question is whether this helps. To see if it does, assume that the pair (X, Y) above consist of (baseline, end-of-treatment) measurements. The previous discussion shows that

$$V(Y) = \sigma_2^2, \qquad V(Y - X) = \sigma_1^2 + \sigma_2^2 - 2\rho\sigma_1\sigma_2,$$

from which it follows that $V(Y - X) < V(Y)$ precisely when $\rho > \sigma_2/2\sigma_1$. It is therefore the correlation that determines whether we gain anything from looking at the change from baseline or not. The cut-off point for this is a correlation of 0.5 when $\sigma_1 = \sigma_2$, in which case we can write the variances as $V(Y) = \sigma_b^2 + \sigma_w^2$ and $V(Y - X) = 2\sigma_w^2$. Another way to express the criterion that $V(Y) > V(Y - X)$ is therefore that the within-subject variability is smaller than the between-subject variability. There is, however, a better way to approach the problem, but before we discuss this, we need to understand why, and when, we can replace the end-of-treatment value with the change from baseline and still get a valid estimate of Δ.

Call the two groups A and B, and let superscript denote group membership, while subscripts denote which variable we compute the mean of. It is a basic property of the mean that

$$\Delta = m_Y^B - m_X^A = (m_{Y-X}^B - m_{Y-X}^A) + (m_X^B - m_X^A),$$

because the mean of a difference is the difference of the means. The first parenthesis on the right is what we measure when we analyze the change from baseline. This will estimate Δ if and only if the second parenthesis is zero. If this is not the case, the two analyses estimate different things. One way to guarantee equality at baseline is to randomize the study.

The answer to the question whether to choose Y or $Y - X$ as outcome variable is that we should choose neither. To see why, first replace the change from baseline with the more general variable $Y - bX$ for some arbitrary constant b. If the groups have equal distributions at baseline, this variable also estimates Δ, irrespective of our choice of b. To get the most efficient outcome variable we should therefore use $Y - bX$ instead, where b is chosen to minimize

$$V(Y - bX) = \sigma_2^2 + b^2\sigma_1^2 - 2b\rho\sigma_1\sigma_2.$$

This is a quadratic function in b, which has its minimum when $b = \rho\sigma_2/\sigma_1$. This may appear difficult to apply in practice, since it looks like we need to have a good knowledge of the distributional parameters in order to be able to compute b, but we will see that this is what is done in an analysis of covariance.

The obvious next question is why we should content ourselves with taking a linear predictor for Y of the form bX – why not use some more general function of X? If we want to predict Y from an observation x of X using a function $g(x)$, how should we choose $g(x)$? Suppose that we assess how good a particular prediction is by measuring the mean squared error of the prediction $E((Y - g(X))^2)$; then the function $g(x)$ that minimizes this is the conditional expectation

$$g(x) = E(Y|X = x).$$

The importance of this function is obvious; by viewing X as an explanatory variable for Y, we can reduce the uncertainty about Y by observing X. The increase in precision is expressed in the mathematical observation that

$$V(Y) = E(V(Y|X)) + V(E(Y|X)).$$

In words: the total variance is the average of the conditional variances (essentially the unexplained variance) plus how much the predictor varies. We may also note that the overall mean is the mean of all conditional means:

$$E(Y) = E(E(Y|X)).$$

(This formula is sometimes called the tower formula for the mean.) The problem here is that we may not know the exact bivariate distribution of (X, Y), and therefore not be able to compute the true predictor $g(x)$. We may, however, know it up to some unknown parameters, so that we have a family of functions $g(\theta, x)$ from which we want to pick one by obtaining an estimate of θ from data. Alternatively, we can simply assume such a functional form, without having any distributional knowledge to justify it. We then try to estimate θ from data by minimizing the function $Q(\theta) = E((Y - g(\theta, X))^2)$. This is the least squares method for the estimation of parameters, the method Gauss used to find Ceres. It will be the subject of Chapter 9.

The simplest choice for a function here, seen below to be appropriate for the bivariate Gaussian distribution, is to use a linear function $g(x) = a + b(x - m_1)$, which is the best linear predictor. As before m_1 is the mean of X. This case is what we considered above, and what we have seen is that $a = m_2$, the mean of Y, that $b = C(X, Y)/V(X) = \rho\sigma_2/\sigma_1$, and that the variance of the residuals is

$$V(Y - g(X)) = V(Y) - V(g(X)) = \sigma_2^2(1 - \rho^2).$$

This is what still remains to be explained in terms of other covariates.

7.4 Bivariate Gaussian distributions

The bivariate counterpart to a standardized Gaussian distribution $N(0, 1)$ should have $m_1 = m_2 = 0$ and $\sigma_1 = \sigma_2 = 1$, but there still are many of them, each corresponding to a different degree of correlation between the components. One way to derive these distributions uses equation (7.2) in its differentiated form, which shows that the density should be given by

$$\varphi_2(x, y; \rho) = \frac{1}{\sqrt{1 - \rho^2}} \varphi\left(\frac{y - \rho x}{\sqrt{1 - \rho^2}}\right) \varphi(x). \tag{7.3}$$

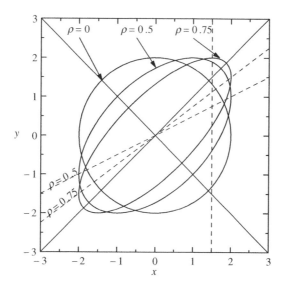

Figure 7.3   The shape of the height contours for standardized bivariate Gaussian distributions for different correlation coefficients.

Here ρ is the Pearson correlation coefficient between the components. Writing out the expression and completing squares, we find that this is

$$\varphi_2(x, y; \rho) = \frac{1}{2\pi\sqrt{1 - \rho^2}} \exp\left(-\frac{x^2 - 2\rho xy + y^2}{2(1 - \rho^2)}\right).$$ (7.4)

We denote this distribution by $N_2(0, 1, \rho)$ and its CDF by $\Phi_2(x, y; \rho)$.

To describe the effect of ρ on the shape of the bivariate Gaussian distribution geometrically, look at Figure 7.3. It shows a contour[1] for the density function for three different choices of the correlation coefficient. Specifically, the curves are the ellipses $x^2 - 2\rho xy + y^2 = 1 - \rho^2$ for the choices $\rho = 0$, 0.5 and 0.75. (These are on the same relative level, relative to the height of the density.) Note the oblique cross in the graph, which corresponds to the two lines $x = y$ and $x = -y$ in the original coordinate system. For all choices of ρ these lines represent the appropriate coordinate system in which to view data (when $\rho = 0$ there is no preferred coordinate system). If we rewrite the equation for the ellipse in this rotated coordinate system (45° counterclockwise) with coordinates s, t, the equation becomes

$$(1 - \rho)s^2 + (1 + \rho)t^2 = 1 - \rho^2 \quad \Leftrightarrow \quad \frac{s^2}{1 + \rho} + \frac{t^2}{1 - \rho} = 1.$$

---

[1]Functions of two variables can be graphically illustrated in the same way as nature topography is illustrated on maps, by indicating height levels as curves in the plane.

This expression confirms what we can see in Figure 7.3, namely that if we increase ρ from 0 toward 1, the ellipse becomes more elongated in the $x = y$ direction and narrower in the $x = -y$ direction. In other words, the closer the correlation is to one, the more concentrated the distribution is around the line $x = y$. Similarly, if the correlation is close to -1, the distribution is concentrated around the line $x = -y$.

On occasion we may have a 2×2 table in which both factors actually have continuous distributions, but data collection has been dichotomized into yes/no categories only. We can then analyze the table in such a way that we actually get a correlation measure between the two factors. This is illustrated in the following example.

Example 7.3 The table below describes a cross-classification according to whether or not a group of workers were exposed to a particular environmental hazard, and whether or not they showed symptoms of bronchitis:

	Not exposed	Exposed
Bronchitis	89	123
No bronchitis	453	318

Each of these two factors can be considered to vary in strength: there may be different degrees of exposure and the bronchitis symptom may vary in severity. However, based on a cut-off point for each, as in a medical test, each subject has found his place in this table. In this way we may have a dichotomization of factors that are really continuous variables, but we have no further recorded information on these. We therefore assume that the underlying continuous data have a standardized bivariate Gaussian distribution, and that the table is obtained using cut-offs a and b for exposure and bronchitis, respectively. As noted earlier, in order to fit the data to such a distribution we need to invoke the correlation coefficient. In all there are three parameters to fit, and three conditions:

$$453/983 = \Phi_2(a, b; \rho), \quad (89 + 453)/983 = \Phi(a), \quad (453 + 318)/983 = \Phi(b).$$

The solution to this system is $a = 0.129$, $b = 0.787$, $\rho = 0.241$, where the last parameter is a description of association. The correlation coefficient in this hypothetical model is called the tetrachoric correlation coefficient.

The bivariate Gaussian is a distribution for which the best linear predictor is also the best of all predictors. To see this, note that equation (7.3) implies that the conditional distribution of Y given that $X = x$ (whose density is $\varphi_2(x, y; \rho)/\varphi(x)$) is also Gaussian, namely $N(\rho x, 1 - \rho^2)$. If we base our understanding of the bivariate Gaussian on the geometric description in Figure 7.3, it may be a little surprising to find that the mean is ρx and not x, since this graph shows that the distribution of the standardized bivariate normal (when $\rho > 0$) is concentrated along the line $y = x$. To see why the mean is ρx, look at the vertical dashed line over the point $x = 1.5$ in Figure 7.3. It intersects the axes of the ellipse in an oblique manner (actually $45°$), so more of it is below the line $y = x$ than is above it. The mean is the mid-point on this line within the ellipse, and the geometrically inclined reader can convince himself that the line of means is precisely the line $y = \rho x$.

The general bivariate Gaussian distribution is defined by the requirement that the mean should be $m = (m_1, m_2)$ and its variance matrix should be $\Sigma = \begin{pmatrix} \sigma_1^2 & \sigma_{12} \\ \sigma_{12} & \sigma_2^2 \end{pmatrix}$. This distribution is denoted by $N(m, \Sigma)$ and is defined by the requirement that

$$\left(\frac{X - m_1}{\sigma_1}, \frac{Y - m_2}{\sigma_2} \right) \in N_2(0, 1, \rho),$$

where $\rho = \sigma_{12}/\sigma_1\sigma_2$. For this distribution the discussion above shows that the distribution of Y conditional on $X = x$ is

$$N\left(m_2 + \rho\frac{\sigma_2}{\sigma_1}(x - m_1), \sigma_2^2(1 - \rho^2) \right).$$

This implies that for a bivariate Gaussian distribution, the conditional mean of Y, given that $X = x$, is the linear predictor

$$E(Y|X = x) = m_2 + \rho\frac{\sigma_2}{\sigma_1}(x - m_1). \tag{7.5}$$

This last equation has some interesting consequences, which will be the subject of the next section.

7.5   Regression to the mean

We will now use the fact that, for the bivariate Gaussian distribution, the best predictor is the linear predictor, to illustrate an interesting and important point about the world and human affairs. It is concerned with the situation where we make two consecutive observations of something, and want to explain the effect we see by a causal relation. There is a specific pattern to what happens in the long run, which leads to certain biases in causal explanations. One example is found in sports, where being on the cover of a *Sports Illustrated* journal, or being awarded the 'Ballon d'Or' in soccer, seems to imply poorer performance in the future. Other examples include the following:

- in education, where special treatment of poor performers all but guarantees their improvement;

- in road safety policies, where actions taken at an accident-prone intersection almost invariably have positive effects;

- the observation that highly intelligent/successful parents often have less intelligent/successful children.

The common denominator for all these examples is a law of nature called *regression to the mean*, which was originally discovered by Francis Galton (see Box 7.3). He studied the relationship between the height of a child and that of its parents. At his disposal he had human data on the heights of 930 children and their 205 parents, which he sex-adjusted to 'male' heights by multiplying all women's heights by 1.08. A quick summary of his result is that:

Box 7.3 On Francis Galton, correlation and regression

The concepts of regression and correlation, as well as the bivariate Gaussian distribution, were all introduced in the second part of the nineteenth century by an English polymath, Sir Francis Galton.

Galton was nephew to Charles Darwin and was intrigued by *The Origin of Species* and its identification of the *survival of the fittest* as the evolutionary force. Inspired by this idea, Galton studied how talent ran in a number of families, including his own and that of the Bernoulli (see Section 11.5). Based on this study Galton became in the 1860s a fervent advocate of positive eugenics, a term he coined in 1883, defining it as 'the study of all agencies under human control which can improve or impair the racial quality of future generations'. ('Positive' refers to the fact that his suggested method was to encourage marriage for the superior, instead of sterilizing the inferior as in the US and some other countries.)

However, there was a disturbing pattern in the data, which was further clarified in a series of investigations into different sizes of seeds of the same species of sweet pea. It had to do with two successive generations. When he finally had sorted it out and published, he chose a different outcome variable, height in successive human generations, in order to drive home the message to a wider audience. This publication, which also introduced the bivariate Gaussian distribution, had as its key message the notion of regression to the mean: for standardized variables we have $E(Y|X = x) = \rho x$, so we should on average expect some regression toward mediocrity. (Galton did not standardize by mean and standard deviation, instead he used the median and half the interquartile range.) This concept, regression, describes how the outcome of one variable predicts the value of another.

What he did not note at the time was that the reverse is also true, that $E(X|Y = y) = \rho y$. The implication of this is that ρ is a more symmetric number related to the two variables. It was not until later, when Galton worked in anthropology on relations like that between a particular bone and the overall length of the body, that he realized this symmetry and invented correlation, both the concept and the word, which appeared in 1888 in a paper titled 'Correlations and Their Measurement Chiefly from Anthropometric Data'. Correlation is a spelling of the word 'co-relation' which was in common use at the time.

- when mid-parents (Galton's term for the average parent height for each child) are taller than mediocrity (Galton's word for the median, which is the mean for the Gaussian distribution), their children tend to be shorter than they are; and

- when mid-parents are shorter than mediocrity, their children tend to be taller than they are.

This summary is probably as good as any description of this phenomenon, in conjunction with the tale in Box 7.4.

Galton's observation is deduced from equation (7.5). To simplify the discussion, assume that the marginal distribution of each generation is the same (same mean and variance). Then

$$E(Y|X = x) - m = \rho(x - m),$$

Box 7.4 A tale about regression to the mean

Daniel Kahnemann is a psychologist who in 2002 received the Nobel prize in Economics (actually, the Bank of Sweden Prize in honor of Alfred Nobel) for his work in behavioral economics. In his speech of thanks he spoke about the following experience.

'I had the most satisfying Eureka experience of my career while attempting to teach flight instructors that praise is more effective than punishment for promoting skill-learning. When I had finished my enthusiastic speech, one of the most seasoned instructors in the audience raised his hand and made his own short speech, which began by conceding that positive reinforcement might be good for the birds, but went on to deny that it was optimal for flight cadets. He said,

"On many occasions I have praised flight cadets for clean execution of some aerobatic maneuver, and in general when they try it again, they do worse. On the other hand, I have often screamed at cadets for bad execution, and in general they do better the next time. So please don't tell us that reinforcement works and punishment does not, because the opposite is the case."

This was a joyous moment, in which I understood an important truth about the world: because we tend to reward others when they do well and punish them when they do badly, and because there is regression to the mean, it is part of the human condition that we are statistically punished for rewarding others and rewarded for punishing them. I immediately arranged a demonstration in which each participant tossed two coins at a target behind his back, without any feedback. We measured the distances from the target and could see that those who had done best the first time had mostly deteriorated on their second try, and vice versa. But I knew that this demonstration would not undo the effects of lifelong exposure to a perverse contingency.'

---

and, since $-1 < \rho < 1$, this means that the expected difference from the mean for the son (i.e., $E(Y|X = x) - m$) is smaller (closer to zero) than that for the mid-parent (i.e., $x - m$). This is what Galton's summary amounts to. The immediate consequence is that if we select only parents with an above average mid-parent height, and look at their children, we will find that these are on average shorter than their parents. Still tall, but shorter. It works the other way around also, in that if we look at tall children, their parents are expected to be, on average, shorter than them, though still tall:

$$E(X|Y = y) - m = \rho(y - m).$$

Actually the son/father example is a better illustration of the special case than the mid-parent/son example of Galton. This is because in Galton's case the variances for X and Y are not equal, and we therefore need to use the more general equation (7.5). Instead the mid-parent height is an average of two heights, so we are probably closer to having $\sigma_1 = \sigma_2/\sqrt{2}$ than to having $\sigma_1 = \sigma_2$. This means that the regression curve in this case is $E(Y|X = x) - m = \sqrt{2}\rho(x - m)$, and it is only when $\rho < 1/\sqrt{2} = 0.71$ that we have regression to the mean (the estimate for ρ in Galton's data was 0.497).

Regression to the mean explains why we should avoid drawing conclusions about an intervention from a change that has been observed over time. To expand on this, note that we can rewrite equation (7.5) as

$$E(Y|X = x) - x = m_2 - m_1 - \left(1 - \rho\frac{\sigma_2}{\sigma_1}\right)(x - m_1).$$

If we let this represent a situation with a pre-test and a post-test for which we assume a common variance, and with an intervention in-between, the relation becomes

$$E(Y|X = x) - x = \Delta - (1 - \rho)(x - m),$$

where m is the pre-test mean and Δ the *true* mean intervention effect. This expression gives the *apparent* intervention effect, which is made up of two terms, the true intervention effect Δ and a term which defines the regression to the mean effect. In a study this becomes a statement about the bias in the expected value of the mean increase due to a difference to the mean at baseline:

$$E(\bar{Y} - \bar{X}|\bar{X} = \bar{x}) = \Delta - (1 - \rho)(\bar{x} - m). \tag{7.6}$$

We see that if we for some reason have $\bar{x} < m$, we will overestimate the effect, whereas the reverse is true if $\bar{x} > m$. This observation becomes important when we compare two groups and build our analysis on the change from baseline. In studies numerical differences at baseline will lead to numerical differences at the end. We therefore need to build an analytic model which adjusts for this, which is what is done in the analysis of covariance methodology (which is actually an analysis of means) to be discussed in Section 8.8.

So far the discussion has been based on the assumption that the underlying distribution is a bivariate Gaussian distribution. A natural question is to ask to what extent these observations hold true for other distributions. To investigate one particular extension, we build on the decomposition

$$X = Z + \xi, \quad Y = Z + \eta,$$

which is one way we can derive the bivariate Gaussian distribution from independent Gaussian components, following the discussion on page 184. This assumption means that each individual has a true value of the response variable, given by Z, but that value is measured with (independent) errors on two occasions. Now drop the condition that Z has a Gaussian distribution, but keep this assumption on ξ and η. Let $g(x)$ denote the probability density for the common marginal distribution of X and Y. We can then compute the following expression for the conditional mean (the derivation of which is outlined in Appendix 7.A.1):

$$E(Y|X = x) = x + \sigma^2(\ln g(x))'. \tag{7.7}$$

This equation is also applicable in the related situation where we have a stochastic variable $(X = Z)$ distributed in a population, which we measure with an error $(Y = Z + \eta)$, and we wish to understand what we can say about the true value based on the observation. This is related to an (in)famous observation in statistics, the Stein effect (see Box 7.5).

Box 7.5 The Stein effect

Assume that there is a characteristic Z in a population which has a $N(m, \eta^2)$ distribution. When we measure Z, we do it with an error ξ which we assume has a $N(0, \sigma^2)$ distribution, so that what we measure, $X = Z + \xi$, has a $N(m, \eta^2 + \sigma^2)$ distribution. Given that we have observed the measurement x in an individual, what can we say about the true value for this individual? The regression to the mean equation shows that

$$E(Z|X = x) = m + \left(1 - \frac{\sigma^2}{\eta^2 + \sigma^2}\right)(x - m),$$

which means that there is a shrinkage toward the mean, which is greater the larger σ^2 is relative to η^2.

This observation is related to a famous result by Charles Stein from 1955. We do not know m, but instead have a single observation x_i (measured with error) on each of n individuals. The intuitive answer to the question how we should estimate the true value Z_i for subject i, is that we should take $Z_i = x_i$. What Stein was able to prove, in a more general context, was that we should really use an estimate $\hat{Z}_i = \bar{x} + c(x_i - \bar{x})$ for a constant $c < 1$.

Comparing this with the observation in the first paragraph, we see that we have replaced m with $\bar{x}$ and that we should have something like $c = 1 - \sigma^2/s^2$ for the constant, where s^2 is the sample variance of the observations. This is not quite true, but if we redefine s^2 by dividing by $n - 3$, instead of $n - 1$, it becomes so. This is done to ensure unbiasedness, and was discussed in Box 6.4.

For the special case with the bivariate normal distribution, the σ^2 in equation (7.7) is the within-subject σ_w^2 and $g(x) = \sigma^{-1}\varphi((x - m)/\sigma)$, where the new σ^2 is the total variance, $\sigma^2 = \sigma_b^2 + \sigma_w^2$. In this case it follows that $(\ln g(x))' = -(x - m)/\sigma^2$, so equation (7.7) becomes

$$E(Y|X = x) = x - \frac{\sigma_w^2}{\sigma_b^2 + \sigma_w^2}(x - m) = x - (1 - \rho)(x - m).$$

We then rederive our regression to the mean equation.

What equation (7.7) tells us is that the correction term is increasing as long as $g(x)$ is increasing, and decreasing when $g(x)$ is decreasing, which means regression toward the mode of the density $g(x)$, that is, the value x that defines the maximum of the function $g(x)$ (a point given by the solution to the equation $g'(x) = 0$). The mean, median and mode just happen to be the same for the Gaussian distribution. However, this is only one particular model extension of the bivariate Gaussian distribution.

Making statements about one outcome variable conditional on the observed values of other variables is what we do when we consider the so-called linear models, which will be the subject of Chapter 9. From a clinical trial perspective it is also important to understand what it means to do something conditional on a condition of the type $a < X < b$. If we, as before, consider X to be the baseline measurement of a clinical variable, a clinical protocol in general defines a particular subpopulation of patients to study, which we hope will make

---

Box 7.6 Conditioning in the bivariate Gaussian distribution

For a general (continuous) bivariate distribution, the conditional density of Y given a condition such s $a < X < b$ ($a = -\infty$ and $b = \infty$ are allowed), is given by

$$f_{Y \mid a < X < b}(y) = \frac{\int_a^b f_{X,Y}(x, y)dx}{P(a < X < b)}.$$

It follows that the expected value $E(Y \mid a < X < b)$ is given by

$$\int_{-\infty}^{\infty} y \int_a^b f_{X,Y}(x, y)dxdy = \int_a^b E(Y \mid X = x)dF_X(x),$$

divided by $F_X(b) - F_X(a)$. Note that

$$E(X \mid a < X < b) = \int_a^b xdF_X(x)/(F_X(b) - F_X(a)) = \mu(a, b).$$

In the situation, discussed in the main text, where we observe a stochastic variable Z twice with independent errors, equation (7.7) (what we denoted $g(x)$ there is $f_X(x)$ here) shows that

$$E(Y - X \mid a < X < b) = \frac{\sigma^2 \int_a^b (\ln f_X(x))' f_X(x)dx}{F_X(b) - F_X(a)} = \sigma^2 \eta(a, b), \qquad (7.8)$$

where $\eta(a, b) = (f_X(b) - f_X(a))/(F_X(b) - F_X(a))$, and also that

$$V(Y \mid a < X < b) = \sigma^2(2 + \sigma^2(\lambda(a, b) - \eta(a, b)^2)), \quad \lambda(a, b) = \frac{f'_X(b) - f'_X(a)}{F_X(b) - F_X(a)}.$$

---

it easier to demonstrate an effect. Such inclusion criteria will imply some kind of restriction on the baseline measurements. For example, in asthma trials we may want to include only patients who take a certain amount of rescue medication per day, and as outcome measurement we may have a lung function measurement. We expect a symptomatic patient to have a more compromised lung function than a non-symptomatic one, so the study population will essentially be selected on a criterion of the form $X \leq a$, though we have not specified a in the protocol. How does regression to the mean manifest itself in such situations?

We consider again the bivariate Gaussian distribution for the case where the pre-treatment and the post-treatment measurement have correlation ρ and the same variance σ^2. We assume that the population mean values are m and $m + \Delta$, respectively, for the two variables so that the true treatment effect is Δ. Our interest is in the estimation of this Δ, which refers to the whole population. Now assume that the net effect of the inclusion criteria is that we only include subjects that fulfill the criterion $a < X < b$ for some a and b, possibly infinitely small or large, respectively. From the discussion in Box 7.6 we have that

$$E(Y - X \mid a < X < b) = \Delta + (1 - \rho)\sigma \frac{\varphi(\frac{b-m}{\sigma}) - \varphi(\frac{a-m}{\sigma})}{\Phi(\frac{b-m}{\sigma}) - \Phi(\frac{a-m}{\sigma})}.$$

The second term on the right shows the bias introduced in the estimate for the treatment effect by having the inclusion criteria (assuming, of course, that our ultimate objective is to find the Δ in the whole population). There is no bias if $\rho = 1$ (which is obvious), and when a and b are taken symmetrically around the mean m (because of the symmetry of $\varphi(x)$). Note that the bias we get in the mean difference $\bar{Y} - \bar{X}$ is not affected by sample size; it is determined by the cut-off points alone.

Example 7.4   Assume that we work with an assay which, like all assays, measures with an error. When measurements are (suspiciously) high, it may be tempting to reanalyze the same sample and use the new measurement instead. What is the consequence of this? When the true value in the sample is Z our first measurement will be $X = Z + \xi$ and the second will be $Y = Z + \eta$, where ξ and η are independent measurement errors. Y is only observed when $X > a$. The expected value for our measurement will be

$$E(X) + E(Y - X|X > a)(1 - F(a)) = E(X) - \sigma^2 g(a),$$

from which we see that we have a negative bias of size $\sigma^2 g(a)$.

7.6   Statistical analysis of bivariate Gaussian data

We now wish to find the two-variable analogue of the (univariate) t-test, which is about obtaining (simultaneous) knowledge about (the components of) the mean vector in a bivariate Gaussian distribution. The key observation is that for a variable Z with a bivariate Gaussian distribution with mean $m = (m_1, m_2)$ and covariance matrix $\Sigma = \begin{pmatrix} \sigma_1^2 & \sigma_{12} \\ \sigma_{12} & \sigma_2^2 \end{pmatrix}$, we have that

$$(Z - m)^t \Sigma^{-1} (Z - m) \in \chi^2(2).$$

Precisely as in the univariate case we usually do not know the variance matrix Σ, and need to estimate it with the sample variance matrix S. The theory that works in one dimension, and was outlined in Appendix 6.A.1, can be generalized at the expense of some slightly more complicated mathematics (see Appendix 7.A.2). The key information is that, for a sample of size n,

$$\frac{n(n - 2)}{2(n - 1)}(\bar{z} - m)^t S^{-1}(\bar{z} - m) \in F(2, n - 2).$$

This is a generalization of the univariate t-test and is called Hotelling's T^2-test. It is used to obtain confidence statements about the mean vector m by use of the confidence function

$$C(m) = F_{2,n-2}\left(\frac{n(n - 2)}{2(n - 1)}(\bar{z} - m)^t S^{-1}(\bar{z} - m)\right), \tag{7.9}$$

where $F_{a,b}(x)$ denotes the CDF of the $F(a, b)$ distribution. This is a function of two variables, which means that what was a confidence interval for the univariate test now becomes a confidence region in the plane. The next example illustrates this.

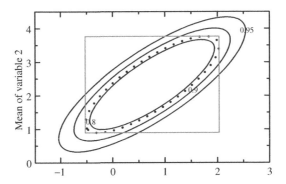

Figure 7.4   The solid contours define simultaneous confidence regions for a Hotelling's T^2-test. Also shown are the univariate confidence intervals and the univariate confidence region as a dotted contour. See text for details.

Example 7.5   The Cushny and Peebles data mentioned in Example 6.5 consisted of 10 pairs of data points, data which we assume come from a bivariate Gaussian distribution. Previously we analyzed the mean difference of the two variables; now we wish to describe our confidence in the pairs of means. The mean vector is estimated as $(0.75, 2.33)$, and the estimated covariance matrix is $\begin{pmatrix} 3.20 & 2.85 \\ 2.85 & 4.01 \end{pmatrix}$. Together these are all we need to calculate the confidence function in equation (7.9), which is described graphically by a contour plot in Figure 7.4. This graph shows three contours of equal value for the function $C(m)$, namely those corresponding to the 0.8, 0.9 and 0.95 confidence levels, respectively. The region enclosed by the last of these levels corresponds to a confidence region of (simultaneous) confidence level 95% for the pair of means. The corresponding univariate 95% confidence intervals for the individual means are indicated by the sides of the rectangle in the graph. This, however, does not enclose a region with (simultaneous) confidence level 95% for the mean vector, but by using the Bonferroni correction we can show that the confidence level is at least 90%.

Based on Figure 7.4 we can derive simultaneous confidence limits for different combinations of the two means. This means that we can compute as many functions as we wish without losing credibility (in other words, no multiplicity correction is needed). Before we proceed, however, look again at the confidence function defined by equation (7.9). It is computed by a two-step procedure. First we compute the value of the quadratic form in the argument, from which we derive the confidence by applying a particular CDF to it. The choice $F_{2,n-2}(x)$ as this CDF gives us simultaneous confidence statements. If we want univariate statements, that is, the limits of a single confidence interval, we can use the $F(1, n-1)$ distribution instead (which is the square of the $t(n-1)$ distribution). In other words, we use the confidence function

$$C(m) = F_{1,n-1}((\bar{u} - m)^t S^{-1}(\bar{u} - m)). \tag{7.10}$$

(Strictly speaking this may only hold true if the statement is based on a linear combination of the components of m, as is the case below, because a linear combination of Gaussian variables is a univariate Gaussian variable; see Appendix 7.A.3.) The 95% region obtained

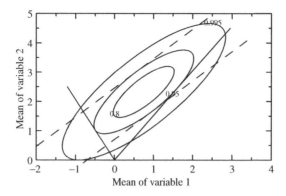

Figure 7.5   How to obtain univariate confidence intervals for a mean difference and a mean ratio from the bivariate confidence function, defined by the t distribution.

using this function is shown in Figure 7.4 as the dotted contour. It is the largest such contour that fits into the rectangle defined by univariate confidence intervals; the curve is tangent to this rectangle at four points. The rectangle is therefore obtained by projecting the dotted contour on to the respective axis.

Example 7.6   Continuing with the previous example, we now wish to derive univariate confidence intervals for two particular functions of the two means, using the (univariate) confidence function in equation (7.10). Three contours for this confidence function, corresponding to the confidence levels 0.8, 0.95 and 0.995, respectively, are shown in Figure 7.5. The reason why we choose these particular levels will soon become apparent.

We wish to identify the confidence intervals both for the mean difference $m_2 - m_1$ and for the mean ratio m_1/m_2 from this graph. To obtain the 95% confidence interval for the mean difference we consider the different lines $m_2 - m_1 = \theta$, and identify those θ for which this line is tangential to the 95% contour of $C(m)$. These are indicated in the graph as the two parallel dashed lines, corresponding to the two θ-values 0.70 and 2.46, respectively. These two numbers are therefore the 95% confidence limits for the mean difference, which agrees with what we found in Example 6.5. Similarly, we obtain the confidence limits for the ratio m_1/m_2 by considering where the line $m_1 = \theta m_2$ is tangent to the 95% contour of $C(m)$. These two lines are illustrated in Figure 7.5 as solid straight lines, and the slopes of these provide us with the limits -0.48 and 0.66, respectively, for the 95% confidence interval.

There is also the contour for the level 0.995 shown in Figure 7.5. This is there to illustrate a particular problem: the line corresponding to the lower confidence limit for the ratio with this confidence level is horizontal, which means an infinitely small lower confidence limit. If we require more confidence than 0.995 we will therefore not get a finite interval. This will be further explored below.

The intervals for the ratio of two means that we obtained above are called Fieller intervals and are analytically described in Box 7.7. We see that two conditions need to be fulfilled in order for this to provide finite intervals, and if we ask for enough confidence, these conditions will be violated for any data. In the next example we explore this further.

Box 7.7 Fieller intervals for the ratio of means

Assume that we have a bivariate stochastic variable (X, Y) with a Gaussian distribution with mean (m_1, m_2) and covariance matrix $\Sigma = (\sigma_{ij})$, and consider the ratio $\theta = m_2/m_1$. We then have that

$$Y - \theta X \in N(0, \sigma^2(\theta)), \quad \text{where } \sigma^2(\theta) = \sigma_{22} - 2\theta\sigma_{12} + \theta^2\sigma_{11}.$$

If Σ is known, this means that the function

$$C(\theta) = \Phi\left(\frac{y - \theta x}{\sigma(\theta)}\right)$$

is a confidence function for the ratio of the means. When we do not know Σ we may estimate it with something that has a Wishart distribution, such as the sample variance, in which case the estimate of $\sigma^2(\theta)$ follows a χ^2 distribution, and we replace the Gaussian CDF with the appropriate t distribution CDF.

There are conditions when such Fieller intervals (of confidence level $1 - 2\alpha$) do not produce finite intervals for the point estimate m_2/m_1, conditions that occur for all data if we ask for sufficiently high confidence level (small α). The interval is defined as those θ for which we have that $(y - \theta x)^2 \leq t_\alpha^2 \sigma^2(\theta)$, which can be rewritten as

$$(t_\alpha^2 \sigma_{22} - y^2) - 2\theta(t_\alpha^2 \sigma_{12} - xy) + \theta^2(t_\alpha^2 \sigma_{11} - x^2) \geq 0.$$

The coefficient on θ^2 has to be negative in order for this to define an interval, and the condition for this is that $|x|/\sqrt{\sigma_{11}} > t_\alpha$, which means that for the denominator m_1 there is enough evidence that it is different from zero at the required confidence level. The other condition is that the discriminant is positive, which can be written

$$t_\alpha^2(z^t \Sigma^{-1} z - t_\alpha^2) \det \Sigma > 0, \quad z = \begin{pmatrix} x \\ y \end{pmatrix}.$$

The first of these criteria is violated first, and when it is, the interval will contain infinity since it allows for division by zero. When both criteria are violated, the inequality holds true for all θ, because this case allows for the indeterminate number $0/0$.

Example 7.7   The top part of the two-sided confidence function for the mean ratio for the Cushny and Peebles data is shown in Figure 7.6. The graph shows that the confidence intervals are somewhat 'fishy' at very high confidence levels. In fact, in order to have a finite interval we must have a confidence level less than 0.995. This level is shown as the lower dashed line in the graph. It is the asymptote of the confidence function in both infinities. For a confidence level above this, but less than 0.9986, the confidence region excludes a finite interval (and includes infinity), whereas for even higher confidence levels all possible values constitute the confidence region; the data are insufficient to provide any information about the parameter with such high confidence.

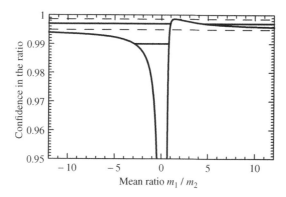

Figure 7.6   The upper part of the two-sided confidence function for the ratio of means for the Cushny–Peebles data.

If instead we had drawn the one-sided confidence function we would not have obtained an increasing function from zero to one. Such a function cannot therefore work as a CDF, which means that this is an example where we cannot interpret the confidence function as a distribution function, and which gave Fisher a headache with his fiducial concept.

7.7   Simultaneous analysis of two binomial proportions

The final section of this chapter is rather mathematical in nature. It is about the general problem of how we do inference on the parameters of interest in the presence of nuisance parameters. The discussion to come, in the context of binomial parameters, is mainly explained in terms of graphics, and it may be worthwhile for the less mathematically interested not to ignore it. The ideas are the same as in the previous section.

In Chapter 5 we discussed how to compare binomial proportions from two different groups. The assumption was that we have an observation x_1 from a $\text{Bin}(n_1, p_1)$ distribution, and an observation x_2 from a $\text{Bin}(n_2, p_2)$ distribution which is independent of the first. We are interested in confidence statements about various combinations of p_1 and p_2. In the discussion in Section 5.3 we noted that the approach taken there failed to take proper account of all uncertainty in the data. Building on the ideas presented in the previous section, we will now remedy this and obtain a simultaneous confidence region for the pair (p_1, p_2). Since it is a simultaneous analysis, we can derive from this as many confidence intervals for combinations of p_1 and p_2 as we wish, without losing control of the overall significance level. The starting point for our investigation will be an observation we made in Section 5.4.1, namely that

$$C(p_1, p_2) = \chi_2 \left(\frac{n_1(p_1^* - p_1)^2}{p_1(1 - p_1)} + \frac{n_2(p_2^* - p_2)^2}{p_2(1 - p_2)} \right) \qquad (7.11)$$

(where p_i^* is the estimate of p_i) is an approximate confidence function for $p = (p_1, p_2)$.

For illustration we again use the first Hodgkin's lymphoma data set, for which we have $p_1^* = 67/101$ and $p_2^* = 43/107$. A few contours for the confidence function $C(p_1, p_2)$ are shown in Figure 7.7, the outermost of which is a 95% confidence region for the pair (p_1, p_2). The other contours represent other confidence levels. We will use the 95%

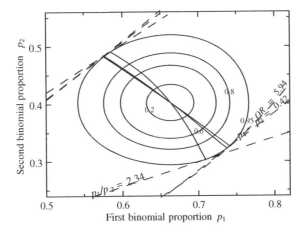

Figure 7.7   Ninety-five percent confidence region for two independent binomial proportions, illustrating how confidence intervals for functions of these are obtained.

contour to derive simultaneous confidence intervals for the most important triad of derived parameters: the difference $\Delta = p_1 - p_2$, the ratio $RR = p_1/p_2$, and the odds ratio $OR = p_1(1 - p_2)/[p_2(1 - p_1)]$.

Confidence intervals are obtained geometrically as follows. A particular parameter is a function of the parameters $f(p_1, p_2)$, for which we consider the curves $f(p_1, p_2) = \theta$ for different, but fixed, θs. Such curves are illustrated in Figure 7.7 as dashed curves (found outside the confidence region). We vary θ until we find the values for which this curve is tangential to the 95% contour, as we did for functions of the means in Example 7.5. These choices of θ determine the confidence interval for the parameter $\theta = f(p_1, p_2)$. Only the curves in the lower part of the graph have been labeled, but they all have their counterparts in the upper portion of the graph. In the graph we see:

- two straight lines $p_1/p_2 = \theta$, for $\theta = 1.19$ and 2.34, that are tangential to the 95% confidence region and therefore define the confidence interval for the risk ratio;

- two straight lines $p_1 - p_2 = \theta$, for $\theta = 0.09$ and 0.42, that are tangential to the confidence region and define the confidence limits for the risk difference;

- two nonlinear curves $p_1(1 - p_2)/[p_2(1 - p_1)] = \theta$, for $\theta = 1.45$ and 5.94, that are tangential to the confidence region and define the confidence limits for the odds ratio.

This approach is geometrically simple, but is more complicated to implement numerically. To get a more convenient numerical method, we note that for a given curve $\theta = f(p_1, p_2)$ we look for the smallest value of the function $C(p_1, p_2)$ on it. Once we have done so for each θ, we look for the θs that intersect with the appropriate confidence contour. Mathematically this means that we define the function

$$C^*(\theta) = \min\{C(p_1, p_2); f(p_1, p_2) = \theta\},$$

from which we obtain the confidence interval as $\{\theta; C^*(\theta) \leq 1 - \alpha\}$. If this was a maximization, instead of a minimization, problem (so that higher surface peaks hide lower ones) this would mean graphically that if we look at the surface $z = C(p_1, p_2)$ in the coordinates (θ, p_2), the function $C^*(\theta)$ would be the profile of the surface as seen from the θ-axis. The method is therefore called profiling. The procedure is also illustrated in Figure 7.7; for each θ the points (p_1, p_2) that define $C^*(\theta)$ are shown as the solid curves within the confidence region. The intersections of these curves with the 95% level contour define the same confidence intervals as above. As before, if we want the univariate confidence interval for a particular θ, we replace the CDF for the $\chi^2(2)$ distribution with the CDF for the $\chi^2(1)$ distribution in the definition of the confidence function.

We can also produce a p-value for the (single) test of the hypothesis that $p_1 = p_2$. For this we first compute the number (corresponding to finding the best fit to this assumption)

$$K = \min_{0<p<1} \left\{ \frac{n_1(p_1^* - p)^2 + n_2(p_2^* - p)^2}{p(1 - p)} \right\} = 14.26,$$

from which the p-value is computed as $1 - \chi_1(K) = 0.00016$. In our case, the p-value is the same as the corresponding p-value in Section 5.4, which was one-sided and therefore should be multiplied by 2 to compare. So in this particular case, accounting for the uncertainty in the reference proportion did not make any difference to our conclusions. We can find this geometrically as the level at which the identity line is tangential to the univariate confidence surface. We may note the analogy between this p-value for equality and the Breslow–Day test in Section 5.5, as illustrated in Figure 7.8 for the data in Example 5.5.

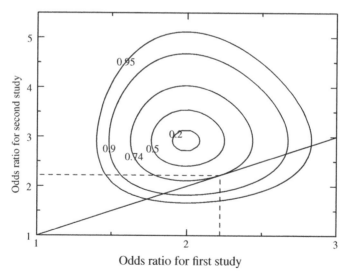

Figure 7.8   If we use different odds ratios in the two terms of Breslow–Day quadratic form in Example 5.5 and apply the $\chi^2(1)$ CDF to it, we get a confidence function with the contours in the graph. The straight line is the line $\theta_1 = \theta_2$, which is tangential to the 0.74 level of the confidence function. This means that the p-value for homogeneity is $p = 0.26$, as was obtained earlier.

Box 7.8 Estimation of the interquartile range

To obtain asymptotic confidence limits for the interquartile range of a distribution, or, more generally, the difference between any two percentiles, we can proceed as follows. Let $x_1 < x_2$ be two percentiles, corresponding to levels p_1 and p_2 respectively, and let $\Delta = x_2 - x_1$. Introduce the notation

$$u_n(x_1, \Delta) = \begin{pmatrix} F_n(x_1) - p_1 \\ F_n(x_1 + \Delta) - p_2 \end{pmatrix}, \quad \Sigma = \begin{pmatrix} p_1(1 - p_1) & p_1 - p_1 p_2 \\ p_1 - p_1 p_2 & p_2(1 - p_2) \end{pmatrix}.$$

Then we have that

$$u_n(x_1, \Delta)^t \Sigma^{-1} u_n(x_1, \Delta) \in As\chi^2(2).$$

To obtain confidence limits for Δ we proceed in the same way as in Section 7.7: for given Δ, estimate $x_1 = x_1(\Delta)$ by minimizing the function above. This gives us a new sample statistic, $P_n(\Delta) = u_n(x_1(\Delta), \Delta)^t \Sigma^{-1} u_n(x_1(\Delta), \Delta)$, and (univariate) knowledge about Δ can now be obtained from the confidence function $C(\Delta) = \chi_1(P_n(\Delta))$.

This whole approach can be adapted to other situations. As one further example we outline how to obtain confidence intervals for the interquartile range in a CDF $F(x)$. This requires us to compare two ordered proportions for the same CDF, such as the upper and lower quartiles, and for this we need to find the appropriate confidence function for this situation. This is found in Appendix 6.A.3, which shows that

$$n(F_n(x) - F(x), F_n(y) - F(y))\Sigma^{-1} \begin{pmatrix} F_n(x) - F(x) \\ F_n(y) - F(x) \end{pmatrix} \in As\chi^2(2),$$

where the variance matrix Σ is given by equation A.3 on page 175. Using this, we can then proceed with calculations in the same way as above to obtain confidence intervals for the interquartile range of a distribution. Some more details can be found in Box 7.8.

Finally, for the analysis of two parameters in independent binomial distributions, the objective function in equation (7.11) is not the only option. In fact, for most statisticians a more natural choice is based on the log-likelihood, which means that we compute the probability of the outcome (p_1^*, p_2^*) from two independent binomial distributions, and take the logarithm of this. In principle we can analyze this using the binomial distributions (and not large-sample approximations), but the mid-P problem in the estimation problem becomes somewhat more complex. However, as a test for the equality of the binomial proportions this exact test is used, and is called Barnard's test. What it amounts to is that we compute the probability $H(p)$ for an outcome at least as extreme as the one we got, which means that we sum the probabilities for all outcomes (x_1, x_2) such that $x_1/n_1 - x_2/n_2 \geq 67/101 - 43/107$ (for a one-sided p-value). In this case the function $H(p)$ is bimodal (has two local maxima), with the largest value occurring for $p = 0.54$. The value of the function at this point is $H(0.54) = 0.00015$, which therefore is the p-value for the hypothesis of equal binomial proportions. This may be considered to be the exact p-value for the test of equality. Alternatively, we can apply large-sample approximations (to be discussed in Chapter 13) to the log-likelihood, which provides almost identical results to those we obtained above.

7.8 Comments and further reading

When understood, the phenomenon of regression to the mean is almost obvious, but it is still consistently misunderstood and considered to be a fallacy, or a paradox, by many. The history of its origin with Galton is described by Stigler (1997) in an issue of *Statistical Methods in Medical Research* devoted to the subject. More details on other aspects of this concept can also be found there. The same author has also given an overview of how Galton came up with the correlation concept (Stigler, 1989). The original publication is Galton (1886).

In the next to last section we introduced the ideas behind one-sample multivariate analysis, based on Gaussian distributions. We will discuss the two-sample case in the next chapter, and a few words about the general theory of multidimensional Gaussian distributions are given in Appendix 7.A.2 below. More comprehensive introductions to this area of statistics can be found in numerous textbooks, including classics such as Anderson (1984) and more practically oriented ones such as Srivastava and Carter (1983).

The Fieller interval was discussed by Edgar Fieller, an early statistician in the pharmaceutical industry. It was developed in connection with work on insulin in the Boots company during the Second World War, but was described later (Fieller, 1954) as a special case of the problem of obtaining confidence limits (though he talked about fiducial limits) for the solution to a polynomial equation with coefficients that have a joint Gaussian distribution. Like us, he used the Cushny–Peebles data as an illustration for the linear case.

The use of the confidence function in equation (7.10) to obtain univariate confidence intervals for a particular combination of the means may only be accurate if we use straight lines in the graphical method (and therefore both for a mean difference and ratio). It is then a consequence of the fact that a linear combination of Gaussian variables (also when dependent) is a univariate Gaussian variable, together with a similar property for the Wishart distribution. For nonlinear functions no accurate general method seems available, but we can use the method outlined to get an approximation, which is expected to be better the more linear the functions are. It allows us at least a large-sample justification for the method for nonlinear parameter functions. For more details, see Appendix 7.A.3 below. The simultaneous approach using the confidence function in (7.9) is always valid, also for nonlinear functions of the parameters.

Stein's paradox is really a paradox in estimation theory (Lehmann and Casella, 1998, Chapter 5) and more general than we have indicated. A popular introduction can be found in an article by Efron and Morris (1977), whereas Stigler (1990) gives a discussion more along our lines.

References

Anderson, T.W. (1984) *An Introduction to Multivariate Statistical Analysis* second edn. John Wiley & Sons.

Das, P. and Mulder, P.G.H. (1983) Regression to the mode. *Statistica Neerlandica*, **37**, 15–21.

Efron, B. and Morris, C. (1977) Stein's paradox in statistics. *Scientific American*, **236**(5), 119–127.

Fieller, E.C. (1954) Some problems in interval estimation. *Journal of the Royal Statistical Society, Series B*, **16**(2), 175–185.

Galton, F. (1886) Regression towards mediocrity in hereditary stature. *Journal of the Anthropological Institute*, **15**, 246–263.

Lehmann, E.L. and Casella, G. (1998) *Theory of Point Estimation* Springer Texts in Statistics 2nd edn. New York: Springer.

Senn, S. (1990) Regression: A new mode for an old meaning. *American Statistician*, **44**(2), 181–183.

Srivastava, M.S. and Carter, E.M. (1983) *An Introduction to Applied Multivariate Statistics*. New York: North-Holland.

Stigler, S.M. (1989) Francis Galton's account of the invention of correlation. *Statistical Science*, **4**(2), 73–86.

Stigler, S.M. (1990) The 1988 neyman memorial lecture: A galtonian perspective on shrinkage estimators. *Statistical Science*, **8**(1), 147–155.

Stigler, S.M. (1997) Regression towards the mean, historically considered. *Statistical Methods in Medical Research*, **6**(2), 103–114.

7.A   Appendix: Some technical comments

7.A.1   The regression to the mode equation

Let a (true) subject characteristic Z have a distribution described by the probability density $h(x)$. When we observe it, there is a measurement error ξ, described by the density $\psi(x)$. In other words, what we observe is $X = Z + \xi$, for which we have the probability density

$$g(x) = \int_{-\infty}^{\infty} \psi(x - u)h(u)du.$$

(We assume the measurement error is independent of Z.) If we assume that the error term follows a Gaussian distribution with density

$$\psi(x) = \varphi_\sigma(x) = \sigma^{-1}\varphi(x/\sigma),$$

what can we say about Z from the observation of X? We know that $\varphi'(x) = -x\varphi(x)$, which means that $\sigma^2\psi'(x) = -x\psi(x)$.

The bivariate distribution for (Z, X) has density $h(z)\psi(x - z)$, so the density for the conditional distribution of Z given that $X = x$ is this, divided by the density $g(x)$ for X. It follows that

$$E(Z - x|X = x) = \frac{1}{g(x)} \int_{-\infty}^{\infty} (z - x)\psi(x - z)h(z)dz = \frac{\sigma^2}{g(x)} \int_{-\infty}^{\infty} \psi'(x - z)h(z)dz,$$

from which we deduce that $E(Z|X = x) = x + \sigma^2(\ln g(x))'$. We can also compute the conditional variance. For this we use the equations

$$V(Z|X = x) = E((Z - x)^2|X = x) - E(Z - x|X = x)^2$$

and

$$\int_{-\infty}^{\infty} (x - z)^2 \psi(x - z)h(z)dz = \int_{-\infty}^{\infty} (\sigma^2\psi(x - z) + \sigma^4\psi''(x - z))h(z)dz.$$

Together these observations imply that

$$V(Z|X = x) = (\sigma^2 + \sigma^4 g''(x)/g(x)) - (\sigma^2 g'(x)/g(x))^2 = \sigma^2 + \sigma^4(\ln g(x))''.$$

In the Gaussian case, where $\sigma = \sigma_w$ and $g(x) = \sigma^{-1}\varphi((x - m)/\sigma)$, $\sigma^2 = \sigma_b^2 + \sigma_w^2$, we have $(\ln g(x))'' = -\sigma^{-2}$ and the variance becomes

$$V(Z|X = x) = \sigma_w^2 - \frac{\sigma_w^4}{\sigma_b^2 + \sigma_w^2} = \sigma_w^2\rho.$$

This analysis is easily extended to cover the situation discussed in Section 7.5, in which $X = Z + \xi$ and $Y = Z + \eta$, where ξ, η are independent with the common probability density $\psi(x)$ above. The distributions of X and Y both have the density $g(x)$ and since we have that $E(Y|X = x) = E(Z|X = x)$, equation (7.7) follows from the analysis above. Moreover

$$V(Y|X = x) = V(Z|X = x) + V(\eta) = 2\sigma^2 + \sigma^4(\ln g(x))'',$$

which in the Gaussian case becomes $\sigma_w^2(1 + \rho) = \sigma^2(1 - \rho^2)$.

Based on these results, Das and Mulder (1983) suggested changing the description of the phenomenon from regression to the mean to regression to the mode, a suggestion not welcomed by everyone (Senn, 1990).

7.A.2   Analysis of data from the multivariate Gaussian distribution

The multivariate Gaussian distribution was introduced at the end of Appendix 6.A.1 as defined by densities that are proportional to the exponential of a quadratic form. In this chapter we have looked more closely into the bivariate case. The extended univariate Gaussian theory that was discussed in the same appendix has its immediate counterpart for multivariate data, though the mathematics now relies more heavily on matrix algebra and therefore is more complicated. Here we only highlight the absolute bare essentials, and refer the reader who wishes to know more to Anderson (1984)

In the same way that the univariate Gaussian distribution is associated with the χ^2, t and F distributions, there is a family of distributions related to the general Gaussian distributions; a family which in the case $p = 1$ reduces to what we discussed in Appendix 6.A.1. In the univariate case the $\chi^2(f)$ distribution was obtained as the distribution of $X^t X = \sum_{i=1}^{f} X_i^2$, where the $X_i \in N(0, 1)$ are all independent, and is a special case of the gamma distribution (namely $\Gamma(f/2, 1/2)$). In the multivariate case the corresponding variable $S = X^t X$ is a $p \times p$ positive definite matrix and therefore has a distribution on the manifold of such matrices. The natural definition of the p-dimensional gamma distribution $\Gamma_p(f, I/2)$ is the distribution on this manifold that has the density that is proportional to

$$(\det S)^{f-(p+1)/2} e^{-\frac{1}{2} \operatorname{Tr} S}.$$

When $p = 1$ this reduces to the definition of the corresponding univariate gamma distribution (the space of positive definite matrices in this case is the space $x > 0$). The general Wishart distribution is defined by the requirement that $S \in \Gamma_p(f, \Sigma)$ precisely when $\Sigma^{-1} S \in \Gamma_p(f, I)$. One of its important properties is that if A is a matrix of correct dimensions, we have that $A S A^t \in \Gamma_p(f, A \Sigma A^t)$ if $S \in \Gamma_p(f, \Sigma)$ (a fact used in Box 7.7). Another is that if $X \in N_p(0, \Sigma)$ and $f S \in \Gamma_p(f/2, \Sigma/2)$ are independent, then

$$\frac{f - p + 1}{p f} X^t S^{-1} X \in F(p, f - p + 1).$$

This was used in the beginning of Section 7.6, for $p = 2$ and $f = n - 1$, and will be used again in Section 8.8.

A further important observation is the generalization of the regression to the mean phenomenon. If we decompose a multivariate Gaussian variable X into two parts

$$(X_1 X_2) \in N_{r+s}\left((m_1 m_2), \begin{pmatrix} \Sigma_{11} & \Sigma_{12} \\ \Sigma_{12}^t & \Sigma_{22} \end{pmatrix}\right),$$

it is easy to show that the conditional distribution of X_1 given that $X_2 = x_2$ has the multivariate Gaussian distribution

$$N_r(m_1 + \Sigma_{12} \Sigma_{22}^{-1}(x_2 - m_2), \Sigma_{1.2}), \quad \Sigma_{1.2} = \Sigma_{11} - \Sigma_{12} \Sigma_{22}^{-1} \Sigma_{12}^t.$$

7.A.3   On the geometric approach to univariate confidence limits

In this appendix we will explain mathematically why the geometric approach to univariate confidence limits of functions of two parameters works, and thereby explain why it gives accurate results for linear combinations of the means. Suppose we want to derive confidence intervals for the parameter $\theta = f(m)$. The geometric criterion, as discussed on page 196, is that the curve $\{m; f(m) = \theta\}$ is tangent to a contour $(\bar{u} - m)^t S^{-1}(\bar{u} - m) = C$, which means that they have proportional normals. In other words, there is a λ such that $S^{-1}(\bar{u} - m) = \lambda f'(m)$. Inserting the expression $\bar{u} - m = \lambda S f'(m)$ into the confidence contour shows that λ is defined by $\lambda^2 f'(m)^t S f'(m) = C$. Next we note that $f(\bar{u}) - f(m) \approx f'(m)^t(\bar{u} - m) = \lambda f'(m)^t S f'(m)$, and if we combine this with the expression above we see that

$$\frac{f'(m)^t(\bar{u} - m)}{\sqrt{f'(m)^t S f'(m)}} = \sqrt{C}.$$

This is immediately applicable to the mean difference $f(m) = m_2 - m_1$, and shows that for a nonlinear function the extent to which the confidence levels are accurate depends on how linear the function $f(m)$ is in the region of interest. That this works for the relations $f(m) = m_1 - \theta m_2 = 0$, defining Fieller intervals, comes from a trivial modification of the argument.

8

How to compare the outcome in two groups

8.1 Introduction

This chapter is a continuation of Chapter 6, with a discussion on how we can compare two CDFs using the corresponding e-CDFs. The two CDFs represent the distributions of an outcome variable in two different groups, and we look for ways to describe how these differ. There are two basic approaches to the comparison of two CDFs, horizontal and vertical. In the horizontal approach we look at how much the CDFs differ horizontally, which can be done using percentiles or the mean, all of which are location measures. If the full distributions are (at least approximately) horizontal shifts of each other, it does not matter whether we use the median (or some other percentile) or the mean to estimate this shift, but the corresponding tests use the data in different ways. The mean, which uses all the data, would be expected to be the best when there is a true shift, whereas the median, which ignores the actual values with low or high ranks, might be better at describing the center of the distribution, if our data contains 'outliers', that is, data in the tails with a heavy influence on the means.

The classical test in the horizontal approach is the t-test, whereas tests that are done using the vertical approach correspond to classical non-parametric tests. For the latter, we will focus our attention on the Wilcoxon test, which is very much the mother of such tests. We will describe how it can be used for different purposes, including parameter estimation. To many proponents of non-parametric testing, the presentation here is probably unnecessarily mathematical. My defense is that this discussion gives us some useful insights into the nature of these tests which are not provided by most standard textbooks. The Wilcoxon test as such also prompts a discussion about ways to compute p-values.

At that stage in this chapter there is no need to have read the previous chapter about the bivariate normal distribution and multivariate analysis. However, we also wish to compare the two groups in situations where we have multivariate data. More precisely, we want to understand how we can use a baseline measurement to improve upon an estimated treatment

Understanding Biostatistics, First Edition. Anders Källén.
© 2011 John Wiley & Sons, Ltd. Published 2011 by John Wiley & Sons, Ltd.

difference, which to a large extent is a continuation of the discussion we had in Section 7.3. This discussion also serves as an introduction to Chapter 9, where general linear models are discussed.

8.2 Simple models that compare two distributions

We will use the data discussed in Section 6.6 for illustration purposes, but now we assume that they come from two independent groups. These data are shown as the two e-CDFs in Figure 8.1. The claim we wish to make is about the true (population) CDFs, which are indicated as dashed curves. We assume that the black curve shows the data of the outcome variable obtained from patients who have been treated with a new drug (also referred to as the active group), while the gray curve shows the distribution for patients who have been given a placebo instead. The CDF for the placebo group is denoted by $F(x)$ and that for the active group by $G(x)$. In most cases we want a simple description of the difference between these CDFs. A single p-value for the hypothesis $G(x) = F(x)$ (for all x, which will not be pointed out below) is, however, not very informative. It gives us the confidence to claim that treatment with the new drug has an effect, but tells us nothing about what this effect looks like. What we wish to demonstrate with the data displayed in Figure 8.1 is that the treatment moves the CDF to the left, which means that small values are more likely in the active group than in the placebo group. The logical extreme of this is to assume that the only difference between $G(x)$ and $F(x)$ is that the former is a shifted version of the latter, so that

$$G(x) = F(x - \theta) \tag{8.1}$$

for some shift parameter θ. In this case $\theta < 0$, since the $G(x)$ curve lies to the left of the $F(x)$ curve. Of course, this does not mean that the corresponding e-CDFs have this relation. The e-CDFs are obtained from the CDFs by adding noise, and that noise may well blur this property.

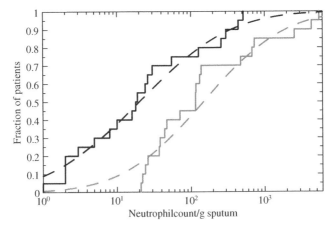

Figure 8.1 The e-CDFs for sputum cell count data for two groups, one (gray) on placebo and one (black) on a treatment. The dashed curves show the corresponding true CDFs.

The assumption in equation (8.1) implies that the location measures, the mean and the percentiles, including the median, are all shifted by the same amount θ, whereas the variances for $F(x)$ and $G(x)$ are the same. This means that whether we estimate the difference between the means or some percentiles, we would be estimating θ. We therefore base our statistical approach on one such measure, but it may not be irrelevant which one we choose. Firstly, equation (8.1) is a model assumption. It is only an approximation of the real world, and may not hold true in the tails of the distributions, but may still be a useful approximation to reality. Different parameters are sensitive to different extents to what goes on in the tails. The mean, for example, is more sensitive than the median.

The data in Figure 8.1 do not really support a parallel shift of $F(x)$, since the plot is on logged data. If we think we see a parallel shift in in Figure 8.1, the model for the sputum data is really of the form

$$G(x) = F(x/\theta), \tag{8.2}$$

a type of model that may be appropriate only if we have a variable that takes only positive values. This model is called the accelerated failure time (AFT) model when it appears in the context of survival data. As discussed in Section 6.6, with this model we derive results for ratios of (geometric) means or ratios of percentiles.

Another model that compares two CDFs is the proportional odds model

$$\frac{G^c(x)}{G(x)} = \theta \frac{F^c(x)}{F(x)}, \tag{8.3}$$

which is used primarily for ordered categorical data. In fact, consider the simple case of binomial data, where the probability for 0 (no event) is $1 - p$ and the probability for 1 (event) is p. Then $F^c(0)/F(0) = p/(1 - p)$ is the odds, which means that θ becomes the odds ratio for the two groups. Related to this is the proportional hazards model $G^c(x) = F^c(x)^\theta$, which is more important for survival data, and will be discussed in that context.

Figure 8.2 illustrates what the function $G(x)$ looks like when it is obtained from $F(x)$ as described by some of these models. We start with the standard normal $F(x) = \Phi(x)$, the

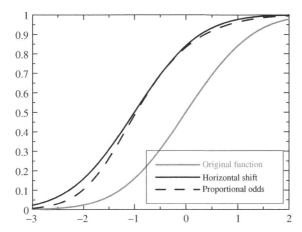

Figure 8.2   Illustration of how some one-parameter models change the location and shape of the CDF. The gray curve shows the original function $F(x)$.

gray curve, which we first shift one step to the left (we take $G(x) = \Phi(x + 1)$), which is the solid black curve. The dashed curve is the CDF from the proportional odds model with factor $\theta = 5$. We see that both these curves are to the left of the original, and that the proportional odds model has also changed the shape of the CDF.

We will use the following notation in this chapter. The sample taken from $F(x)$ is denoted by $x_1, \ldots, x_n$ and that from $G(x)$ is denoted by $y_1, \ldots, y_m$. Combining these two samples gives a sample from a population whose CDF is given by

$$\Psi(x) = rF(x) + (1 - r)G(x).$$

Here r is the probability that a randomly chosen individual is from the first group. If we have two real groups, such as men and women, it is the fraction of the first group within the total, whereas in a randomized clinical trial situation, r is defined by the allocation ratio to the two groups. With a balanced randomization we have $r = 1/2$. The samples from the two groups define two e-CDFs, $F_n(x)$ and $G_m(x)$ respectively, and the e-CDF for the total sample is

$$\Psi_{nm}(x) = \frac{n}{n + m} F_n(x) + \frac{m}{n + m} G_m(x).$$

Note that the assumption that this is an estimate of $\Psi(x)$ rests on the assumption that the sample sizes actually reflect the underlying fraction r. Note also that the rank of a particular observation x in the combined sample is given by

$$R_{nm}(x) = nF_n(x) + mG_m(x);$$

in other words, $\Psi_{nm}(x) = R_{nm}(x)/(n + m)$. In a randomized study we usually make all statistical analysis conditionally on the randomization outcome, in which case we take $r = n/(n + m)$ in the definition of $\Psi(x)$.

8.3   Comparison done the horizontal way

We will now discuss how we can compare the distributions $F(x)$ and $G(x)$ horizontally – how to describe the difference between the two CDFs in terms of location measures. The most important of these location measures are certain percentiles, in particular the median, together with the mean, and the discussion is closely related to the discussion on understanding a single CDF in Section 6.5.

We start with the mean. If the two distributions have means and standard deviations (μ_1, σ_1) and (μ_2, σ_2), respectively, the CLT tells us that

$$\bar{Y} - \bar{X} \in AsN\left(\Delta, \frac{\sigma_1^2}{n} + \frac{\sigma_2^2}{m}\right),$$

with $\Delta = \mu_2 - \mu_1$. To be a true horizontal shift, we need to have equal variances σ^2, and then we have that

$$\sqrt{\frac{nm}{n + m}}(\bar{Y} - \bar{X} - \Delta) \in AsN(0, \sigma^2).$$

Box 8.1 Correlation of an outcome with a group variable

It is of some interest to understand what correlation measures turn into when we correlate an outcome variable Y with a group variable X. For this we need numerical values for the group variable, and for convenience we choose $X = 1$ for the first group and $X = -1$ for the second. Let the fraction from the first group be r, so that the mean value of X is $2r - 1$. With this notation we have that $P(Y \leq y|X = 1) = F(y)$ and $P(Y \leq y|X = -1) = G(y)$, and the covariance is

$$\int y dF(y)(1 - (2r - 1))r + \int y dG(y)(-1 - (2r - 1))(1 - r) = 2r(1 - r)(m_1 - m_2).$$

The Pearson correlation coefficient is therefore half the mean difference normalized to the standard deviation of Y in the whole sample, since $V(X) = 4r(1 - r)$.

In order to compute Kendall's tau defined in Box 7.1, we need to compute the probability of events of the form $(Y_1 - Y_2)(X_1 - X_2) > 0$, a criterion we see is equivalent to the outcome variable being larger for a subject from group 1 than for group 2. This means that if we sample one subject from each group and let Z_i be the outcome for group i, then

$$\tau = P(Z_1 > Z_2) - P(Z_1 < Z_2).$$

This relates Kendall's tau to the important non-parametric Wilcoxon test as discussed in the main text. If we have more than one group, ordered in some way (e.g., according to which dose of a drug a particular group is given), Kendall's tau leads us to a related non-parametric test, the Jonckheere–Terpstra test, which is sometimes used to establish dose response.

This actually reduces this discussion to the results obtained for the one-sample t-test in Section 6.5. All we need is an estimate of σ^2, for which we use the pooled sample variance defined by

$$s^2 = \frac{(n - 1)s_1^2 + (m - 1)s_2^2}{n + m - 2},$$

which leads us to the approximate confidence function

$$C(\Delta) = \Phi\left(\frac{\Delta - (\bar{y} - \bar{x})}{s\sqrt{1/n + 1/m}}\right).$$

If our original data are actually described by shifted Gaussian distributions, we obtain exact inference if we replace $\Phi(x)$ with the CDF for the $t(n + m - 2)$ distribution, because under that assumption the numerator of s^2 is a sum of two independent χ^2 distributions with $n - 1$ and $m - 1$ degrees of freedom, respectively. As for the single mean parameter, the Bayesian approach to these data allows us to view $C(\Delta)$ as a posterior probability function, with carefully chosen (non-probabilistic) priors for the parameters (the same as in Section 6.5). We can also use this observation to derive the *a posteriori* density distribution from an informative

---

Box 8.2 Analysis of a percentile difference for two groups

Here we outline how we can obtain confidence claims for the difference for a particular percentile for two independent distributions $F(x)$ and $G(x)$. Consider any particular percentile, x_p, of $F(x)$ (so that $F(x_p) = p$). Let the corresponding percentile for $G(x)$ be written $x_p + \theta$, so that $G(x_p + \theta) = p$. In order to obtain knowledge about θ we use the approximation

$$\frac{(F_n(x_p) - p)^2}{V(F_n(x_p))} \in \chi^2(1), \quad \text{where } V(F_n(x_p)) = p(1 - p)/n,$$

for $F(x)$, and similarly for $G(x)$. Since the test statistics for $F(x_p)$ and $G(x_p + \theta)$ are independent, we can sum two quadratic forms to get

$$\frac{(F_n(x_p) - p)^2}{V(F_n(x_p))} + \frac{(G_m(x_p + \theta) - p)^2}{V(G_m(x_p + \theta))} \in \chi^2(2).$$

From this we derive a confidence function for (x_p, θ), and in order to obtain knowledge about θ we now profile x_p out of this, as in Section 7.7. This means that for given θ we estimate $x_p = x_p(\theta)$ by minimization, which gives us a function $P_{mn}(\theta)$ of θ alone. The end result is the two-sided confidence function $C(\theta) = \chi_1(P_{mn}(\theta))$, from which we can obtain asymptotically correct knowledge about θ. Properly modified, this approach also allows us to compare the two percentiles in other ways, such as their ratio.

---

a priori distribution for the mean difference, but we will follow the biostatistical tradition and not discuss this any further.

In order to apply the discussion above to the data in Figure 8.1 we estimate the group mean values of the logarithmic data to 4.99 and 3.00, respectively, which gives us a mean difference of 1.99 with 95% confidence interval (0.80, 3.17). This is the (estimated) size of the shift in Figure 8.1. However, to get the shift back to the original measurement scale we back-transform by exponentiation. This gives us a ratio of geometric means for treated versus placebo of $e^{-1.99}$, which is 14%, with 95% confidence interval (4.2, 45)%.

Next we consider the median. A key difference between the mean and the median is that whereas the difference of two means is the mean of the difference, this need not be true for medians. So the approach is more complicated, building on ideas used in Section 7.7 when we analyzed two independent binomial parameters. An outline is given in Box 8.2. When we carry out this analysis for our sputum data on the log scale, we get an estimated median difference of 1.83 with 95% confidence interval (0.26, 3.39), which we can back-transform to a statement about the ratio of the medians for the two distributions as being 16% with 95% confidence interval (3.4, 77)%. For these data there is therefore a close agreement between the mean and median estimates of a possible horizontal shift. The confidence interval for the median is wider than that for the mean, because the mean value analysis uses more of the information in the data than the median analysis does.

Now we return to the original cell count scale, instead of their log values. For the mean values we have the estimates 752 and 93 for the two groups, giving a mean difference estimate of 659 with 95% confidence interval (−34, 1351). In this computation we have assumed equal

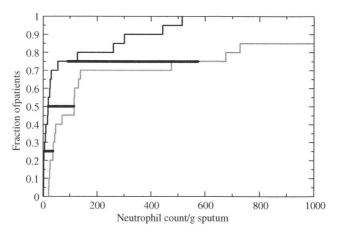

Figure 8.3 The percentile difference increases with level for the sputum data on the original scale.

standard deviations and used the t distribution. But the standard deviations are estimated as 1521 and 157 respectively, so it is probably not a valid assumption to make. Alternatively, we can analyze the difference for some percentiles, as described in Box 8.2. The result for the median and the two quartiles is shown in the following table:

Parameter	Estimate	95% CI
25%	−33.5	(−69, −8)
50%	−96.5	(−459, −14)
75%	−485	(−4260, 312)

Figure 8.3 shows the percentile differences as horizontal lines connecting the corresponding percentiles for the two e-CDFs. Clearly the differences vary considerably between levels, strongly indicating that it is not appropriate to assume that the two distributions are simple shifts of each other.

A little mathematics explains this observation. For two Gaussian distributions $N(\mu_1, \sigma_1^2)$ and $N(\mu_2, \sigma_2^2)$, the percentile difference is given by

$$\theta_p = \mu_1 - \mu_2 + (\sigma_1 - \sigma_2)z_p,$$

where z_p is the pth percentile for the standardized Gaussian distribution. In particular, this shows that we have a horizontal shift precisely when the variances are equal; only then is θ_p independent of p. For lognormal distributions we have the percentiles $x_p = e^{\mu + \sigma z_p}$, and when we have equal σ, the percentile difference becomes

$$\theta_p = e^{\mu_1 + \sigma z_p} - e^{\mu_2 + \sigma z_p} = e^{\sigma z_p}(e^{\mu_1} - e^{\mu_2}),$$

which increases with p if $\mu_1 \neq \mu_2$. This is, to a reasonable approximation, what happens in our data.

8.4 Analysis done the vertical way

This section is about the famous Wilcoxon two-group test, which may come as a surprise, since a casual look at the following pages shows a considerable number of mathematical expressions involving integrals. The Wilcoxon test is usually presented as a very simple test: rank-transform data and do a few simple calculations. However, the framework within which we derive it has a number of important extensions built into it. It becomes more than a simple test for equality; it also provides a method to do parameter estimation. The price for this is a little more mathematics.

The vertical approach for comparing two independent CDFs $F(x)$ and $G(x)$ looks for a vertical shift $\Delta(x) = F(x) - G(x)$ of the two CDFs at different points x. In this section we will tacitly assume that the true CDFs are both continuous, a restriction that will be removed in Section 8.6. Essentially there are two major ways to derive scalar quantities from $\Delta(x)$ that we can use to test the null hypothesis that $\Delta(x) = 0$ for all x. The obvious way may be to look for the maximal difference, which we estimate by

$$D_{nm} = \max_x |F_n(x) - G_m(x)|.$$

This leads us to the Kolmogorov–Smirnov test, which is discussed in Appendix 6.A.2. Alternatively we may take a weighted average $\int_{-\infty}^{\infty} \Delta(x)dw(x)$ of the $\Delta(x)$ with some weight function $w(x)$. A partial integration and some algebra shows that this is proportional to

$$\int_{-\infty}^{\infty} w(x)d(F(x) - \Psi(x)), \tag{8.4}$$

which is the integral we want to concentrate on. It is zero if the distributions are equal, whatever weights we choose. We typically want to take the weights where the data are, and one way to ensure this is to use a function of the CDF for the combined sample, so that $w(x) = a(\Psi(x))$. This means that the requirement for the integral to be zero is that

$$\int_{-\infty}^{\infty} a(\Psi(x))dF(x) = \int_0^1 a(u)du,$$

where we have changed to the variable $u = \Psi(x)$ in the integral $\int a(\Psi)d\Psi$ (we consider only continuous distributions at present). The corresponding test statistic is obtained if we insert e-CDFs instead of the CDFs:

$$\int_{-\infty}^{\infty} a(\Psi_{nm}(x))dF_n(x).$$

We can use this to test the hypothesis of equality and also to estimate a parameter in some simple models. We will illustrate this in some detail by looking closely at the simplest choice of weight function, which is to take $a(u) = u$, so that $w(x) = \Psi(x)$. This gives us the test statistic

$$\int_{-\infty}^{\infty} \frac{R_{mn}(x)}{n+m}dF_n(x) = \frac{1}{n(n+m)} \sum_{i=1}^{n} R_{mn}(x_i),$$

so this test amounts to the analysis of the total rank sum of the first sample and is the well-known *Wilcoxon rank sum statistic*. Under the null hypothesis $F(x) = G(x)$, the expected value of the rank sum is $n(n + m + 1)/2$, so we test the null hypothesis of equality by comparing the rank

---

Box 8.3 The Wilcoxon test and the rank-transformed t-test

To compute the p-value for the Wilcoxon test based on the statistic $W = \sum_1^n R(x_i)$, we first compute its mean and variance under the null hypothesis (we assume no ties):

$$E(W) = \frac{n(N + 1)}{2}, \quad V(W) = \frac{nm(N + 1)}{12},$$

where $N = n + m$. Appealing to large-sample theory, we can show that

$$T = \frac{W - E(W)}{\sqrt{V(W)}} \in AsN(0, 1),$$

from which p-values can be computed.

This is closely related to the following procedure: first rank-transform all the data and then apply the t-test to the resulting data. For this test the mean difference and pooled sample variance are given by

$$\hat{\Delta}_R = \frac{N}{nm}\left(W - \frac{n(N + 1)}{2}\right) \quad \text{and} \quad s_R^2 = \frac{N}{(N - 2)nm}V(W)(N - 1 - T^2),$$

respectively. The corresponding t-statistic is therefore given by

$$t_R = \frac{\hat{\Delta}_R}{s_R\sqrt{\frac{1}{n} + \frac{1}{m}}} = \frac{T}{\sqrt{1 + \frac{1-T^2}{N-2}}}.$$

This relationship holds also in the presence of ties, and shows that for all practical purposes we can compute the Wilcoxon test by applying the t-test to rank-transformed data, at least when the sample size is large enough to allow for the asymptotic p-values (conventionally considered to be when $N \geq 40$).

There is also a relation to the Wilcoxon probability P_W. In fact, the relation between Mann–Whitney scores and the rank sum discussed in the text shows that the true mean difference is $\Delta_R = N(P_W - 1/2)$. We can therefore use the information obtained from the t-test on ranks to obtain an estimate and approximate confidence interval for P_W, when the sample is large.

---

sum to this value. This test is frequently used in biostatistics as an alternative to Student's t-test to compare two independent distributions when data are distinctly non-Gaussian in nature. The Wilcoxon rank sum test is sometimes referred to as a test of medians, which it is not. It is a test of the mean rank and has no more to do with medians than the t-test has. In fact, it is essentially a t-test on the ranks of the observations, instead of the original observations, as outlined in Box 8.3. To put it another way, we apply the t-test to the variable $\Psi(X)$ instead of X (which is a non-linear transformation of data), a variable which we know has a uniform distribution on $(0, 1)$ when the null hypothesis of equal distributions for the two groups holds. When it comes to computing p-values for the Wilcoxon test, this can be done in different ways, which we will discuss in the next section. For the rest of this section we will instead try to better understand the nature of the test. It is to a large extent a mathematical investigation of relationships and is about how we transform the test into a parameter estimation method.

Box 8.4 An alternative derivation of the Hodges–Lehmann estimator

---

An alternative derivation of equation (8.7) is based on the squared horizontal difference between the two CDFs,

$$L_2 = \int_{-\infty}^{\infty} (F(x) - G(x))^2 \, dx$$

as the measure of the extent to which they differ. We can use this to estimate the θ in the shift model defined by equation (8.1) by considering the function

$$L_2(\theta) = \int_{-\infty}^{\infty} (F(x - \theta) - G(x))^2 \, dx.$$

For this function we have that $L_2(0) = L_2$; also, when $F(x) = G(x)$, we have that $L_2(\theta)$ has its minimum (which is zero) when $\theta = 0$. Therefore, if the minimum of $L_2(\theta)$ is at a point other than zero, we cannot have equality of the two distribution functions. A short calculation shows that θ is the point where $L_2(\theta)$ is minimal precisely when equation (8.7) is fulfilled.

This characterization of the Hodges–Lehmann estimator suggests various related parameters obtained by minimizing the integral

$$\int_{-\infty}^{\infty} w(x)(F(x - \theta) - G(x))^2 \, dx$$

as a function of θ, for some specified weight function, preferably of the form $w(x) = a(\Psi(x))$.

We have seen that the Wilcoxon test is derived from the relationship

$$\int_{-\infty}^{\infty} F(x) dF(x) = \frac{1}{2}, \tag{8.5}$$

which is true for every continuous CDF $F(x)$. This statement says that the mean value of $F(X)$ is 1/2, when $F(x)$ is the CDF for the stochastic variable X. When $G(x) = F(x)$ there are a few equivalent statements, including

$$\int_{-\infty}^{\infty} G(x) dF(x) = \int_{-\infty}^{\infty} \Psi(x) dF(x) = \int_{-\infty}^{\infty} \Psi(x) d\Psi(x) = \frac{1}{2}.$$

For each of these integrals there is a test statistic obtained by replacing the CDFs with e-CDFs, and each such test statistic will provide the Wilcoxon test. We will pick the first of these, because this will simplify variance computation. We will also generalize it, so that we let the function $G(x)$ depend on a single parameter θ, which we write $G(x, \theta)$. Providing evidence that the equation

$$\int_{-\infty}^{\infty} G(x, \theta) dF(x) = \frac{1}{2} \tag{8.6}$$

does not hold then constitutes evidence against the null hypothesis that $G(x, \theta) = F(x)$. More importantly, we can use this equation to estimate and obtain knowledge of θ.

The classical example of parameter estimation here is when we do it on the shift model $G(x) = F(x - \theta)$. We then have that $F(x) = G(x + \theta)$, and the θ we require is the parameter value that solves the equation

$$\int_{-\infty}^{\infty} G(x + \theta) dF(x) = \frac{1}{2}. \tag{8.7}$$

This equation means that θ is the median of the CDF

$$H(z) = \int_{-\infty}^{\infty} G(x + z) dF(x) = \int \int_{y - x \leq z} dF(x) dG(y).$$

This is the CDF for the stochastic variable $Z = Y - X$, where X and Y are independent with CDFs $F(x)$ and $G(x)$, respectively. When $F(x) = G(x)$ we have that $H(0) = 1/2$, where $H(0)$ is the probability $P(Y < X)$ that Y is less than X, which we call the Wilcoxon probability and denote by P_W. This parameter must be clearly distinguished from the p-value obtained from an application of the Wilcoxon test to test the null hypothesis.

From this discussion we can immediately derive two ways to disprove the null hypothesis that $G(x) = F(x)$. We can either prove that the median θ of $H(z)$ is not zero, or that the Wilcoxon probability

$$P_W = \int_{-\infty}^{\infty} G(x) dF(x)$$

is not $1/2$. Whichever method we use, we need to estimate $H(z)$. For this we replace the CDFs with the corresponding e-CDFs and obtain the estimate

$$H_{mn}(z) = \int_{-\infty}^{\infty} G_m(x + z) dF_n(x).$$

This is a sum, and if we expand it we find that it can be written as

$$H_{mn}(z) = \frac{1}{n} \sum_j G_m(z + x_i) = \frac{1}{nm} \sum_{i,j} I(y_j - x_i \leq z),$$

so this function is obtained as the e-CDF for all the nm differences $z_{ij} = y_j - x_i$. The sample median of $H_{mn}(z)$ is called the Hodges–Lehmann estimate of location. It therefore serves the same purpose of separating two distributions as the mean or median group difference. Also note that $H_{mn}(0)$, which estimates P_W, can be written as $H_{mn}(0) = U_{mn}/mn$, where U_{mn} is called the Mann–Whitney statistic and is defined as

$$U_{nm} = \sum_{ij} \frac{1}{2}(1 + Z_{ij}), \quad \text{where } Z_{ij} = \begin{cases} 1 & \text{if } y_j > x_i, \\ -1 & \text{if } y_j < x_i. \end{cases}$$

The Z_{ij} are called the Mann–Whitney scores, and we may note that the sum of these, divided by the product nm, estimates the relative difference

$$RD = P(Y > X) - P(Y < X) = 2P_W - 1.$$

This is a transformation of the Wilcoxon probability that relates the comparison of continuous data for two groups to the corresponding description of binary data. It also relates the Wilcoxon

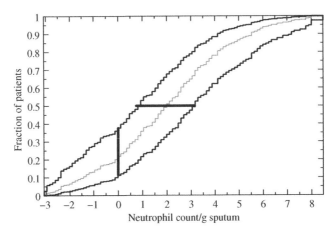

Figure 8.4   The e-CDF for the treatment difference together with pointwise confidence limits, also showing estimates and confidence limits for the Hodges–Lehmann estimate (horizontal line) of location and the Wilcoxon probability (vertical line).

probability to Kendall's τ (see Box 8.1). Note further that, for continuous data, the odds $P_W/(1 - P_W)$ can be interpreted as an odds ratio comparing the two groups, (see page 127; this is the generalized odds mentioned in Box 7.1). If we compare Gaussian variables with a mean difference Δ and a common variance σ^2, this log-odds ratio is essentially proportional to the standardized difference Δ/σ. (This can be seen by approximating the Gaussian distribution with a logistic one; see Box 9.4.) Finally, the Mann–Whitney statistic is equivalent to the Wilcoxon rank sum, because $n(n + 1)/2 + U_{mn}$ is equal to

$$n^2 \int_{-\infty}^{\infty} F_n(x)dF_n(x) + nm \int_{-\infty}^{\infty} G_m(x)dF_n(x) = n \int_{-\infty}^{\infty} R_{mn}(x)dF_n(x),$$

where the right-hand side is the rank sum of the first group.

Figure 8.4 shows the function $H_{mn}(z)$ for the logarithm of the sputum cell count data as the gray staircase function in the middle. It is surrounded by pointwise 95% confidence limits, such that for a given z, the vertical interval determined by the black curves defines the pointwise confidence interval for $H(z)$. How these confidence limits are calculated will be discussed below. In the graph there are also two confidence intervals shown as solid lines: the horizontal one at the level $1/2$ shows that of the Hodges–Lehmann estimator, which is estimated as 1.97 with 95% confidence interval (0.74, 3.15). The vertical line represents a confidence interval for the Wilcoxon probability P_W for which we have the estimate 0.22 with 95% confidence interval (0.11, 0.38). Both confidence intervals convey proof, at the conventional two-sided 5% significance level, of a treatment effect – the first excludes 0 and the second excludes $1/2$.

We have tacitly analyzed the log-count of the sputum data. If we choose to analyze the count directly, the only thing that changes is the shape of $H_{mn}(z)$ and the Hodges–Lehmann estimate. The Wilcoxon test and the Wilcoxon probability are both independent of the scale on which we analyze the data. On the original scale we obtain a Hodges–Lehmann location estimate of 60 with 95% confidence interval (19, 129). This is the median of the differences, as opposed to the difference in medians which was analyzed earlier. The estimate 1.97 discussed above refers to the median of the differences $\ln X - \ln Y = \ln X/Y$, and by exponentiation we find that the drug effect is estimated as 14% with 95% confidence interval $(4.3, 48)\%$ for active versus placebo.

The reader may have noted that the confidence interval for the Hodges–Lehmann estimate in Figure 8.4 does not quite fit with the pointwise confidence interval curves. That is because the confidence interval for the Hodges–Lehmann estimate in this case was obtained in the way it is usually computed and not the correct way. To understand the difference we need to take a closer look at how the pointwise confidence intervals for $H(z)$ are constructed.

We first take $z = 0$ and focus on the Wilcoxon probability. The basic fact that underlies the computation of the confidence interval for P_W is the observation that the stochastic variable $H_{mn} = \int_{-\infty}^{\infty} G_m(x)dF_n(x)$ has an asymptotic Gaussian distribution with mean P_W and with the variance $\sigma^2(P_W)$, with the function $\sigma^2(P)$ defined by

$$
\begin{aligned}
nm\,\sigma^2(P) =& (m-1)\left(\int_{-\infty}^{\infty} G(x)^2 dF(x) - P^2\right)\\
&+ (n-1)\left(\int_{-\infty}^{\infty} F(x)^2 dG(x) - (1-P)^2\right) + P(1-P).
\end{aligned}
\tag{8.8}
$$

One derivation of this formula is given in Appendix 8.A, in which the Wilcoxon probability parameter is discussed as a U-statistic, and where it is also shown that the distribution of the estimator is asymptotically Gaussian. From this we derive the (approximate) confidence function

$$
C(P_W) = \Phi\left(\frac{P_W - H_{nm}}{\sigma(P_W)}\right),
$$

which we can use to obtain knowledge about P_W. To get the corresponding expression for $H(z)$ we only replace $G(x)$ by $G(x + z)$ and P_W by $H(z)$ in the integrals above. Denote the corresponding standard deviation by $\sigma(H(z), z)$.

In order to get the confidence limits for the Hodges–Lehmann estimate θ, note that for θ to be the median means that it solves the equation $H(\theta) = 1/2$, which in turn implies that we obtain knowledge about θ from the confidence function

$$
C(\theta) = \Phi\left(\frac{1/2 - H_{nm}(\theta)}{\sigma(1/2, \theta)}\right).
$$

This is the correct way to compute the confidence limits for the Hodges–Lehmann estimate. The standard way, referred to above, is to use the approximation obtained by taking $\theta = 0$ and $G(x) = F(x)$ when we compute the variance. Some calculations shows that the variance $\sigma^2(1/2, 0)$ equals $(n + m + 1)/(12nm)$, so the confidence limits for θ derived from this would be the solutions to the equations

$$
H_{nm}(\theta) = \frac{1}{2} \pm z_{\alpha/2}\sqrt{\frac{n+m+1}{12nm}}.
$$

There is one further geometric interpretation of P_W, shown in Figure 8.5. The dashed curve in this graph is the path of CDF pairs $(F(x), G(x))$ inside the unit square, obtained by varying x. In this graph the value of the Wilcoxon probability parameter P_W is the area above this curve. We can assess the difference between the two distributions from how much this

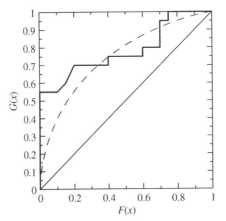

Figure 8.5   The geometric interpretation of the Wilcoxon parameter as an area. P_W is the area above the dashed line, which is estimated by the area above the polygon.

curve departs from the equality line, and a summary measure of this is to see how much P_W deviates from 1/2. Another way to state this is as follows. The function $K(x) = F(G^{-1}(x))$ is a distribution function on $(0, 1)$, and P_W is the mean of this function. If $F(x) = G(x)$ we have $K(x) = x$ as the uniform distribution on $(0, 1)$, which has mean 1/2. We encountered this type of curve in the form of ROC curves in connection with diagnostic tests.

The parameter estimation version of the Wilcoxon test (equation (8.6)) is more general than the shift model and can be applied to all kinds of one-parameter models for the two distributions. If we assume the AFT model $G(x) = F(x/\theta)$ instead, we have that θ should solve the equation

$$\int_{-\infty}^{\infty} G(\theta x) dF(x) = \frac{1}{2}.$$

A short calculation shows that this means that θ is the median of the distribution for the variable $Z = Y/X$, with X and Y as before. For the proportional odds model in equation (8.3) the corresponding equation for θ is

$$\int_{-\infty}^{\infty} \frac{\theta G(x) dF(x)}{\theta G(x) + G^c(x)} = \frac{1}{2}.$$

In either model, we replace CDFs with e-CDFs in order to obtain an estimating equation for θ from data.

Example 8.1   Figure 8.6 illustrates the different models discussed above, applied to the sputum data, except that the translation model is written as $G(x) = F(x + \theta)$ and the AFT model is written as $G(x) = F(\theta x)$, in order to make θ positive in the first case and greater than one in the second. Figure 8.6(a) shows the integrals $\int_{-\infty}^{\infty} G_m(x, \theta) dF_n(x)$ as functions of θ. The different estimates of θ are obtained from where these curves intersect the horizontal line at level 1/2. For the shift model we find the same estimate as before, the Hodges–Lehmann estimate. For the AFT model we find the estimate 7.18, which is the same estimate

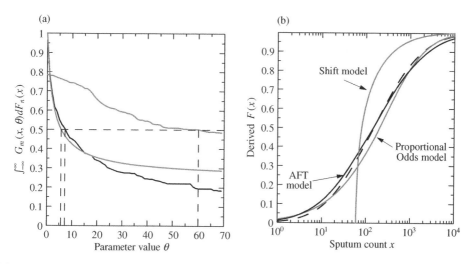

Figure 8.6 Fitting different models to the sputum count data, based on the Wilcoxon test. In (a) the integrals involved are plotted as functions of θ to illustrate how we obtain the estimate; in (b) the predicted $F(x)$ from the models is compared to the original (dashed curve).

as previously obtained for the Hodges–Lehmann estimate on logged sputum data. Finally, we find the estimate 5.64 for the proportional odds model.

Using these estimates of θ, we can reconstruct $F(x)$ from $G(x)$ for each of the models and compare them to the original $F(x)$ (which is the dashed curve in the graph). The result is shown in Figure 8.6(b). We see, as expected, that the pure shift model is poor, but it is hard to separate the other two models from the original. What really should drive the choice of model is an understanding of what one wishes to describe.

In order to obtain knowledge about the model parameters we repeat steps we have already outlined. The general estimating equation is

$$U(\theta) = \int_{-\infty}^{\infty} G_m(x, \theta)dF_n(x) - \frac{1}{2} = 0,$$

and we appeal to the CLT to argue that

$$U(\theta) \in AsN(0, V_\theta(U(\theta))).$$

It only remains to determine the variance. But this is given by equation (8.8): all we need to do is put $P = 1/2$ and replace $G(x)$ with $G(x, \theta)$. We therefore find that $V_\theta(U(\theta))$ is $1/nm$ times

$$(m-1)\int_{-\infty}^{\infty} G(x, \theta)^2 dF(x) + (n-1)\int_{-\infty}^{\infty} F(x)^2 dG(x, \theta) - \frac{1}{4}(m+n-3),$$

which we estimate by inserting e-CDFs instead of CDFs. The confidence function for θ is then defined by

$$C(\theta) = \Phi\left(\frac{U(\theta)}{\sqrt{\hat{V}_\theta(U(\theta))}}\right).$$

We conclude this section by observing that the Wilcoxon test is only a role model for non-parametric tests, albeit the most important example of such tests. We can define other related two-group tests from almost any function $a(u)$ defined on the interval $[0, 1]$ by using the relation

$$\int_{-\infty}^{\infty} a(G(x))dF(x) = \int_0^1 a(u)du,$$

which is valid for continuous distributions when $G(x) = F(x)$ (we need the integral on the right to be finite). If we insert $\Psi(x)$ instead of $G(x)$ in the argument for $a(u)$, this provides us with the framework for many classical linear rank tests. However, using $G(x)$ instead has its own merits, because it is then clear how to compute the variance and how to do parameter estimation.

8.5 Some ways to compute p-values

From the discussion so far we can deduce how to compute the p-values for tests of the null hypothesis $F(x) = G(x)$. However, there are some enhancements of this that justify a separate discussion on this particular problem.

We start with the t-test. Under the assumption of Gaussian distributions with a common variance we can compute an exact p-value for the null hypothesis of equal means. This is because the test statistic $(\bar{y} - \bar{x})/s\sqrt{n^{-1} + m^{-1}}$ follows a t distribution, based on which we compute the p-value. In real life the assumptions are never exactly fulfilled and the value of the test statistic is therefore only approximately t-distributed. As a consequence, the p-value we compute is only an approximation of the 'true' p-value. How far off it is in a particular situation is impossible to assess accurately, since we do not know the *exact* distribution of the test statistic. However, we can often argue that the approximation is fit for purpose, because of the mathematical result described by the CLT. How good the approximation is depends on the sample size and on how symmetric the data distribution is – in other words, how Gaussian-like it is. It is the data that are analyzed that matter; sometimes this variable is the change in an outcome measurement, where both the before and after values may have highly skewed distributions, while the change is a reasonably symmetric variable.

This discussion on the t-test does not apply to the Wilcoxon test. Here the situation is different because only a finite number of values are possible for the test statistic, the rank sum, given the sample size. Theoretically we can therefore enumerate all the cases, and label those that lead to a sufficiently extreme value for the test statistic. The p-value becomes a simple fraction. This would be an exact p-value for the null hypothesis, valid whatever the distribution $F(x)$ looks like, and is the reason why they are called non-parametric tests. The problem is that unless sample sizes are relatively small, the numbers involved are huge and time-consuming to compute. As an example, for the sputum data we compute the Wilcoxon rank sum as 523. To compute the p-value, we look at all possible ways of allocating the observed data points to the two groups with the group sizes kept constant. For each such permutation of data we compute the test statistic, and from that we can compute the CDF for the rank sums. By picking out tails, we can compute the p-value. This is however not a convenient method in our two-group case, with 40 data points equally divided between two groups, because there are $\binom{40}{20} \approx 1.4 \cdot 10^{11}$ such permutations. Because of the sheer size of this, we do not compute this p-value here, but there are algorithms developed that also allow us to compute the exact p-values for moderately large data sets.

Unless the sample sizes are small, the CLT provides us with a method to compute approximations to the p-value for the Wilcoxon test, which has been described above. To apply it, we need to compute the expected rank sum and its variance under the assumption of $G(x) = F(x)$, which is a pure combinatorial problem, the result of which was given in Box 8.3. Note that the accuracy depends only on sample sizes, and not on the actual distribution $F(x)$, as was the case for the t-test.

There is a version of the exact computation that can be applied to the t-test as well, which we describe in the next example.

Example 8.2   For the t-test, the distribution of the test statistic was deduced from mathematical operations based on distributional assumptions. An alternative approach is to deduce the distribution of the test statistic by estimating the combined CDF $\Psi(x)$ with its e-CDF $\Psi_{mn}(x)$, and enumerate all possible assignments of groups to this data. For each such assignment we have one value of the test statistic, and consequently the process gives us the CDF for the test statistic under these conditions, from which a p-value can be computed as a simple fraction.

A test of the kind described in Example 8.2 is called a *randomization test* (or permutation test). The p-values obtained from such tests are often referred to as exact, which they are if the (combined) sample obtained is the only possible sample. In other words, the p-value obtained in this way is from a conditional test, where we take the values of the observations in that particular experiment as given. As an unconditional test, however, it is not exact – how accurate the p-value is depends on how well the e-CDF of the combined sample approximates the true CDF. If we apply this procedure to the Wilcoxon rank sum statistic, we derive the combinatorial p-value described above, since it is the same procedure as was described for the computation of the exact p-value for the Wilcoxon test. This also implies that its computational hurdles apply to the randomization tests as well.

A related method, which is also feasible for large samples, is to use bootstrapping. This method was described in Section 6.7 as a method to estimate the CDF for the test statistic by repeated resampling. With this method we perform resampling a large number of times and for each of these samples we compute the test statistic. This gives us an estimate of the distribution of the test statistic for the actual data, which puts us in a position to compute the appropriate p-value. This method will be approximate both for the t-test and for the Wilcoxon test. How good the approximation is depends on how well the two e-CDFs describe their respective CDFs, and also on how many samples we take.

Example 8.3   For the sputum data we computed the Wilcoxon test statistic as 523. The two-sided p-value $2(1 - F(523))$ can then be computed from a bootstrap-derived estimate of the distribution $F(x)$ of the rank test based on 10 000 samples, as 0.0021, but it varies slightly between different runs since tail probabilities need many resamplings. In this case the CDF we obtain by bootstrapping is for all practical purposes identical to the Gaussian CDF whose mean is the expected value of the Wilcoxon ranks statistic, here 410, and whose standard deviation is the standard deviation of the Wilcoxon rank statistic, here 37. Using the Gaussian distribution we compute the p-value as 0.0022. This is the conventional way to compute the p-value for the Wilcoxon test when the sample sizes are not too small.

8.6   The discrete Wilcoxon test

In our discussion of the Wilcoxon test we have so far assumed that we have continuous distributions, so that the defining relation (8.5) holds. We will now modify this to take into account discontinuity points. The general equation, which is true for all types of CDFs, is

$$\int_{-\infty}^{\infty} \frac{F(x-) + F(x)}{2} dF(x) = \frac{1}{2}. \qquad (8.9)$$

To see why this is, take the example of a purely discrete distribution with jumps at the points $x_1, x_2, \ldots$ Then the integral is the sum

$$\sum_k \frac{F(x_k) + F(x_{k-1})}{2}(F(x_k) - F(x_{k-1})) = \frac{1}{2}\sum_k (F(x_k)^2 - F(x_{k-1})^2) = \frac{1}{2}.$$

This means that the original Wilcoxon relation in equation (8.5) holds true, only if we define the CDF $F(x)$ by the mid-point at points of discontinuity (see page 152). If we do so, we can apply the previous discussion to all distributions. The only change is that there is a tie-correction in formula 8.8 for the variance at jump points, in that we need to subtract a term $\frac{1}{4}\sum_x \Delta G(x)\Delta F(x)$ from $P_W(1 - P_W)$, a fact that is demonstrated in Appendix 8.A.

In the example below we investigate what the Wilcoxon test corresponds to when we compare two independent binomial distributions. Recall that for the mid-point CDF $F(x)$ of a Bin$(1, p)$ distribution, graphically shown in Figure 6.3, we have that

$$F(0) = \frac{1-p}{2}, \quad F(1) = 1 - \frac{p}{2}, \quad \Delta F(0) = 1 - p, \quad \Delta F(1) = p,$$

information which form the basis for the computations in the next example. In this section we will assume that all CDFs are defined by the mid-point value at jumps.

Example 8.4   Let $F(x)$ be the (mid-point) CDF for a Bin$(1, p)$ distribution, and $G(x)$ the CDF for a Bin$(1, q)$ distribution. We then have that

$$P_W = \int_{-\infty}^{\infty} G(x)dF(x) = G(0)\Delta F(0) + G(1)\Delta F(1) = \frac{1}{2}(1 + p - q),$$

so that $P_W = 1/2$ precisely when $p = q$. The corresponding estimator is $(1 + \hat{p} - \hat{q})/2$, which means that the Wilcoxon test is precisely the standard binomial test. Computing the variance as outlined we get what we expect: $(p(1 - p)/n + q(1 - q)/m)/4$.

We can extend this to other models for binomial distributions. If we take the proportional odds model instead, we use the function $G(x, \theta) = \theta G(x)/(G^c(x) + \theta G(x))$, which is the CDF for a Bin$(1, Q)$ distribution with $Q = q/(q + \theta(1 - q))$. It follows that

$$\int_{-\infty}^{\infty} G(x, \theta) dF(x) = \frac{1}{2} + \frac{1}{2} \left(p - \frac{q}{q + \theta(1 - q)} \right),$$

which is $1/2$ precisely when $\theta = q(1 - p)/p(1 - q)$. In 2×2 table notation the estimating equation is

$$x_{11} - \frac{n_1 x_{21}}{x_{21} + \theta x_{22}} = 0,$$

which therefore provides us with the empirical odds ratio as the estimate for θ. Applying this to the table in the first Hodgkin's lymphoma trial in Section 4.5.1, we obtain the 90% confidence interval $(1.86, 4.74)$, which agrees well with what we obtained in Figure 5.2.

This example naturally extends to ordered data with more than two categories. This is illustrated in the following example.

Example 8.5   The following table describes the outcome of a (hypothetical) clinical trial comparing an active treatment to a placebo, in patients who have suffered a stroke. The outcome variable is a disability score which goes from 1 (full ability) to 6 (completely disabled). The data obtained (number of patients) are displayed in Figure 8.7 and tabulated as follows:

Group	Disability score 1	2	3	4	5	6	Sum
Placebo	93	170	99	108	175	204	849
Active	332	262	83	65	62	49	853

Figure 8.7(a) shows the (ordinary) e-CDFs for the two groups, which are the cumulative probabilities of the rows in the table. The e-CDF for the active group lies to the left of that of the placebo group which indicates an improvement, since low number categories indicate

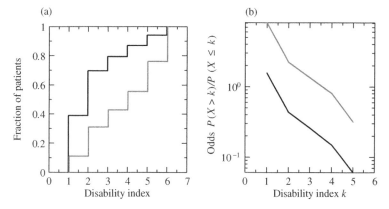

Figure 8.7   Description of the odds ratio for a disability index: (a) shows the CDFs for the two groups, (b) the category odds for each group.

Box 8.5 The relation between a proportional odds model and the Wilcoxon test

---

The relation between the proportional odds model and the Wilcoxon test goes deeper than for other models in that the Wilcoxon test is also intimately related to the likelihood estimation of the parameter. For continuous data the proportional odds model implies that

$$dF(x) = \frac{\theta dG(x)}{(\theta G(x) + G^c(x))^2},$$

which means that the log-likelihood is

$$n \ln \theta - 2 \sum_{i=1}^{n} \ln(\theta G(x_i) + G^c(x_i)),$$

ignoring terms that do not involve θ. Upon differentiation we obtain the estimating equation

$$\frac{n}{\theta} - 2 \sum_{i=1}^{n} \frac{G(x_i)}{\theta G(x_i) + G^c(x_i)} = 0,$$

a relation which we can also write as

$$\int_{-\infty}^{\infty} \frac{\theta G(x) dF_n(x)}{\theta G(x) + G^c(x)} = \frac{1}{2}.$$

At a discontinuity point we instead have

$$\Delta F(x) = \frac{\theta \Delta G(x)}{(\theta G(x-) + G^c(x-))(\theta G(x) + G^c(x))}$$

and the function to integrate becomes the average of the expression at $x-$ and x. In both cases, if we let n go to infinity, we see that in the limit the score equation is the same as equation (8.6). The Wilcoxon method of estimation is to replace $G(x)$ with $G_m(x)$, which essentially means replacing nuisance parameters with estimates.

---

less disability. However, the treatment effect is not a simple shift to the left, which we cannot expect when we study categorical data, since the jump points are fixed. So another description is called upon.

In Figure 8.7(b) the estimates for the CDF odds $F^c(x)/F(x)$ have been plotted for each category x and for each group. The y-scale is logarithmic, which means that the vertical distance between the two groups gives the logarithm of the corresponding odds ratio for that category. If these two curves are parallel (as seems to be the case), we would have a constant odds ratio, which therefore would be a universal description of the difference of the two treatments; in other words, the proportional odds model. Applying the Wilcoxon test to these data means an extension to a six-category case of the discussion in the previous example for the two-category case. The result is that a common odds ratio is estimated as 5.2 with 95% confidence interval (4.3, 6.2), so the odds of being at most in any particular category are five times larger for the active group than for the placebo group.

Proportional odds models for categorical data are usually discussed in the context of the analysis of multinomial data. The approach is the same as above, but the formulation using the Wilcoxon test is more general and possibly more transparent. We should also note that the odds ratio involved here is different from the generalization of the 2×2 table odds ratio discussed on page 127. In fact, in the previous example, if we denote the outcome for the control group by X and that for the treated group by Y, we have that $P(X < Y) = 0.18$ and $P(Y < X) = 0.67$, which means that the ratio $P(Y < X)/P(X < Y) = 3.70$. This is the generalized odds ratio mentioned in Box 7.1, and is different from the odds ratio θ in the proportional odds model. Whereas θ is a parameter in a precise model for the two distributions, the proportional odds model, the generalized odds ratio makes no such assumptions.

8.7   The two-period crossover trial

In a two-period crossover study we measure an outcome variable in patients under two different conditions, for example, with and without a particular treatment. Contrary to what may be intuitive, the way to analyze such studies is actually to compare two independent groups; in this section we will explain why. We consider an experiment in which an active drug is compared to a placebo, and let X denote the outcome in the placebo period (denoted P), and Y the outcome variable in the period with active drug (denoted A). At first glance it may look like that the appropriate thing to analyze is $Y - X$, so we first need to explain why this is not the case.

To analyze $Y - X$ would be appropriate if X and Y have mean m_1 and m_2, respectively, and we want to estimate $\Delta = m_2 - m_1$. However, for various reasons the distributions of X and Y, in particular their mean values, are expected to depend on a few design factors. We refer here to design factors, not individual demographic or disease factors. Such design factors are related to the fact that X and Y cannot be assessed simultaneously, but need to be measured on different occasions. The mean values in the two periods may for such a study look like the following:

	Period 1	Period 2
AP	0.5	0.4
PA	0.3	0.6

The patients treated with A in the first period have mean value 0.5, whereas the mean is 0.3 for those who are treated with placebo in the same period. In the second period the means for A and P are both increased by 0.1. We interpret this to say that there is a *period effect* of magnitude 0.1. Such period effects can occur for many reasons, reasons that were previously discussed as study effects and included the important regression to the mean phenomenon, as well as learning effects – if the experiment involves some procedure that you can improve on by practice, you may expect better results in the second period. The simple psychological effect of not really knowing what to expect when you do the experiment for the first time, but having some experience of it the second time, may also lead to period effects. Since there are many possible reasons for a period effect, it is not clear *a priori* in which period such an effect occurs, and therefore absolute treatment effects are not really meaningful (does A have mean value 0.5 or 0.6?). It is only the treatment difference that has a true meaning.

Next we consider the following table:

	Period 1	Period 2
AP	0.5	0.4
PA	0.3	0.5

We see here that the mean value of A is the same in both periods, so there is no period effect. However, the mean on P is 0.3 in the first period and $0.3 + 0.1$ in the second period. This is when it is tested after A has been given in the first period. We can express this in many equivalent ways:

1. there is a residual effect of treatment A into the next period;

2. there is a treatment-by-period interaction;

3. there is a carry-over effect in the sequence AP.

Again there are many reasons why this may occur. A simple explanation may be that there is still drug left in the body in the second period in amounts sufficient to have an effect. The simple remedy for this is to have a wash-out period of sufficient length between the two periods. There can also be residual effects of the treatment that are not dependent on drug concentration – the extreme example being that you are actually cured of the disease when treated with the drug. Other possibilities include psychological reasons where expectations in the second period are higher if you received the active drug in the first period than if you received placebo, provided expectations can affect the measured outcome.

Based on this, we can write a table of mean values for a two-period crossover experiment as follows:

	Period 1	Period 2
AP	$\mu + \pi_1 + \tau_A$	$\mu + \pi_2 + \tau_P + \lambda_A$
PA	$\mu + \pi_1 + \tau_P$	$\mu + \pi_2 + \tau_A + \lambda_P$

Here μ is an overall average effect, π_i the effect of period $i = 1, 2$, τ_X the effect of treatment ($X = A, P$) and λ_X the residual effect of treatment X when it was taken in the previous period. In all we have seven parameters, but only four mean values, so we cannot estimate all these parameters.

We therefore need to reduce the number of parameters. We have already noted that it is not clear from the experimental setup alone whether a particular period effect $\pi = \pi_2 - \pi_1$ occurs in the first or in the second period, or perhaps partly in both. It is a matter of defining an absolute reference, which we cannot do. We may therefore decide to put the effect of period 1 into the overall average μ (i.e., assume that $\pi_1 = 0$), so that $\pi_2 = \pi$ becomes the period effect. Similarly, it is not possible to estimate the absolute treatment effects, and we may decide to include τ_P in μ and take $\tau_A = \tau$ to be the treatment difference. As a final step we can write $\lambda = \lambda_A - \lambda_P$ and build λ_P into π, giving us the following table with only four parameters:

	Period 1	Period 2
AP	$\mu + \tau$	$\mu + \pi + \lambda$
PA	μ	$\mu + \pi + \tau$

We now compare this to a table of mean values:

	Period 1	Period 2
AP	a	b
PA	c	d

We find that we have four equations and four unknowns, so we can solve for the parameters:

$$\mu = c, \quad \tau = a - c, \quad \pi = d - a, \quad \lambda = b - c - d + a.$$

This is, however, a disappointing result. It means that when we estimate the treatment effect τ, we only use data from the first period. That is not what we want, because it means that having the second period was a complete waste of time. It was because we wanted to efficiently estimate the treatment effect that we carried out the study, and in order to use the power of the crossover study we must compare the treatment effect within each treatment sequence. In our notation this means that we want to use $a - b$ and $d - c$, and take the average of these numbers as an estimate of τ. However,

$$\frac{a - b + d - c}{2} = \tau - \frac{\lambda}{2},$$

so we can only use all the data to estimate τ if $\lambda = 0$. This observation has profound implications. It means that if we want to use the power of the crossover study in the analysis, we need to make sure that there are no (differential) carry-over effects. The elimination of these must come with the study design – if present we lose all the power of the crossover when it comes to estimate the treatment effect, since we are forced to make the comparison as a parallel group study instead, using first-period data only.

We therefore need to eliminate carry-over effects by design. We can, however, allow for period effects, because we can handle them in the analysis. Or rather, we can perform the analysis in such a way that the treatment effect is estimated independently of any period effect (provided effects are additive). The appropriate analysis of the study is as a parallel group study, with the two groups defined by the sequence of treatments, AP and PA. If we form the variable $Z = Y - X$ and compare it between the two groups, we see that for group AP the mean value is $\tau + \pi$, whereas for group PA it is $\tau - \pi$. It follows that the mean difference between the groups is 2π, so with this procedure we can estimate the size of the period effect. A simple modification of this allows us to estimate the treatment effect: since the sum of the two mean values is 2τ we have that, if we define a new variable as $Z = Y - X$ for group AP but as $Z = X - Y$ for group PA, then the mean difference between the two groups is 2τ. This procedure also gives us an unbiased treatment estimate in the presence of a period effect. We can replace mean value here with any location measurement, and use the appropriate two-group statistical method for the analysis.

Example 8.6   Consider again the paired data discussed in Section 6.6, but this time assume it is obtained in a properly randomized crossover study, in which patients with odd numbers are randomized to the sequence AP and those with even numbers to the sequence PA. The following table shows the mean values of the log-transformed data on the two treatments for each sequence:

Sequence	X	Y	Average
AP	5.12	3.15	4.13
PA	4.87	2.85	3.86
Average:	4.99	3.00	.

To get the treatment difference we compute the difference in the column averages as -1.99, which is also the average of the differences within sequence. In other words, twice the difference is the same as the sum of the mean differences within sequence. So, forming $Y - X$ for the AP group and $X - Y$ for the PA group, and statistically comparing these two groups with a mean value test (in other words, a t-test) on the log scale, we can arrive at the estimate above and 95% confidence limits for it. If we do that, and back-transform, we get a (geometric) treatment mean ratio of 0.137 with 95% confidence interval (0.090, 0.209). Alternatively, we can do a non-parametric estimation using the Hodges–Lehmann estimator for the group difference, still on log-transformed data, to obtain a similar estimate of the treatment ratio, namely 0.134 with 95% confidence interval (0.085, 0.215).

Note that these treatment estimates are adjusted for a possible overall period effect in the experiment. To estimate the mean period effect we note that twice the period effect is given by $5.12 + 2.85 - (3.15 + 4.87) = -0.05$, which means that the potential period effect is estimated to be a 2.5% increase (in geometric mean ratio).

8.8   Multivariate analysis and analysis of covariance

We now want to extend the discussion on how to compare two groups to situations where we have more than one measurement on each subject at our disposal. Suppose there are two measurements, X and Y, and let $F(x, y)$ be their bivariate CDF for the first group, and $G(x, y)$ for the second. How do we now compare groups? There are two different questions here, depending on the situation:

1. It may be that X and Y measure different aspects of a disease and we want to make an effect claim by simultaneously analyzing both these variables.

2. Alternatively, it may be that X and Y measure the same outcome variable, but on two different occasions. If X is a baseline measurement and Y is an end-of-treatment measurement and we have a randomized study, how can we capitalize on the fact that the population distributions of X for the two groups are the same?

Most of our discussion will be concerned with the second case, but we first need to outline how to approach the first.

We have discussed how to extend the one-sample t-test to cover more than one variable in Section 7.6, and the extension of this to the two-group case is done in the same way as for the univariate Gaussian distribution. To fix notation, assume that we have n observations of a p-vector with a $N_p(\mu_1, \Sigma_1)$ distribution, and another set of m observations from an (independent) $N_p(\mu_2, \Sigma_2)$ distribution. Let $\Delta = \mu_1 - \mu_2$ denote the mean difference between the groups (which is a vector). We estimate Δ using the difference $\hat{\Delta}$ in the arithmetic means of the two distributions, for which we have that

$$\hat{\Delta} \in N_p(\Delta, \Sigma_1/n + \Sigma_2/m).$$

If we furthermore assume that the two groups have the same variance matrix, we can estimate this common variance matrix Σ with the pooled sample variance S based on $f = n + m - 2$ degrees of freedom. In other words, $\hat{\Delta} \in N_p(\Delta, \Sigma(1/n + 1/m))$, and the sample variance S is an estimate of Σ such that $fS \in \Gamma_p(f/2, \Sigma/2)$. If we replace S with $S(1/n + 1/m)$, it becomes the variance estimate of $\hat{\Delta}$ and we have (see Appendix 7.A.2) that

$$\frac{f - p + 1}{pf}(\hat{\Delta} - \Delta)^t S^{-1}(\hat{\Delta} - \Delta) \in F(p, f - p + 1).$$

From this we can derive a confidence function for the parameter vector Δ. The case $p = 1$ is the classical two-sample t-test, since $F(1, f) = t^2(f)$ and the test statistic is the square of the t-test statistic in this case.

Having disposed of this case, we now turn to the second problem above: how to use information from a baseline measurement in the analysis of an end-point variable. For the rest of this section we will restrict attention to the bivariate case $p = 2$ with one (univariate) baseline variable X and one (univariate) end-point variable Y. This will also link into the discussion in Section 7.3. By way of illustration we consider a data set of 21 subjects for which a particular outcome variable was measured twice in two groups, called A and B. The table below shows the mean values for the two groups:

Group	X	Y
A	2.36	3.69
B	2.22	3.88
$B - A$	0.14	-0.19

We first construct the two-dimensional confidence function above for the mean group differences $\Delta = (\Delta_1, \Delta_2)$. Its 90% contour is shown in Figure 8.8 as the solid ellipse, and we have indicated the point estimates for the group differences as the intersection of axis-parallel lines. The rectangle in Figure 8.8 is derived from the corresponding univariate 95% confidence intervals, and is a confidence region for Δ with a coverage level of at least 90%. When we construct univariate confidence intervals we do not use this confidence function, but the one we obtain when we replace the $F(2, f - 1)$ distribution with the $F(1, f) = t^2(f)$ distribution. Its 95% contour is shown as the dotted ellipse, and we see that it is tangential to the rectangle close to the lower left and upper right corners.

As discussed in Section 7.3 we have a few options when we want to estimate Δ_2. If we only consider Y, and ignore whatever information may be stored in X, we find the estimate of Δ_2 to be -0.192 with 95% confidence interval equal to $(-0.53, 0.14)$. This corresponds to the vertical side of the rectangle in Figure 8.8. The corresponding two-sided p-value

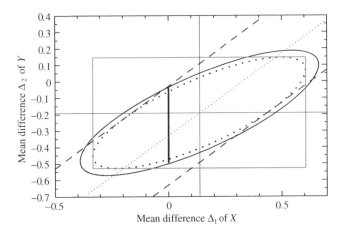

Figure 8.8 The solid curve shows the 90% simultaneous confidence region for the mean differences of two dependent Gaussian variables. The rectangle shows the univariate 95% confidence intervals and the dashed lines the univariate 95% confidence interval for the difference $\Delta_2 - \Delta_1$. For the dotted curve and the vertical line segment, see the text.

for the test of equality of the means is 0.25, and is therefore not even close to being statistically significant at any conventional significance level. Another natural approach is to make the same analysis with the change from baseline, $Y - X$, as outcome variable. This means that we estimate $\Delta_2 - \Delta_1$ instead of Δ_2, but if the study is randomized, we have $\Delta_1 = 0$ by design, so we estimate the same parameter. This analysis gives us the estimate -0.33 with 95% confidence interval $(-0.63, -0.03)$. The corresponding two-sided p-value is 0.033, within the conventional limit of 5% required to declare an effect of the drug. This analysis is graphically illustrated in Figure 8.8 as the three oblique lines. They are given by equations of the form $\Delta_2 = \theta + \Delta_1$, and their intersections with the vertical line $\Delta_1 = 0$ provide us with the point estimate (middle line) for the difference (θ) as well as the 95% confidence limits (outer lines). These lines are also tangential to the dotted univariate 95% contour.

When we replace the analysis of Y with the analysis of $Y - X$ we see that, for these particular data, the size of the drug effect increases considerably. To understand why, note that Y depends both on which group it was measured in and on the corresponding value of X. A subject with a higher value at baseline is expected to have a higher value at the end, compared to one with a lower value at baseline. The two groups differ numerically at baseline and this mean difference in X carries over to the mean values of Y, and since the control group had a higher mean at baseline, the mean difference of the two groups in Y will tend toward no effect. This effect is adjusted for when we look at the increase from baseline. But we must also take into account the effect of regression to the mean. We made the observation in equation (7.6) that looking at the change from baseline alone may provide a bias in the estimate of the treatment difference Δ_2. We know that if the average difference at baseline is Δ_1^*, then the expected mean difference, conditional on the knowledge of the baseline data, is given by $\Delta_2 - (1 - \rho)\Delta_1^*$, where ρ is the correlation between Y and X and we have assumed equal variance. In numbers it means that the true Δ_2 should be estimated from $-0.33 = \Delta_2 - (1 - \rho)0.14$, if we only knew ρ. A crude estimate of ρ

from data (assuming no group difference) is around 0.70, which gives an estimate of Δ_2 of approximately -0.29.

In order to make this mathematically robust, return first to Figure 8.8. There is a vertical interval indicated over $\Delta_1 = 0$. This *is not derived from some intersection with the 90% contour line*, contrary to appearances. The interval shows the confidence interval for Δ_2 in a situation wherew we know that $\Delta_1 = 0$, and we wish to understand how it is derived. The difference between this and the examples above is that there we considered *functions* of Δ, but now we put a *constraint* on Δ, namely that $\Delta_1 = 0$. Instead of being a problem with one parameter of interest and one nuisance parameter, a constraint reduces the problem to a one-parameter problem. In order to see how that changes the picture, recall first that the contour lines are levels of the quadratic form

$$Q(\hat{\Delta}) = (\hat{\Delta} - \Delta)^t S^{-1} (\hat{\Delta} - \Delta),$$

but considered as a function of Δ, with the estimate Δ^* replacing the estimator $\hat{\Delta}$. (Explicitly, $Q(\Delta) = (\Delta - \Delta^*)^t S^{-1} (\Delta - \Delta^*)$, with similar expressions for the components below.) We can decompose $Q(\hat{\Delta})$ as

$$Q(\hat{\Delta}) = Q_1(\hat{\Delta}_1) + Q_{2.1}(\hat{\Delta}_2 | \hat{\Delta}_1),$$

where

$$Q_1(\hat{\Delta}_1) = \frac{(\hat{\Delta}_1 - \Delta_1)^2}{S_{11}}, \quad Q_{2.1}(\hat{\Delta}_2 | \hat{\Delta}_1) = \frac{(\hat{\Delta}_2 - \Delta_2 - \frac{S_{12}}{S_{11}}(\hat{\Delta}_1 - \Delta_1))^2}{S_{2.1}},$$

with $S_{2.1} = S_{22} - S_{12}^2/S_{11}$, a notation we also use for Σ.

We first assume that Σ is known, so that we can take $S = \Sigma$ above. That allows us to highlight what is going on in a first step, before we can address some mathematical complications in the general case. When Σ is known (and we take $S = \Sigma$), $Q(\hat{\Delta})$ has a $\chi^2(2)$ distribution, because it is the sum of two independent $\chi^2(1)$-distributed variables, $Q_1(\hat{\Delta}_1)$ and $Q_{2.1}(\hat{\Delta}_2 | \hat{\Delta}_1)$. Since linear combinations of Gaussian variables, dependent or otherwise, also are Gaussian, we have that

$$\hat{\Delta}_2 - \frac{\Sigma_{12}}{\Sigma_{11}} \hat{\Delta}_1 \in N \left(\Delta_2 - \frac{\Sigma_{12}}{\Sigma_{11}} \Delta_1, \Sigma_{2.1} \right).$$

The counterpart to the confidence function above would now be $\chi_2(Q(\Delta))$. However, if we know that $\Delta_1 = 0$, the term $Q_1(\Delta_1)$ is only noise, and addresses a question we are not interested in. We therefore focus exclusively on the second quadratic form and consider the confidence function $\chi_1(Q_{2.1}(\Delta_2 | \Delta_1))$. When the first quadratic term is pure noise, these two confidence functions should provide similar results, but the latter should be more efficient. This constitutes the overall explanation of the difference between what we see in Figure 8.8 and what we actually do.

However, when Σ is not known but estimated with S as above, there are some further mathematical complications. These are concerned with the actual distribution of $Q_{2.1}(\hat{\Delta}_2 | \hat{\Delta}_1)$, which is now less straightforward. That general theory of multivariate Gaussian distributions tells us the following two facts, both of which refer to conditional distributions given the values of $\hat{\Delta}_1$ and S_{11}:

$$\hat{\Delta}_2 - \frac{S_{12}}{S_{11}}\hat{\Delta}_1 \in N\left(\Delta_2 - \frac{\Sigma_{12}}{\Sigma_{11}}\Delta_1, (1 + Q_1(\hat{\Delta}_1)/f)\Sigma_{2.1}\right) \quad \text{and} \quad \frac{S_{2.1}}{\Sigma_{2.1}} \in \chi^2(f-1).$$

Moreover, conditionally on $\hat{\Delta}_1$ and S_{11}, these are independent, and it follows that

$$\frac{f-1}{f}\frac{Q_{2.1}(\hat{\Delta}_2|\hat{\Delta}_1)}{1 + Q_1(\hat{\Delta}_1)/f} \in F(1, f-1) = t^2(f-1).$$

This corresponds to the $\chi^2(1)$ distribution of $Q_{2.1}(\hat{\Delta}_2|\hat{\Delta}_1)$ above, and gives us the appropriate confidence function for Δ_2. When we use this to obtain a confidence interval for Δ_2, we say that we perform an unconditional test; what the corresponding conditional test is will be described below.

The analysis is illustrated graphically in Figure 8.9. Gray dots refer to the data in group A and black dots to the data in group B, and there are two straight lines describing the estimate of the conditional mean function $E(Y|X = x) = \bar{x} + b(x - \bar{x})$, $b = S_{12}/S_{11}$, for the two groups separately (same grayness for points and line). Since we assumed that the covariance matrix was the same in the two groups, these lines are parallel. It is the vertical distance between these two lines that defines the treatment difference. The dashed lines gives the predictions for the two means of the outcome variable, conditional on the overall baseline average. This illustrates the difference between the result of this analysis and the analysis of the unadjusted means for the two groups, which started this discussion. The unadjusted means are shown as the triangles close to the y-axis, and are not quite on the dashed lines, but if we move these triangles horizontally until they intersect the respective line, and from there go down to the x-axis, we find the baseline means for the two groups. The difference between the triangles and the respective dashed lines is therefore a graphical illustration of how much the mean difference at baseline affects the crude mean difference.

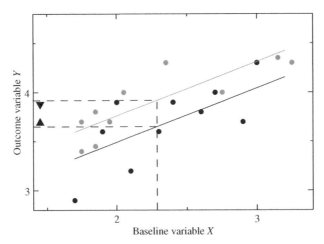

Figure 8.9 How an analysis of covariance works. Black and gray dots represent data from the two groups, and the parallel lines the regression lines from the analysis for each of the groups. The triangles are crude mean values of the outcome variable for each group and the dashed lines show how the adjusted means for each group are obtained.

When we look at Figure 8.9, we may view the problem in a slightly different way. We know that the conditional distribution of Y on X is a linear function as in the graph, because the data are Gaussian. We get an (almost) equivalent analysis if we switch our assumptions so that we assume this linear relation to start with, and only consider the conditional distribution of Y. In doing this we switch to a univariate probability model. Specifically, let z be an indicator variable for group membership, which we choose to be 1 for group A and -1 for group B. The model assumption is then that

$$E(Y|Z = z, X = x) = \theta_1 + \theta_2 z + \theta_3 x,$$

for unknown parameters $\theta = (\theta_1, \theta_2, \theta_3)$, together with an assumption that this conditional distribution of Y has a Gaussian distribution with a common variance for all individuals. To analyze this we can use least squares to estimate θ; this is described in some detail in Example 9.1. What this theory tells us is that we get parameter estimates corresponding to those above, in particular that $\theta_3 = S_{12}/S_{11}$ and $2\theta_2 = \Delta_2^*$. There is a difference, however, because with this model we end up with a confidence function for Δ_2 which is based on the observation (in the notation above) that

$$\frac{f-1}{f} Q_{2.1}(\hat{\Delta}_2 | \hat{\Delta}_1) \in t^2(f - 1).$$

We refer to this test as the conditional test for the mean parameter difference. It is a so-called linear model (see Chapter 9); this particular case is called an analysis of covariance (ANCOVA). The difference from the unconditional test above lies in which residual variation is used in the test. Both tests takes the form

$$\hat{\sigma}^{-2} \left(\hat{\Delta}_2 - \Delta_2 - \frac{S_{12}}{S_{11}} (\hat{\Delta}_1 - \Delta_1) \right)^2 \in t^2(f - 1),$$

where for $\hat{\sigma}^2$ we take $s^2 = f\Sigma_{2.1}/(f - 1)$ for the conditional test, but use the slightly larger variance $s^2(1 + Q_1(\Delta_1^*)/f)$ for the unconditional test. The results of these two analyses of our data are summarized in the following table:

Type	Δ_2^*	95% CI
Conditional	-0.268	$(-0.491, -0.044)$
Unconditional	-0.268	$(-0.497, -0.038)$

The difference between the conditional and unconditional test above is not that great. They are different, since they use different probability models. The unconditional test is a test on a bivariate Gaussian distribution, though with a conditional analysis at the end. The conditional tests is a test of a linear model with the residual assumed to have a (univariate) Gaussian distribution. In the first case the linear predictor is a consequence of the bivariate Gaussian distribution of the mean differences, in the second case it is the basic model assumption.

The unconditional method lends itself rather readily to an ANCOVA without the Gaussian assumption, as long as we accept that the analysis is only approximate. The assumption of equal variance is not necessary, since we can replace $\Sigma(1/n + 1/m)$ with the expression $\Sigma_1/n + \Sigma_2/m$, and estimate this using group-specific sample variances. Appealing to the

CLT to justify a large-sample bivariate Gaussian distribution approximation for $\hat{\Delta}$, we have a relatively easy road to the application of the unconditional method to non-Gaussian data. The appropriateness of the approximation ultimately depends on sample size and how skewed the data are. In this case the linear regression lines will be statements about the arithmetic means only, not about the data; it is the arithmetic means we assume have a bivariate Gaussian distribution.

We have discussed how we can employ the ANCOVA to adjust for observed baseline differences when there is no true difference there, and obtain estimates of treatment effects with higher precision. However, this is not the only use for ANCOVA. In fact, we could argue that if adjustment is what we want to do, we should think of the unconditional method. The conditional method is a modeling approach to data, something we will discuss in much more detail in the next chapter. As an introduction to this discussion we conclude this chapter with an illustration of an ANCOVA in which we do not want to make a group comparison for the same value of the covariate. In fact, quite the opposite.

Example 8.7   The pharmacokinetics of a new drug was investigated in 6 men and 5 women. When clearance was computed and compared head-on between the two groups, no difference could be found. But then the graph in Figure 8.10 was produced. The straight lines are analysis of covariance estimates for each of the sexes, in which the logarithm of clearance is analyzed with an ANCOVA, with the logarithm of body weight taking the place of the baseline measurement. What we see is that, as functions of body weight W, clearance differs between sexes: on average it is $0.82W^{1.26}$ for men and $1.11W^{1.26}$ for women. The ratio, for the same weight, is 1.35. But then, why should we want to compare clearance at the same weight? The important information is that, despite a body weight dependence, men and women clear the drug from plasma equally well. But knowing that clearance depends on body weight may also be important when we choose the dose for a particular patient.

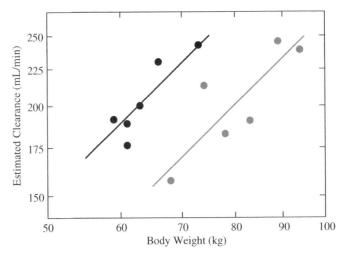

Figure 8.10   Example of an ANCOVA in which we do not want to adjust to equal covariate means. Gray data represent men, black data women.

8.9   Comments and further reading

This chapter has discussed two-group comparisons using what are perhaps the two most often used and best-known tests in medical statistics, Student's t-test and Wilcoxon's rank sum test. The former is an extension of the one-sample t-test we discussed in the previous chapter. The two-group test was actually introduced by R. A. Fisher who, fifteen years after its publication, discovered William Gosset's original publication on the one-sample case and extended it, not only to this two-group situation, but all the way to what became known as analysis of variance (Senn, 2008). The t-test addresses the difference in means and requires complete data, whereas the way to obtain confidence limits for a percentile difference, discussed in Box 8.2, can be extended to situations with censored data (Su and Wei, 1993), by using the appropriate variance estimate.

Lehmann (1998) gives a general introduction to nonparametric tests. The Wilcoxon test was developed as a rank test by Frank Wilcoxon in 1945 and then, later and independently, by Mann and Whitney, using the scores that are named for them. The test is therefore often referred to as the Wilcoxon–Mann–Whitney test, with the score version leading to the Wilcoxon probability (Halperin et al., 1987). The Hodges–Lehmann shift estimator was introduced later (Hodges and Lehmann, 1963) as the median of all the differences. The characterization in Box 8.4 was given in the paper by Fine (1966), though the discussion there is in terms of e-CDFs instead of CDFs. The relation between the t-test and the Wilcoxon test discussed in Box 8.3 was introduced (Conover and Iman, 1981) as a pedagogical technique to use rank transformations as a bridge between parametric and nonparametric statistics (and also as a method to carry out nonparametric statistics in statistical software that may not include such methods). The initial enthusiasm about the prospect of applying this philosophy (to rank data before analysis) to more general problems soon waned, when it was realized that the nonlinear transformation involved did not always produce sensible tests (Thompson, 1991). The generalized odds ratio mentioned in the text was discussed by Agresti (1980), but seems not to have gained much interest in the biostatistics community.

I have found no reference for the description of the various aspects of the nature of the Wilcoxon test discussed in this chapter; in particular about parameter estimation in different models using a simple estimating equation. The particular relation between the proportional odds model and the Wilcoxon test is, however, well known and has been used to determine the power of a study for categorical data (Whitehead, 1993).

The fact that p-values can be computed in different ways, and why the way they are computed matters, was illustrated in (Bergmann et al., 2000), where the outcomes for a Wilcoxon test on a particular data set were compared between 11 PC-based statistical software programs. The p-values varied, depending on whether a large-sample approximation or an exact permutation form of the test was used and, in the former case, whether or not a continuity correction (see page 98) was used. The key message is that you need to understand precisely what your particular software does, before you use it.

The discussion of crossover studies is short, and focused on the design issues and on the analysis of a two-period crossover study as a parallel group study. Similar discussions as well as discussion of how to analyze crossover studies with more than two periods can be found in either Jones and Kenward (1989) or Senn (2002).

For mathematical details on the unconditional test in Section 8.8, see Rao (1967) and references therein, as well as Marden and Perlman (1980). It should be noted that we do not discuss the test statistic $Q(\hat{\Delta}_2|\hat{\Delta}_1)$ itself; this has a different distribution which is described by Rao. For more on the nonparametric analysis of covariance, see Koch et al. (1982).

References

Agresti, A. (1980) Generalized odds ratio for ordinal data. *Biometrics*, **36**(1), 59–67.

Bergmann, R., Ludbrook, J. and Spooren, W. (2000) Different outcomes of the Wilcoxon-Mann-Whitney test from different statistics packages. *American Statistician*, **54**, 72–77.

Conover, W. and Iman, RL. (1981) Rank transformations as a bridge between parametric and nonparametric statistics. *American Statistician* **35**(3), 124–133.

Fine, T. (1966) On the Hodges and Lehmann shift estimator in the two sample problem. *Annals of Mathematical Statistics*, **37**(6), 1814–1818.

Halperin, M., Gilbert P.R. and Lachin J.M. (1987) Distribution-free confidence intervals for $\Pr(X_1 < X_2)$. *Biometrics*, **43**, 71–80.

Hodges, J.L. and Lehmann, E.L. (1963) Estimates of location based on rank tests. *Annals of Mathematical Statistics*, **34**, 598–611.

Jones, B. and Kenward, M.G. (1989) *Design and Analysis of Cross-Over Trials* number 34 in *Monographs on Statistics and Applied Probability*. London:Chapman & Hall.

Koch, G.G., Mara, I.A., Davis, G.W. and Gillings, D.B. (1982) A review of some statistical methods for covariance analysis of categorical data. *Biometrics* **38**(3), 563–595.

Kowalski, J. and Tu, X.M. (2008) *Modern Applied U-Statistics*. Hoboken, NJ: John Wiley & Sons, Inc.

Lehmann, E.L. (1998) *Nonparametrics: Statistical Methods Based on Ranks* revised 1st edn. Upper Saddle River,NJ: Prentice Hall.

Marden, J. and Perlman, M.D. (1980) Invariant tests for means with covariates. *Annals of Statistics*, **8**(1), 25–63.

Rao, C.R. (1967) *Proceedings of the Fifth Berkeley Symposium on Mathematical Statistics and Probability* vol. 1 Berkeley: University of California Press chapter Least squares theory using an estimated dispersion matrix and its application to measurement of signals, pp. 355–372.

Senn, S. (2002) *Cross-over Trials in Clinical Research* Statistics in Practice second edn. Chichester: John Wiley & Sons, Ltd.

Senn, S. (2008) A century of *t*-tests. *Significance*, **5**(1), 37–39.

Su, J.Q. and Wei, L.J. (1993) Nonparametric estimation for the difference or ratio of median failure times. *Biometrics*, **49**(2), 603–607.

Thompson, G. (1991) A note on the rank transform for interactions. *Biometrika*, **78**(3), 697–701.

Whitehead, J. (1993) Sample size calculations for ordered categorical data. *Statistics in Medicine*, **12**, 2257–2271.

8.A   Appendix: About U-statistics

The arithmetic mean, the sample variance and the Wilcoxon probability estimator have in common that they are all U-statistics. Such statistics consists of sums of dependent variables which are asymptotically dominated by a sum of independent variables, so that we can appeal to the CLT for a large-sample theory. Only a brief overview will be given here; for details, see Kowalski and Tu (2008).

Consider a sample $\{x_1, \ldots, x_n\}$ from a CDF $F(x)$. To define a U-statistic of order 1 and degree k we pick a symmetric function $g(x_1, \ldots, x_k)$ of k arguments. This function defines a population parameter $\phi = E(g(X_1, \ldots, X_k))$. There are a total of $\binom{n}{k}$ subsets $\{x_{i_1}, \ldots, x_{i_k}\}$ of the original sample, subsets we denote by x_I, where I is a set of indices. If we denote the collection of all such subsets by Π, we can estimate ϕ by using the average

$$U_n = \frac{1}{\binom{n}{k}} \sum_\Pi g(x_{i_1}, \ldots, x_{i_k}), \tag{8.10}$$

which is an unbiased estimator of ϕ. The members of Π are dependent when they share x_is, which is an important observation when we are to compute the variance of U_n. This definition is not restricted to univariate data; we can allow multivariate distributions where x_is are vectors. Simple examples among the univariate tests are $g(x) = x$, which defines a U-statistic of degree 1, namely the arithmetic mean (estimating the population mean) and $g(x, y) = (x - y)^2/2$, which defines a U-statistic of degree 2, namely the sample variance s_n^2 (see Box 8.6), an unbiased estimate of the population variance σ^2. Among the bivariate tests we find $g(x, y) = (x_1 - y_1)(x_2 - y_2)/2$, which defines the (Pearson) covariance, and $g(x, y) = 2I((x_1 - y_1)(x_2 - y_2) > 0) - 1$, which defines Kendall's τ. The theory of U-statistics therefore provides a unified framework for the analysis of all these parameters.

In order to obtain knowledge about the population parameter ϕ from the corresponding U-statistic we usually appeal to large-sample theory. For this we need to compute the variance $V(U_n) = \binom{n}{k}^{-2} \sum_{I,J} C(g(x_I), g(x_J))$. For a covariance term in this sum to be non-zero, the corresponding index sets I and J need to have a non-empty intersection, and for symmetry reasons the covariance then depends only on the number of indices in this intersection. If we therefore introduce the functions

$$g_d(x_1, \ldots, x_d) = \int \cdots \int g(x_1, \ldots, x_d, y_1, \ldots, y_{k-d}) dF(y_1) \ldots F(y_{k-d}),$$

we see that if the indices have d variables in common the covariance is $C(g(x_I, x_J)) = V(g_d(X_1, \ldots, X_d)) = \sigma_d^2$. A combinatorial argument now shows that

$$V(U_n) = \sum_{d=1}^{k} \pi_d \sigma_d^2, \quad \text{where } \pi_d = \frac{\binom{k}{d}\binom{n-k}{k-d}}{\binom{n}{k}}.$$

If we increase n, the dominant term in this expression is the first term, for which we have that $\pi_1 \sim k^2/n$. For the special case of a U-statistic of degree 2 this is

$$\binom{n}{2} V(U_n) = 2(n - 2)V(g_1(X)) + V(g(X_1, X_2)).$$

Box 8.6 Estimating the variance

We can use U-statistics to derive the sample estimate of σ^2 with the correct degrees of freedom. Let X and Y be independent stochastic variables with the same CDF $F(x)$. Then the difference $X - Y$ has mean zero and variance $2\sigma^2$, which implies that

$$2\sigma^2 = \int\int (x - y)^2 dF(x)dF(y) = 2\int\int_{x<y} (x - y)^2 dF(x)dF(y).$$

We can therefore estimate σ^2 with

$$\frac{1}{\binom{n}{2}}\sum_{i<j}(x_i - x_j)^2 = \frac{1}{n-1}\sum_i (x_i - \bar{x})^2.$$

Here we have used the observation, easily verified by brute force computation, that

$$\sum_i (x_i - \bar{x})^2 = \frac{1}{n}\sum_{i<j}(x_i - x_j)^2.$$

So by finding a formula for σ^2 that does not involve m we get a proper estimator that has the correct expected value σ^2.

Example 8.8   With $g(x, y) = (x - y)^2/2$ we have that

$$g_1(x) = \int \frac{1}{2}(x - y)^2 dF(y) = \frac{x^2 + \sigma^2}{2},$$

whose expected value is σ^2, and for which the variance is given by

$$V(g_1(X)) = \int_{-\infty}^{\infty}\left(\frac{x^2 + \sigma^2}{2} - \sigma^2\right)^2 dF(x) = \frac{1}{4}(\mu_4 - \sigma^4).$$

Since we also have that

$$V(g(X_1, X_2)) = \frac{1}{4}\int (x - y)^4 dF(x)dF(y) = \frac{1}{2}(\mu_4 + 3\sigma^4), \quad \mu_4 = \int_{-\infty}^{\infty}(x - m)^4 dF(x),$$

we find that

$$V(s_n^2) = \frac{2(n-2)(\mu_4 - \sigma^4) + (2\mu_4 + 6\sigma^4)}{4\binom{n}{2}} = \frac{(n-1)\mu_4 - (n-5)\sigma^4}{n(n-1)}.$$

Asymptotically, when n becomes large, this is approximately $(\mu_4 - \sigma^4)/n$, and we have that $\sqrt{n}(s_n^2 - \sigma^2) \in AsN(0, \mu_4 - \sigma^4)$.

The large-sample theory for U-statistics involves two steps: first we show that the $d = 1$ term dominates, and then we apply the CLT to this dominant term. For this we first note that the conditional expectation $E(g(X_I)|X_i = x_i)$ equals $g_1(x_i)$ if $i \in I$, and equals ϕ otherwise, which implies that $E(U_n - \phi|X = x_i)$ equals

$$\binom{n}{k}^{-1}\left(\binom{n-1}{k-1}g_1(x_i) + \left(\binom{n}{k} - \binom{n-1}{k-1} - 1\right)\phi\right) = \frac{k}{n}(g_1(x_i) - \phi).$$

Therefore the best predictor for $U_n - \phi$, which is a sum of functions of only one variable, is

$$\hat{U}_n = \frac{k}{n}\sum_{i=1}^{n}(g_1(x_i) - \phi).$$

Furthermore, $E((U_n - \hat{U}_n)^2) = V(U_n) - V(\hat{U}_n)$ goes to zero when we increase n, because the two terms both equal k^2/n asymptotically, an observation known as Hájek's projection theorem. Appealing to the CLT, we find that $U_n \in AsN(\phi, k^2\sigma_1^2/n)$, provided $\sigma_1^2 = V(g_1(X)) > 0$. In summary, to obtain knowledge about ϕ we use the fact that

$$\frac{(U_n - \phi)^2}{V(U_n)} \in As\chi^2(1).$$

The two-group situation is similar, but involves more complex formulas, and in order to avoid the notational complexity of the general case, we consider only the simple case with U-statistics of degree $(1, 1)$. Such a U-statistic takes the form

$$U_{mn} = \frac{1}{nm}\sum_{i,j}g(x_i, y_j),$$

and estimates the parameter $\phi = E(g(X, Y))$. Denote the CDF for X by $F(x)$ and that for Y by $G(x)$, and note that $mnV(U_{mn})$ is given by

$$V(g(X, Y)) + (n - 1)C(g(X_1, Y), g(X_2, Y)) + (m - 1)C(g(X, Y_1), g(X, Y_2)).$$

If we introduce the two functions

$$g_{10}(x) = \int_{-\infty}^{\infty}g(x, y)dG(y) - \phi, \quad g_{01}(y) = \int_{-\infty}^{\infty}g(x, y)dF(x) - \phi,$$

we have that $C(g(X, Y_1), g(X, Y_2)) = E(g_{10}(X)^2) = \sigma_{10}^2$, with similar expressions for the other covariances. It follows that

$$V(U_{mn}) = \frac{1}{nm}((m - 1)\sigma_{01}^2 + (n - 1)\sigma_{10}^2 + \sigma_{11}^2), \tag{8.11}$$

where $\sigma_{11}^2 = V(g(X, Y))$.

Example 8.9 The Wilcoxon parameter

$$P_W = P(Y < X) + P(X = Y)/2 = \int_{-\infty}^{\infty}G(x)dF(x),$$

with CDFs defined by their mid-point value at jump points, is estimated by the U-statistic of degree $(1, 1)$, defined by the kernel function

$$g(x, y) = I(y < x) + \frac{1}{2}I(x = y).$$

For this kernel function we have that $g_{10}(x) = G(x) - P_W$ and $g_{01}(y) = F^c(y) - P_W = (1 - P_W) - F(y)$. Moreover,

$$g(x, y)^2 = I(x < y) + I(x = y)/4 = g(x, y) - I(x = y)/4,$$

which means that the variance of $g(X, Y)$ equals

$$P_W(1 - P_W) - \frac{1}{4}P(X = Y) = P_W(1 - P_W) - \frac{1}{4}\sum_k \Delta G(x_k)\Delta F(x_k).$$

From this it follows that the variance for this U-statistic is given by

$$nmV(U_{nm}) = (n - 1)\int_{-\infty}^{\infty}(F(x) - (1 - P_W))^2 dG(x)$$

$$+ (m - 1)\int_{-\infty}^{\infty}(G(x) - P_W)^2 dF(x) + P_W(1 - P_W) - \frac{1}{4}\sum_k \Delta G(x_k)\Delta F(x_k).$$

In the continuous case this is formula (8.8) which we used when we discussed the Wilcoxon test. The full formula was used in Section 8.6.

To derive the large-sample theory, we again use the fact that the projection of $U_{mn} - \phi$ onto the space of functions of one variable only, either X or Y, is

$$\hat{U}_{nm} = \frac{1}{m}\sum_{i=1}^{m} g_{10}(X_i) + \frac{1}{n}\sum_{j=1}^{n} g_{01}(Y_j).$$

In particular, we have that $V(\hat{U}_{nm}) = \sigma_{10}^2/m + \sigma_{01}^2/n$. As before, we see that asymptotically U_{mn} and $\hat{U}_{nm}$ have the same Gaussian distribution and therefore that

$$\frac{U_{nm} - \phi}{\sqrt{V(U_{nm})}} \in AsN(0, 1)$$

when $n, m \to \infty$ in such a way that $m/n \to \lambda$, provided one of the following conditions is fulfilled: (1) that $\lambda = 0$ and $\sigma_{10}^2 > 0$; or (2) that $\lambda > 0$ and one of σ_{10}^2 or σ_{01}^2 is positive.

9

Least squares, linear models and beyond

9.1 Introduction

Up to now we have discussed group comparisons and how a statistical analysis of an outcome variable may allow us to conclude that an exposure, possibly a drug, has an effect on this outcome variable. If it has, the exposure is a predictor for such a response, in that we expect differences in the response in individuals who are exposed, compared to those who are not exposed. But we warned early on that there might be more to this than meets the eye – there may be confounders operating in the background. The purpose of the present chapter is to discuss how we take such confounders into account in a model for the data, as a continuation of the discussion that concluded the last chapter on adjustment for baseline information.

In this chapter we will construct models that account for covariates (or explanatory variables) and we will see how we estimate the corresponding model parameters, mostly the mean. Many such models are linear, which means that the values of explanatory variables are simply added. Among such models we find the classical case of analysis of variance (ANOVA), a model in which parameter values are obtained using least squares estimation. We introduced least squares estimation in Section 7.3, a discussion we now will follow up. We will discuss variations of this particular estimation method, and end the chapter by briefly discussing a large family of distributions, which includes many of the important ones in statistics such as the Gaussian, binomial and Poisson distributions, for which these estimation methods also constitute the maximum likelihood theory. This will provide a unified framework for doing statistics in classical ANOVA, logistic regression and other important models.

However, before we do this, we need to discuss heterogeneity. One of the main purposes of linear models is to use covariates to explain heterogeneity in the response in the population, or to demonstrate the importance of covariates in explaining such heterogeneity. Heterogeneity is essentially the problem that there are, still unidentified, covariates that make the model correct, but since we have not been able to include them in the model, the model we actually

Understanding Biostatistics, First Edition. Anders Källén.
© 2011 John Wiley & Sons, Ltd. Published 2011 by John Wiley & Sons, Ltd.

Box 9.1 On different types of mathematical models

A mathematical model is a set of mathematical expressions which constitute a simplification of reality in order to expose some important aspect of it. Very often a key part is the estimation of some parameters describing the model. There is a crude division of models into mechanistic and empirical models (though different names may be used).

A *mechanistic* model is a model which is constructed in a search for the basic mechanisms underlying the processes that are studied. The main purpose is therefore related to *understanding* and strives to describe the structure of the underlying mechanism and the laws governing them in the most general way possible. This typically requires abstraction and idealization in order to eliminate the specific circumstances of the particular situation.

An *empirical* model is used to make predictions that can guide behavior. In this case quality is not measured by how 'true' something is, but by its ability to provide useful predictions in the special circumstances of the situation. These models only need to provide a good approximation of reality in the relevant situation, and may be based on a convenient and flexible family of models from which we select the one that provides the best fit to data. As an example, a linear regression may be perfectly adequate for the problem under consideration when we study only a limited range of options, and when it cannot provide a reasonable fit outside this range.

---

analyze is not correct. What is the effect of such misspecification, of omitting these variables from the analysis? The answers differ. We will see that for some models this is a question of precision in statements, for others more a question about bias.

9.2   The purpose of mathematical models

A mathematical model that relates an outcome y to some input data x is essentially a function $y = f(x)$. The modeling process is the identification of both the vector x and the functional relationship $f(x)$. We assess how good the model is by performing an experiment that provides us with data (x_i, y_i) and looking at the residuals $y_i - f(x_i)$. This is zero for all data points if the model is correct and there are no measurement errors in the data. In real life this may only be true on average, and the residuals may not be precisely zero. In many applications, in particular in physics, the model is defined implicitly, possibly as solutions to differential equations, and the only sources of errors are those from measurements. In biology in general, and medicine in particular, a true functional relationship is usually unattainable, due to biological variation and measurement errors. This is a serious practical complication, but philosophically the problem is the same: can we find data, and a functional relationship for these data, to enable us to predict a particular outcome? Not surprisingly, there are many aspects of this modeling that should be taken into account, one of which is the purpose of the model (Box 9.1), what aspect of the distribution of the outcome it is that we want to model. In this chapter our focus will be on modeling the mean of a distribution.

First, a word about terminology. When we relate the outcome to the input, the purpose of the input is to act as either explanatory variables or predictors. The difference is one of words only. A variable is an explanatory variable if it explains some of the variability we have in the outcome, and it is a predictor if we use its value to get a more refined prediction of the outcome than not knowing it would provide. The modeling is the same whichever word we use, but the prediction terminology helps us better understand what we do. With no predictors available, and forced to guess the outcome, our most likely guess would be the mean. The reason why this is a sensible guess is that it minimizes the expected value of the squared residuals (we could also use the median, which means minimizing the expected value of the absolute residuals, but that is mathematically more complicated). If we have a model that relates the outcome to a particular predictor, and we have measured that predictor, we would instead use the conditional mean of the outcome, given the value of the covariate for prediction. The model is good if this increases the precision in our prediction. Another term we have used in earlier chapters is 'confounder', used primarily in epidemiology. A confounder is an explanatory variable, other than the one that is under investigation, which is a predictor of the response. In a non-epidemiology context the corresponding term is often 'covariate'. In the same way as the exposure we study is not a confounder, the explanatory variable we have designed an experiment to investigate is often not included among the covariates. On other occasions they are. We will mostly use the term 'covariate' in our discussion for any of these concepts, and in a wide sense.

We can divide covariates into those that are fixed, like gender and similar variables, that stay fixed if we repeat the experiment on the same subject, and those that are random, which means that if we take a new measurement of such a covariate, we will probably get a new value. Examples of random covariates include baseline measurements of outcome variables such as blood pressure or lung function. However, in this chapter all analysis will be conditional analysis on observed covariate values, which means that we also consider the random covariates to be fixed. As a consequence we can, strictly speaking, only generalize the results to situations with the same set of covariate values, and must use other means to generalize to the general population. This may be an important point philosophically, but is mostly ignored in practice.

Suppose we are comparing some treatments, or exposures, on an outcome variable with a Gaussian distribution and that we have a number of covariates we may wish to include in a linear statistical model. Why should we contemplate doing so? Here are some reasons.

1. To adjust for inherent differences among comparison groups in order to reduce bias. This is of particular importance in studies where groups cannot be balanced by the use of randomization or matching, as is the case in many observational studies.

2. To generate more powerful statistical tests through variance reduction, which will take place if an appropriate covariate explains some of the variation present. Adjustment for baseline measurements in a randomized experiment is an example of this.

3. To induce equivalence of comparison groups that are generated by randomization. Randomization guarantees approximate equivalence, but statistical adjustment can offset minor imbalances for important predictors for the outcome variable.

4. To clarify to what extent treatment effects are explained by other factors, potentially leading to a change in the interpretation of treatment effects. Conversely, the lack of explanatory factors would help to substantiate the independent existence of treatment effects.

5. To study the degree to which findings are uniform across subpopulations. For example, if treatment effects are considered to be specific to certain age groups, an assessment of the interaction between treatment and age would help to clarify whether an overall treatment effect can be generalized to all ages.

We have emphasized that the setting was that of classical ANOVA (or ANCOVA), which is not fully applicable to many other models, in particular the second item on the list. This is because some of the arguments require a parameter capturing unexplained variance, which can be reduced in size by identifying predictors. Many models do not have such parameters, classical logistic regression being one example. As we will see, the inclusion/exclusion of a covariate then becomes a more complex business, and is more related to bias than precision. But the above list summarizes the main intuitive reasons why we want to include covariates in statistical models.

By way of introduction to the rest of this chapter, which is about estimation methods, we will look a little more deeply into the conditional test in Section 8.8, the ANCOVA.

Example 9.1   For the outcome variable we have two explanatory variables:

1. An indicator variable Z which defines group membership. It takes the value 1 for a subject in group A and the value -1 for a subject in group B.

2. The baseline measurement of the outcome variable, denoted X.

We now introduce the following notation. Let x_1 denote a one for all observations (this is introduced for notational convenience), let x_2 be the indicator for group membership (the observation of Z) and let x_3 be the observed difference of X from the observed average $\bar{x}$ in the whole sample. In this notation the linear model for the conditional mean of the outcome variable is given by the expression

$$E(Y|X = x) = \theta_1 x_1 + \theta_2 x_2 + \theta_3 x_3.$$

To estimate the vector $\theta = (\theta_1, \theta_2, \theta_3)$ we can use least squares, which means that we minimize the quadratic form

$$Q(\theta) = \sum_i (y_i - (\theta_1 x_{i1} + \theta_2 x_{i2} + \theta_3 x_{i3}))^2.$$

A short computation shows that this is equivalent to solving the following three equations:

$$\sum_i x_{ij} y_i = \theta_1 \sum_i x_{i1} x_{ij} + \theta_2 \sum_i x_{i2} x_{ij} + \theta_3 \sum_i x_{i3} x_{ij}, \quad j = 1, 2, 3.$$

We do not write down the explicit solution to this system of equations here but note the following:

1. All the θ_i will be Gaussian variables, since they are linear combinations of the y_i, which were assumed to have Gaussian distributions.

2. The θ_i are correlated.

Once we have found the θ_i we can compute the baseline adjusted group means, which are $\hat{m}_A = \hat{\theta}_1 + \hat{\theta}_2$ for group A and $\hat{m}_B = \hat{\theta}_1 - \hat{\theta}_2$ for group B. What we have now are group

---

Box 9.2 Matrix algebra and linear models

Matrix algebra offers a useful notation for linear models, where the regression function takes the form $f(\theta, x) = x\theta$. In matrix notation this can be written as $f(\theta, x) = A\theta$, where the matrix A consists of the different row vectors x_i of covariate data and θ is a column vector. A is called the *design matrix* for the problem, and the equation that defines the least squares estimator $\hat{\theta}$ becomes

$$A^t(y - A\theta) = 0.$$

Explicitly this means that

$$\hat{\theta} = (A^t A)^{-1} A^t y,$$

from which we have that $E(\hat{\theta}|A) = (A^t A)^{-1} A^t m$, which equals θ_0 if the model $m = A\theta_0$ holds true. Its variance is given by

$$V(\hat{\theta}|A) = (A^t A)^{-1} A^t \Lambda A (A^t A)^{-1},$$

where Λ denotes the diagonal matrix for which the ith element on the diagonal is $\sigma^2(x_i)$. In the important special case where $\sigma^2(x) = \sigma^2$ is independent of x, this simplifies to $V(\hat{\theta}|A) = \sigma^2(A^t A)^{-1}$, and if our outcome data y come from a Gaussian distribution, then

$$\hat{\theta} \in N(\theta, \sigma^2(A^t A)^{-1}).$$

If we instead use weighted least squares (WLS), the estimating equation becomes

$$A^t \Lambda^{-1}(y - A\theta) = 0,$$

and the WLS estimator is given by

$$\hat{\theta} = (A^t \Lambda^{-1} A)^{-1} A^t \Lambda^{-1} y.$$

In this case $\hat{\theta}$ has a Gaussian distribution with mean θ and variance $(A^t \Lambda^{-1} A)^{-1}$.

---

means corresponding to groups that are as equal as our data (and model) allow them to be, except for the treatment. They should therefore provide the fairest test for a treatment effect.

The actual treatment effect can be expressed in different ways. One is the obvious difference $m_A - m_B$, another is the mean ratio m_A/m_B. With the estimates above, and the associated covariance matrix, it is easy to derive confidence intervals (and corresponding p-values) for any of these. For the difference we have that $m_A - m_B = 2\theta_2$, so knowledge about this comes directly from the original parameters. For the ratio we apply Fieller's method (see Box 7.7), after we have identified the bivariate Gaussian distribution for $(\hat{m}_A, \hat{m}_B)$.

The notation in the example becomes much simpler if we use matrix algebra. It is not that the actual computations change, but they can be organized in a way that simplifies the analysis and allows for easy computer implementation of more general statistical models. The basic steps are outlined in Box 9.2.

In Example 9.1 we should note that we go through a two-step process. We first obtain updated mean estimates for the two groups, which are adjusted to the same value of the baseline covariate. We then analyze these updated estimates in a second analysis step. We will illustrate this two-step way of thinking further, first in this chapter with a simple logistic regression example, and then in the next chapter, when we discuss dose–response relationships.

9.3   Different ways to do least squares

Given a set of explanatory variables $X = (X_1, \ldots, X_k)$, we wish to find a function $f(X)$ such that if we have the observation x of X, the value $f(x)$ is a good prediction of what the outcome variable Y will be. In order to determine the function we need to understand how we decide when a predictor is a good predictor. In what sense is it good? Somehow it should mean that the residuals $Y - f(X)$ are 'small', but how do we measure that? One suggestion was given in Section 7.3, namely that we should minimize the expected value of the squared mean residuals, which implies that $f(x) = E(Y|X = x)$, the conditional mean of Y given the information that $X = x$. This is not the only way to define what a good predictor should be, but it is a much used method, and mathematically convenient, so we will stick to it. It means that we utilize the method of least squares which we have seen was developed by Gauss who, in 1823, replaced his first proof (which we outlined in Box 4.10) with a new one, partly because he considered the assumption of a Gaussian distribution too narrow. Instead he justified the method of least squares, assuming a linear function for the predictor, as the method that gives the linear unbiased estimate with the smallest variance.

So, let $f(x) = E(Y|X = x)$ denote the conditional mean, which minimizes the expression $E((Y - f(X))^2)$. Denote the conditional variance $V(Y|X = x)$ by $\sigma^2(x)$. In general we may not know $f(x)$ (we do not know $\sigma^2(x)$ either, but ignore that for a moment) so we replace it with a function $f(\theta, x)$ of some specified type, dependent on some unknown parameter vector θ which is to be estimated. The idea is that there is a true parameter value θ_0 such that

$$E(Y|X = x) = f(\theta_0, x),$$

which in many cases will only be an approximation. By definition, θ_0 minimizes the mean of the squared residuals, which is the function $Q(\theta) = E((Y - f(\theta, X))^2)$, and is therefore a solution to the equation

$$Q'(\theta) = E(f'(\theta, X)^t (Y - f(\theta, X))) = 0.$$

Here f' refers to differentiation with respect to θ. In order to estimate θ_0 from a set of data points $(x_1, y_1), \ldots, (x_n, y_n)$ (where each x_i is a k-tuple), we use the estimating equation

$$\frac{1}{n} \sum_{i=1}^{n} f'(\theta, x_i)^t (y_i - f(\theta, x_i)) = 0,$$

obtained by replacing the CDF in the formula for the expected value with the corresponding e-CDF. The solution of this is the *least squares* (LS) estimate of θ. The simplest case would be to take a linear function in x, that is, $f(\theta, x) = x\theta$ for a parameter vector θ, where the row vector x may start with a one in order to account for an intercept. Such functions can be justified either because they are as simple as they can be, or because we approximate $f(\theta, x)$ with a linear function in x.

Box 9.3 Why is analysis of means called analysis of variance?

The analysis of conditional means of an outcome variable on covariates when the relationship is a linear one (mathematically $E(Y|X = x) = x\theta$), is called analysis of variance if the covariates all are categorial (called factors), and it is called analysis of covariance if at least one of the covariates is continuous and enters the model in such a way that the mean is proportional (the slope) to its value. The reason why it is called an analysis of variance in the former case is that the analysis consists of splitting sums of squares into parts defining the estimation of factor effects, plus a final part of unexplained noise, the residual sums of squares. Sums of squares estimate variances, which explains the terminology. Such methods once led to convenient ways of doing the analysis by hand when the data were balanced, but with present-day technology the matrix formulation indicated in Box 9.2 is more powerful and easily implemented in computer software.

There is also a geometric formulation of the sums of squares approach, applicable for all types of covariates, which allows a compact mathematical theory. In this formulation data are considered to be part of a large Euclidean space, with a metric defined by the covariance structure of the Gaussian model. Conditional means are then orthogonal projections on subspaces representing different sub-models. By expressing such spaces in matrix formulation, calculations with this approach are equivalent to the matrix formulation.

We can also use this method if the variance structure is not constant, but depends on the covariate(s). There is, however, an alternative and better way to approach this. In that method we look for the θ that minimizes the *weighted least squares*

$$E\left(\frac{(Y - f(\theta, X))^2}{\sigma^2(X)}\right).$$

An estimate of θ from data is obtained by minimizing the estimate of this expression:

$$Q(\theta) = \frac{1}{n}\sum_{i=1}^{n}\sigma(x_i)^{-2}(y_i - f(\theta, x_i))^2.$$

At this point we digress slightly to make a comment on the choice of weights. In certain applications (modeling in pharmacokinetics is one), a least squares analysis is sometimes done by weighting not on a conditional variance $\sigma^2(x_i)$, but on either the value y_i of the outcome variable, or its square y_i^2. Consider the latter case. Since $\ln x_2 - \ln x_1 \approx x_1^{-1}(x_2 - x_1)$ we have that

$$\sum_{i=1}^{n} y_i^{-2}(y_i - f(\theta, x_i))^2 \approx \sum_{i=1}^{n}(\ln(y_i) - \ln(f(\theta, x_i)))^2.$$

It is therefore more natural to do the actual estimation on a log scale instead in this case. This may be because the original distribution resembles a lognormal rather than a normal distribution. Similarly, a variance proportional to y_i would mean that we should do the analysis on the square root of data instead of on the raw data. This is the case for Poisson data, for which the square root transformation has a 'normalizing' effect. End of digression.

The notation above indicates that we know what the variance is as a function of the covariates. In general this may not be true; there may be additional unknown parameters ϕ in the variance, which we therefore write as $\sigma^2(x, \phi)$. The standard ANOVA serves as an example, where the variance is assumed constant but unknown, so that $\sigma^2(x, \phi) = \phi$ (which is σ^2). When ϕ enters as a proportionality factor like this, its value does not affect the estimation of θ, but in other cases it will. However, we can never ignore ϕ altogether; it contributes to the variance of the estimator, so when we want to describe our confidence in θ from data we need to take it into account.

The situation is different if there is an overlap between θ and ϕ, which occurs, for example, when the variance depends on the mean. This is the case for both binomial and Poisson data. Since those parts of ϕ that are not in θ do not really play any role, we omit them from notation and assume that the conditional variance of Y given that $X = x$ is given by an expression $\sigma^2(\theta, x)$. The quadratic form $Q(\theta)$ then reads

$$Q(\theta) = \frac{1}{n} \sum_{i=1}^{n} \sigma(\theta, x_i)^{-2} (y_i - f(\theta, x_i))^2.$$

This is a more complex situation to handle in full generality, because the derivative of $Q(\theta)$ is not nearly as simple as before. If we differentiate $Q(\theta)$ we find that the derivative is the sum of two terms, namely

$$U(\theta) = \frac{1}{n} \sum_{i=1}^{n} \sigma(\theta, x_i)^{-2} f'(\theta, x_i)^t (y_i - f(\theta, x_i)) \tag{9.1}$$

and a term involving the derivative of the variance. When the overriding purpose of the analysis is to get a good fit of y to $f(\theta, x)$, it is tempting to replace the estimating equation $Q'(\theta) = 0$ with the equation $U(\theta) = 0$. The solution to this equation is called the *generalized least squares* (GLS) estimate. It is not as arbitrary as it may seem at first glance, because there is a whole family of estimation problems for which this is the 'right thing' to do, which we will discuss later in this chapter.

As an endnote to this, it is not difficult to obtain confidence statements about the GLS estimate above, because the variance of the stochastic variable $\sigma(\theta, x)^{-2} f'(\theta, x)(Y - f(\theta, x))$ is

$$\sigma(\theta, x)^{-4} f'(\theta, x)^t V(Y|X = x) f'(\theta, x) = \sigma(\theta, x)^{-2} f'(\theta, x)^t f'(\theta, x),$$

where we have used the fact that $V(Y|X = x) = \sigma^2(\theta, x)$. From this the variance of $U(\theta)$ is easily derived, and we get the confidence function $C(\theta) = \Phi(U(\theta)/\sqrt{V(U(\theta))})$, based on large-sample theory.

9.4   Logistic regression, with variations

To illustrate the general discussion, we look at the example of accounting for covariates in a model involving binomial distributions. We will do this with a view to estimating not probabilities, but odds ratios. The definition of the odds ratio, $OR = p_1(1 - p_2)/[p_2(1 - p_1)]$,

Box 9.4 The logistic distribution

Logistic regression is associated with the logistic function

$$\Psi(x) = \frac{1}{1 + e^{-x}},$$

which defines a distribution function with some interesting properties. One such property is that it satisfies the *logistic equation*

$$\Psi'(x) = \Psi(x)(1 - \Psi(x)). \qquad (9.2)$$

This equation is much used in ecology to describe a population that undergoes exponential growth until it experiences crowding effects that limit its ultimate size (here normalized to one) and is derived from the equation for exponential growth, $\Psi'(x) = r\Psi(x)$, with a non-constant growth rate $r = 1 - \Psi(x)$. It was introduced by Verhulst in 1845, though the name came into general use 30 years later and is derived from the French *logistique*, referring to the lodgement of troops.

The logistic distribution closely resembles a Gaussian distribution, in fact, $\Psi(x)$ is almost indistinguishable from the Gaussian distribution $\Phi(ax)$, where $a = 16\sqrt{3}/15\pi \approx 0.59$, as the following graph shows:

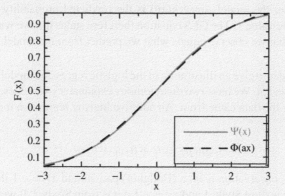

The family of logistic distributions defined by $\Psi_{\lambda,\gamma}(x) = \Psi(-\ln(\lambda) + \gamma x)$ is closed under the proportional odds model, because

$$\theta \frac{\Psi^c_{\lambda,\gamma}(x)}{\Psi_{\lambda,\gamma}(x)} = \theta\lambda e^{-\gamma x} = \frac{\Psi^c_{\theta\lambda,\gamma}(x)}{\Psi_{\theta\lambda,\gamma}(x)}.$$

This explains why we perform a logistic regression when we want to analyze odds ratios.

---

means precisely that

$$\ln OR = \ln \frac{p_1}{1 - p_1} - \ln \frac{p_2}{1 - p_2}.$$

Introduce the notation

$$\text{logit}(p) = \ln \frac{p}{1-p}$$

for the log-odds and let $\theta = \ln OR$ be the log-odds ratio. What we have written above can then also be written $\text{logit}(p) = a + x\theta$, where x is $+1$ for group 1, and 0 for group 2, and a is the log-odds for group 2 (we can also write it in other ways, for example with x taking values ± 1 instead, which changes the meaning of the parameters a and θ but gives the same result if we account for this change in meaning). This type of model is called a (linear) *logistic* model. The general assumption is that the value of a binomial parameter p in a population depends on observed covariate values x in such a way that $\text{logit}(p) = x\theta$. The solution to this equation can be written as $p(\theta, x) = h(x\theta)$, where $h(u) = 1/(1 + e^{-u})$ is called the logistic function and is briefly presented in Box 9.4. The odds ratio θ is estimated by solving the GLS equation

$$U(\theta) = \frac{1}{n} \sum_{i=1}^{n} x_i(y_i - h(x_i\theta)) = 0.$$

The sum here is the difference between two sums. The first is $\sum x_i y_i / n$ which is, since y_i is a 0/1 variable, the mean value of the covariate values for the records with $y_i = 1$, times the fraction of these among the total number of subjects. The second sum is the corresponding predicted value from the model, since $h(x\theta)$ is the predicted probability of an event for a record with covariate value x. The GLS equation therefore states that we wish to find the odds ratio θ for which what we observe equals what we predict from the model using θ.

Example 9.2 In order to give an illustration of the logistic regression model we revisit the data tabulated in Example 5.4. We have two dichotomous explanatory variables, the tonsillectomy status and the study the data came from. An additive logistic regression model for these data can be written

$$\text{logit}(p) = \theta_1 + \theta_2 x_1 + \theta_3 x_2,$$

where $x_1 = 1$ if the observation is on a (Hodgkin) patient and $x_1 = -1$ if it is on a control, whereas $x_2 = 1$ if it is from Study 1 and $x_2 = -1$ if it is from Study 2. If we insert this into the estimating equation and solve it, we find the parameter estimates $\theta_1 = -0.060$, $\theta_2 = 0.192$ and $\theta_3 = 0.400$. Figure 9.1 compares the predictions of the model, which are the (circle) symbols connected with lines, with the corresponding empirical log-odds, which are the free-floating (triangle) symbols. That the two lines are parallel is precisely what the model specifies and the vertical distance between patients and controls defines the common logged odds ratio. This means that the odds ratio is given by $e^{2\theta_2}$, from which we derive the estimate 2.22 with 95% confidence interval (1.73, 2.86). This is almost the same as the Mantel–Haenszel method gave in Section 5.3.

We can also introduce an interaction term into the model. This means that we add one more parameter θ_4, such that

$$\text{logit}(p) = \theta_1 + \theta_2 x_1 + \theta_3 x_2 + \theta_4 x_1 x_2.$$

This extra parameter measures the interaction between studies and groups; it allows us to adjust the two lines in Figure 9.1 so that they connect the observed log-odds (we say we

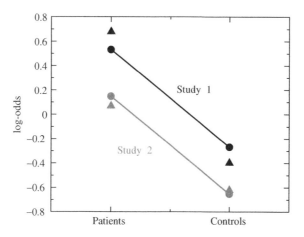

Figure 9.1 Data and model fit for a logistic regression for a meta-analysis of two studies on Hodgkin's lymphoma.

have a saturated model) and are no longer parallel. Testing the hypothesis $\theta_4 = 0$ that there is no interaction between study and patient group corresponds to the Breslow–Day test in Section 5.5, and gives the same p-value. A little algebra shows that $4\theta_4$ is geometrically the difference between the difference in the log-odds for patients and the difference for the controls.

The results in the example are obtained assuming binomial distributions for data, but were found to be very similar to what the Mantel–Haenszel approach gives, which is based on the hypergeometric distribution, a conditional distribution for two independent binomials. This conditional analysis is mainly used for case–control studies, since the meaningful parameter is then the odds ratio. To expand on this, consider the situation where we have a case–control study examining a particular exposure. We also have some additional confounders that we want to adjust for in the analysis. Just as we can obtain Fisher's exact test from a 2×2 table of two binomials by constructing a conditional test, we can derive a similar test for this situation from the logistic regression model. To see how, we rederive the exact test in a more general notation. A logistic regression model is defined by the probability function $p(y)$ defined by

$$\prod_i \binom{n_i}{y_i} \frac{e^{y_i(\alpha + x_i \beta)}}{(1 + e^{(\alpha + x_i \beta)})^{n_i}} = e^{\alpha y_+ + yx\beta} \prod_i \frac{\binom{n_i}{y_i}}{(1 + e^{(\alpha + x_i \beta)})^{n_i}}, \tag{9.3}$$

where $yx = \sum_i y_i x_i$ as before. Here we have assumed that there are n_i observations with covariate vector x_i and we have written the linear term as $\alpha + x\beta$ in order to single out the intercept α (x and β do not need to be univariate). The conditional distribution of Y on the condition $Y_+ = r$ eliminates α, but is somewhat involved from a notational perspective. The key observation is that the probability for the event $Y_+ = r$ is given by the sum of the probabilities $p(z)$ for all z such that $z_+ = r$, showing that the conditional distribution of Y

given that $Y_+ = r$ is given by

$$\frac{e^{yx\beta} \prod_i \binom{n_i}{y_i}}{\sum_{\{z:z_+=r\}} \prod_i e^{zx\beta} \binom{n_i}{z_i}}.$$

In the special case where we have no additional covariates except exposure, and therefore only one explanatory variable, with values 0 and 1, this probability function describes a hypergeometric distribution.

Regression analysis of β using the general distribution above is referred to as conditional logistic regression. To compute the estimating equation for β derived from this, using maximum likelihood, is rather computer-intensive, because we need to identify and calculate the sum in the denominator. Once this is done, however, we can choose to proceed by appealing to large-sample likelihood theory to derive approximate inference about β. Alternatively, we can use combinatorial arguments.

However, this does not correspond to the Mantel–Haenszel test. We have used the variable study as a covariate, whereas Mantel–Haenszel uses it to define a stratification. But we can generalize to that situation too. Suppose that we want to assess the effect of an exposure in a case–control study, and have at our disposal a series of strata, within each of which we have a 2×2 table with exposure-response numbers. The logistic regression model for this is

$$\text{logit}(p_i) = \alpha_i + x\beta,$$

where p_i is the probability for a case in stratum i and x a single 0/1 variable. In addition, we have a series of stratum-specific intercepts. It is only β that is of interest here; all the α_i are nuisance parameters. When we compute the probability function for this case, we note that each 2×2 table contributes two factors in the product formula in equation (9.3), one for each of the two values of x. The probability function is therefore the following product over strata:

$$\prod_i \frac{\binom{n_{i1}}{y_{i1}} \binom{n_{i2}}{y_{i2}} e^{\alpha_i y_{i+} + \beta y_{i1}}}{(1 + e^{\alpha_i + \beta})^{n_i}}.$$

To eliminate the α_i we compute the conditional distribution on all the conditions $y_{i+} = r_i$ corresponding to what we do for Fisher's exact test for each individual table. With the same argument as above, we find that the conditional distribution is

$$\prod_i \frac{\binom{n_{i1}}{y_{i1}} \binom{n_{i2}}{r_i - y_{i1}} e^{\beta y_{i1}}}{\sum_k \binom{n_{i1}}{k} \binom{n_{i2}}{r_i - k} e^{\beta k}},$$

which means (not surprisingly) that this is the product of a series of non-central hypergeometric probabilities. The conditional logistic regression is therefore equivalent to the Mantel–Haenszel method in terms of model; what differs slightly is the way the actual parameter estimation is performed. To solve for β in the expression above, we consider the logarithm of the probability function as a function of β and differentiate it, to obtain the estimating equation $\sum_i (y_{i1} - E_\beta(y_{i1})) = 0$. In terms of estimation, the conditional logistic regression approach in this case is therefore the same as the use of the Mantel–Haenszel quadratic form $Q_{\text{MH}}(e^\beta)$ in Section 5.5.

The logistic model is not the only possible model for binomial data. Instead of using the logistic function as $h(u)$ in the relation $p(\theta, x) = h(x\theta)$ we can use other functions. The corresponding estimating equation becomes a weighted version of what constitutes the estimating equation for the logistic model:

$$U(\theta) = \sum_{i=1}^{n} w_i(x_i\theta)x_i(y_i - h(x_i\theta)) = 0, \qquad w_i(u) = \frac{h'(u)}{h(u)(1 - h(u))}. \qquad (9.4)$$

We have seen in Box 9.4 that the logistic function is almost indistinguishable from the function $\Phi(au)$, where $a = 0.59$. To use the cumulative Gaussian as response function instead is therefore an almost indistinguishable alternative, though the GLS equation becomes more complicated. Such a model is called a probit model. Translating regression coefficients between the two models involves the use of the constant a.

There is one very good aspect of the probit model, compared to the logistic model. Suppose that we have two binary outcomes, each of which we want to model with the linear model $a + bx$, allowing for different regression coefficients for the two outcomes. Suppose further that the two outcomes are correlated in some way. We can then simultaneously analyze the two outcomes by replacing the univariate response function $\Phi(u)$ with the bivariate standardized Gaussian CDF $\Phi_2(u, v; \rho)$. If we organize the outcomes for each covariate value in a 2×2 table, we can model this in a way similar to what we did in Example 7.4.1 when we introduced the tetrachoric correlation coefficient. The model in question is

$$p(1, 1|x) = \Phi_2(a_1 + b_1x, a_2 + b_2x; \rho),$$
$$p(1, 0|x) = \Phi(a_1 + b_1x) - \Phi_2(a_1 + b_1x, a_2 + b_2x; \rho),$$
$$p(0, 1|x) = \Phi(a_2 + b_2x) - \Phi_2(a_1 + b_1x, a_2 + b_2x; \rho).$$

A similar extension does not come naturally for the logistic model.

Another useful response function for binomial data is $h(u) = 1 - e^{-e^u}$. This is the function we should use when we are primarily interested in rates rather than proportions (see Section 2.7). The logistic regression model is the appropriate model for the analysis of odds ratios, but this model is the appropriate model to use when we analyze hazard rates, recording only whether or not the event has occurred during the observation period. A regression model for the hazard would then typically be of the form $\lambda(x)T_i = e^{\ln T_i + x\beta}$, where T_i is the observation time for subject i. This means that the variable $\ln T_i$ is part of the linear model, but with a known coefficient, namely one. Such a variable is called an *offset* variable.

9.5 The two-step modeling approach

When we discussed the weighted least squares approach to estimation above we assumed that the n observations in the data set were independent. We can, however, easily extend to the case when the observations are dependent, as long as we account for the correlation between the observations. This provides the building block for the two-step approach to modeling that was indicated earlier and will revisited in the next chapter. The basic idea is that if the observations y_i are gathered into an n-vector y, $f(\theta)$ denotes the n-vector with components previously denoted $f(\theta, x_i)$ and $V(\theta)$ is the variance matrix of the outcome variable (possibly dependent not only on θ, but also auxiliary parameters not included in the notation and assumed known in the present discussion), then the GLS estimate of θ is obtained by solving the equation

$$f'(\theta)^t V(\theta)^{-1}(y - f(\theta)) = 0.$$

To illustrate the two-step approach to analysis, suppose we have p groups and a p-vector $m = (m_1, \ldots, m_p)$ of true group means, and assume that we have an estimator $\hat{m}$ of m such that $\hat{m} \in N(m, \Sigma)$, where the variance matrix Σ is known. That we may get correlations when this is the output of a non-trivial first analysis is to be expected, but that Σ should be known is unrealistic. It is a convenient assumption for the moment, soon to be relaxed. Next we impose a linear model on m, so that $m = B\theta$. This is not an equation with a unique solution, because there are usually fewer parameters in θ than in m, so the design matrix B has fewer columns than rows. The GLS equation (actually WLS in this case) for θ is given by

$$B^t \Sigma^{-1}(\hat{m} - B\theta) = 0,$$

which is a linear system with solution

$$\hat{\theta} = (B^t \Sigma^{-1} B)^{-1} B^t \Sigma^{-1} \hat{m} \in N(\theta, (B^t \Sigma^{-1} B)^{-1}).$$

All this is a summary of some matrix algebra discussed earlier. What is different now is that we will have the estimate of the m-vector as the result of a preliminary analysis, and what we have just discussed constitutes a second step in the analysis. Why we want to do this may be somewhat clarified if we revisit Example 9.2.

Example 9.3   In a two-step analysis of the data in Example 9.2 we first compute for each of the four groups the (empirical) log-odds and its standard deviation (see Section 4.6.1). These are shown in the following table (standard deviation in parenthesis; all correlations are zero by design).

Patients, Study 1	Controls, Study 1	Patients, Study 2	Controls, Study 2
0.678 (0.211)	0.069 (0.152)	−0.398 (0.197)	−0.621 (0.097)

The standard errors are only estimates, but if we assume that they are exact, we can apply the analysis discussed above. The model, which is illustrated by the straight lines in Figure 9.1, is obtained if we apply the model $m = \theta_1 + \theta_2 x_1 + \theta_3 x_2$ to the log-odds m. (As before, x_1 is 1 for patients and −1 for controls and x_2 takes value 1 for Study 1 and −1 for Study 2.) In other words, putting the log-odds in a vector in the order of appearance in the table above, we define the model $m = B\theta$, where

$$B = \begin{pmatrix} 1 & 1 & 1 \\ 1 & -1 & 1 \\ 1 & 1 & -1 \\ 1 & -1 & -1 \end{pmatrix}.$$

It follows that the estimate of $\theta = (\theta_1, \theta_2, \theta_3)$ is given by

$$\hat{\theta} = (B^t \Sigma^{-1} B)^{-1} B^t \Sigma^{-1} \begin{pmatrix} 0.678 \\ 0.069 \\ -0.398 \\ -0.621 \end{pmatrix} = \begin{pmatrix} -0.061 \\ 0.192 \\ 0.399 \end{pmatrix},$$

where Σ is the diagonal matrix with the squared standard errors on the diagonal. We see that the result is almost identical to what we found in Example 9.2. The slight difference is explained by the difference between the Σ we use here, which is the one obtained from individual groups, and the corresponding Σ used in the iterative process of the original model, which is based on the model.

We may note in passing that if we add another column to B, consisting of zeros except for the last element which is a one, and redo the analysis, we reproduce the saturated model in Example 9.2. In this case this is an invertible transformation from the four group mean log-odds to four new parameters, and for this case the variance matrix estimate is precisely the same in the two analyses, and the results therefore identical.

In the first step of this example we reduced all data to estimates of a few parameters, accompanied by an estimate of the variance matrix for these estimates. This is what is summarized in the table, and is a data reduction step. At this stage we could have taken the opportunity to adjust the estimates for some covariates describing, for example, patient characteristics. Whatever model we use in the first step, the second step is the same.

There may appear to be a price to pay when we do this analysis, since we need to use an estimate of the variance of $\hat{m}$. However, the analysis is (almost) equivalent to the customary analysis presented in Example 9.2, which also uses this approximation. It is not 100% correct, but it is (almost) the standard large-sample approximation. But there are situations when we do not even need to do it approximately. If our original data are from a linear model with Gaussian data with a common variance σ^2, then the analysis in Box 9.2 shows that $\hat{m} \in N(m, \sigma^2 \Lambda)$, where $\Lambda = (A^t A)^{-1}$ is a known matrix derived from the design matrix A. This means that if we use the estimate s^2 of σ^2 from the first model, we get an exact analysis, provided we use the t distribution in the analysis (with the degrees of freedom from the first-step analysis).

Example 9.4   Suppose that we have an estimator $\hat{m}$ which has a bivariate $N_2(m, \Sigma)$ distribution, and that we want to obtain information about m_2, given that we know that $m_1 = 0$. This is the ANCOVA problem, and it is an example of the analysis above. In fact, the model is $m = B\theta$ where $B = \begin{pmatrix} 0 \\ 1 \end{pmatrix}$. The least squares estimate of θ is now given by $\hat{\theta} = (B^t \Sigma^{-1} B)^{-1} B^t \Sigma^{-1} \hat{m}$. The formula for the inverse of Σ is

$$\Sigma^{-1} = (\det \Sigma)^{-1} \begin{pmatrix} \Sigma_{22} & -\Sigma_{12} \\ -\Sigma_{12} & \Sigma_{11} \end{pmatrix},$$

which means that

$$(B^t \Sigma^{-1} B)^{-1} = (\det \Sigma)/\Sigma_{11} = \Sigma_{22} - \Sigma_{12}^2/\Sigma_{11} = \Sigma_{2.1}.$$

Inserting this in the formula above, we find that the estimate of θ is given by $\hat{\theta} = \hat{m}_2 - \Sigma_{12}\hat{m}_1/\Sigma_{11}$. This is what we obtained in Section 8.8.

The two-step method is most useful in cases where there is a nonlinear relationship to be described. From the discussion above we know that the equation to solve in order to estimate θ for the model $m = f(\theta)$ is the GLS equation

$$f'(\theta)^t \Sigma^{-1}(\hat{m} - f(\theta)) = 0.$$

In order to obtain knowledge about θ we can either compute the variance of the stochastic variable on the left here, or use the fact that the original assumptions mean that

$$Q(\theta) = (\hat{m} - f(\theta))^t \Sigma^{-1}(\hat{m} - f(\theta)) \in \chi^2(p).$$

This gives us a confidence function to work with. As above, there are modifications in the detail, depending on what information we have on Σ. If we only have an estimate of Σ we need to modify the distribution appropriately, or appeal to large-sample theory and use the estimated one as fixed to obtain reasonable approximations. Here $\hat{m}$ has been presented as adjusted means from a first-step analysis, but what it really consists of are estimates of parameters from the model of this first analysis.

So far we have discussed how to get simultaneous confidence in all the model parameters. What if we want a simultaneous confidence region for only a subset of these, with the remaining ones considered nuisance parameters? More generally, what if we want a confidence interval for a particular combination of the model parameters, and not a complex confidence region for them all? This question was discussed, albeit in a special case, in Section 7.7, and a very brief and superficial description of the method used, profiling, is as follows. Suppose we want confidence information for a parameter $\eta = g(\theta)$.[1] Instead of minimizing $Q(\theta)$ to get an estimate of θ, we then minimize it under constraints $g(\theta) = \eta$ for different η. This gives us a function $Q(\eta)$ of the new parameter alone, and we can obtain univariate confidence statements about this by using the confidence function $\chi_1(Q(\eta))$. This is a general method that produces reasonable results on most occasions, at least if the sample is not too small.

9.6 The effect of missing covariates

The classical ANOVA model and the logistic model, together with the other models for binomial data discussed above, are members of a larger family of regression models, called generalized linear models (GLMs). One common property of these is that the mean of the outcome variable is modeled in terms of covariates, as an expression of the form $E(Y|X = x) = h(x\beta)$, for some function $h(u)$ called a *response function*. An important question which then arises is the following. Suppose that there is a true model containing one set of covariates, but that we have omitted to measure some of these and model the mean using only the ones we have measured. What are the consequences? In the ANOVA situation we get more unexplained residual variance, but no other effect on what we estimate. In the general case, including the logistic model, the situation is more complicated.

---

[1] This does not have to be one parameter; the function $g(\theta)$ can define more than one parameter, but it is simpler if we think of it as a single parameter.

Specifically, let Y be the outcome variable, X the measured covariates and Z the missing one (we combine any set of omitted covariates into a single one for this discussion). Assume that $E(Y|X = x, Z = z) = h(x\beta + z)$ for some response function $h(u)$ and that the population CDF of Z is given by $G(z)$. If there is an intercept parameter β_0 in the linear model $x\beta$, we assume that $G(z)$ has mean zero, since any non-zero mean can be incorporated into the intercept. Since we have not measured Z, the mean we observe is not $h(x\beta + z)$, but instead

$$E(Y|X = x) = \int h(x\beta + z)dG(z).$$

This new response function may differ considerably in shape from the original. A graphical illustration of this, in a slightly different (but equivalent) setting, is given in Figure 10.4 on page 270 in the next chapter. The model has therefore changed and this poses the following immediate question: if we fit the reduced data to the original model (with the response function $h(u)$; remember that we do not know $G(z)$, so there is not much else we can do), how much do the regression coefficients from that model differ from the original one? We want to compare β to the β^* which makes the function $h(x\beta^*)$ the best approximation to $E(Y|X = x)$, and we want to understand what the difference between β and β^* is. The answer depends on the choice of response function and on the distribution $G(z)$.

Before we look at a few examples, let us link this up with the discussion on individual risks and population risks. We never know what the true individual risk is, the best we can do is to obtain predictive covariates and use the conditional mean in the appropriate subpopulation as the prediction for this. But this varies with how many, and which, covariates we include. There is one prediction $E(Y|X = x, Z = z)$ if we know both X and Z, and another, $E(Y|X = x)$ if we only know the former. The purpose of estimating the regression coefficients β is to understand how sensitive the outcome is to X, and we most often think of that as individual sensitivity. If we use the same model (i.e., response function) in the two cases, we have two inconsistent models, and it should not be a surprise that the regression coefficients shift their interpretation, and therefore their value. They do not do so if the response function is the identity ($h(u) = u$), so in that case, as for ANOVA, the meaning is independent of how many covariates we include in the model.

Example 9.5   Consider the case of an exponential response function, $h(u) = e^u$, and assume that Z has a Gaussian distribution, so that

$$E(Y|X = x) = e^{x\beta} \int_{-\infty}^{\infty} e^z d\Phi(z/\sigma) = e^{\sigma^2/2} e^{x\beta}.$$

This means that patient heterogeneity inflates the conditional mean of Y compared with a homogeneous population. Compared to the equation $E(Y|X = x) = e^{x\beta^*}$, we find that β and β^* coincide except for the constant, for which we have the relation $\beta_0 + \sigma^2/2 = \beta_0^*$. This is the mildest effect we get on the mean when we omit covariates, except for the identity function, when there is no effect.

Other response functions have other effects. The following example gives the effect of omitting covariates for the important case of the logistic regression when the omitted variable is Gaussian.

Example 9.6 Consider first the probit model for which $h(u) = \Phi(u)$ and assume that Z has the Gaussian distribution with mean zero and variance σ^2. Under that assumption,

$$E(Y|X = x) = \int_{-\infty}^{\infty} \Phi(x\beta + \sigma z) d\Phi(z) = \Phi\left(\frac{x\beta}{\sqrt{1 + \sigma^2}}\right).$$

(To evaluate the integral in the middle, note that it is the CDF for the difference between independent $N(x\beta, \sigma^2)$ and $N(0, 1)$ variables.) It follows that $\beta^* = \beta/\sqrt{1 + \sigma^2}$, so all coefficients are affected and regressed toward zero. An approximation for the logistic regression follows from this by appealing to Box 9.4; it gives us $\beta^* \approx \beta/\sqrt{1 + a^2\sigma^2}$, where $a = 0.59$.

Note that the Z we have discussed above is actually γZ, where γ is the true regression coefficient for the unknown covariate, and it is really $\gamma^2\sigma^2$ that enters the expressions above. The effect is therefore the combined effect of the heterogeneity in Z, as measured by the variance, and the predictive power, as measured by γ, of the covariate.

Example 9.7 Consider a 2×2 table describing an exposure-response relationship. The model we have is $\text{logit}(p(x, \xi)) = \xi + \beta x$, where $x = 1$ for an exposed individual and $x = 0$ for a control, and $\theta = e^\beta$ is the odds ratio we require. The ξs are allowed to vary between individuals with a CDF $P(\xi)$. The odds ratio calculated from the population data is then

$$\psi = \frac{P_1(1 - P_0)}{(1 - P_1)P_0}, \quad \text{where} \quad P_x = \int_{-\infty}^{\infty} \frac{e^{\xi+\beta x}}{1 + e^{\xi+\beta x}} dP(\xi).$$

If we approximate the logistic distribution with the Gaussian CDF and assume that the distribution for ξ is $N(\alpha, \sigma^2)$, we see that (approximately)

$$\ln\left(\frac{P_x}{1 - P_x}\right) = \frac{\alpha + \beta x}{\sqrt{1 + a^2\sigma^2}},$$

from which it follows that $\psi \approx \theta^\nu$, where $\nu = 1/\sqrt{1 + a^2\sigma^2} < 1$. This means that, because of the heterogeneity, the population odds ratio we calculate from data (ψ) is expected to be biased toward one, compared with the odds ratio that is relevant for the individual (θ).

This is of particular relevance to a one-to-one matched study, for which we have the probability table (see page 98)

		Controls(C^c)	
		E	E^c
Cases	E	P_{11}	P_{10}
(C)	E^c	P_{01}	P_{00}

where

$$P_{xy} = e^{\beta x} \int_{-\infty}^{\infty} \frac{e^{(x+y)\xi}}{(1 + e^{\xi+\beta x})(1 + e^\xi)} dP(\xi).$$

The assumption here is that the value ξ is common to the case and the control (that is what the matching tries to achieve). We see that $P_{10}/P_{01} = \theta$, so this estimate of the odds ratio is not influenced by the heterogeneity. This explains why, for the second Hodgkin's lymphoma study in Section 4.5.1, we obtained a smaller estimate from the first analysis than from the second. The relation is $1.47 = 2.14^{v}$, from which we get an indication of the heterogeneity by solving for v. However, there is no concomitant increase in precision, the 90% confidence interval for ψ is $(0.88, 2.46)$, whereas that for θ from the matched analysis is $(1.03, 4.47)$ (derived from the Wilson interval for a single binomial parameter by transforming to the odds). This is consistent with the discussion above, and also reflects the fact that fewer observations are used in the second analysis.

The difference between β and β^* may be considered some kind of misspecification bias, because we analyze the wrong model. This may not always be a good use of the word 'bias', because it really only reflects the fact that the overall effect seen in a population depends on how much heterogeneity is left unaccounted for. The more predictive covariates we include, the smaller is this residual heterogeneity. The general observation is that the larger the heterogeneity, the smaller is the effect we see in the population (we have $|\beta^*| \leq |\beta|$ in the notation above, with the difference increasing with the heterogeneity). The exception here are the identity and exponential response functions. A simple regression model in a heterogeneous environment provides estimates, for example treatment estimates, that may be reasonably accurate from a population perspective, but wrong when interpreted as individual effects. This distinction between the population perspective and the individual perspective will play a major role in our discussion on survival data and the Cox model later. It is also further discussed in the next chapter, where we discuss the difference between a subject-specific and a population averaged approach to the description of dose response.

We may also note that because of the general observation that the variance can be decomposed as $V(Y) = E(V(Y|Z)) + V(E(Y|Z))$ (applied to the conditional distribution of Y given that $X = x$), the (conditional) variance of Y always increases with omitted covariates, including the otherwise harmless case with the identity response function. However, this does not imply that the precision in the estimated regression coefficients has to increase when we include more predictive covariates, if they get a new meaning.

9.7   The exponential family of distributions

The rest of this chapter is more mathematical in character. It is about a particular drive in mathematics – the wish to generalize and systematize, to see what is common to a number of particular cases and find a general formulation which treats these as special cases. We seek a general theory, including all the proofs necessary, which we can apply to the particular cases, without the need for individual proofs. This is something that appeals to mathematicians, and statistics is a subdiscipline of mathematics. We therefore wish to take this opportunity to formulate as part of a general framework the regression theories so far encountered. This will give us the tools to find variations of these, applicable to specific problems (not that we will make much use of these tools, but at least we will be able to if we wish).

We will use the following definition. A distribution is said to belong to the *exponential family* of distributions if its CDF can be written in the form

$$dF_\phi(x, \theta) = e^{(x\theta - \kappa(\theta))/\phi} dF_\phi(x) \tag{9.5}$$

for a parameter vector θ and a positive scalar ϕ. Here $\kappa(\theta)$ is a function of θ alone and the reference function $F_\phi(x)$ (which does not have to be a CDF) does not depend on θ (but is allowed to depend on ϕ). The parameter ϕ is called the *dispersion parameter* of the distribution. The first and obvious example is the case where we take the dispersion parameter to be one, $dF_\phi(x)$ to be one when $x > 0$ and zero otherwise, and $\theta = -\lambda$. The result is the probability density function $dF(x, \lambda) = \lambda e^{-\lambda x}$, $x > 0$, which is the exponential distribution.

The definition above is not the only definition possible for the exponential family. The most common definition is probably to write its density as

$$a(\theta)e^{Q(\theta)T(x)}dF(x)$$

for some reference function $F(x)$. The form in equation (9.5) is the special case when we (1) use $Q(\theta)$ as parameter instead of θ, (2) consider the distribution of $T(X)$ instead of that of the original variable X, and (3) introduce the additional extra dispersion parameter ϕ. The form in equation (9.5) is called the canonical form for the family and the parameter θ is called the natural (or canonical) parameter for the distribution. If we understand the canonical form for the family, we understand the general exponential family. Some key examples of distribution families found within the exponential family are listed in Box 9.5. Of particular interest here is that for the binomial distribution the natural parameter θ is not the proportion p, but the log-odds $\ln(p/(1 - p))$.

The examples in Box 9.5 constitute only a sample, but not all distributions encountered so far belong to the exponential family. The t distribution is one exception, another is the logistic distribution, but the logistic function plays a fundamental role for the binomial distribution, since it maps the natural parameter (the log-odds) to the binomial proportion.

All these examples are univariate distributions. For any of them it is the case that if we have a sample of n independent observations x_i of the same distribution, the multivariate distribution is proportional to

$$e^{n(\bar{x}\theta - \kappa(\theta))/\phi},$$

and the proportionality factor does not depend on θ (but may depend on ϕ). This means that the CDF of a sample of n independent observations is summarized by the arithmetic average $\bar{x}$, which has (essentially) the same distribution as the components, with the dispersion parameter ϕ replaced by ϕ/n.

The expression for the CDF that defines the exponential family leads to a simple and explicit form for the mean and variance of such a distribution, determined by the function $\kappa(\theta)$. This also provides us with a method to estimate the natural parameter (though not ϕ, which we consider known in this discussion). These formulas are obtained by differentiating the relation

$$e^{\kappa(\theta)/\phi} = \int e^{x\theta/\phi}dF_\phi(x),$$

with respect to θ. The first formula we obtain is a formula for the mean:

$$E_\theta(X) = \kappa'(\theta).$$

Another differentiation and we find a similar equation for the variance:

$$V_\theta(X) = \phi\kappa''(\theta).$$

Box 9.5 Some important subfamilies of the exponential family

Many of the important distributions we have encountered so far belong to the exponential family.

The probability function for the *binomial distribution* can be written

$$\binom{n}{x} p^x (1-p)^{n-x} = \binom{n}{x} (1-p)^n e^{\theta x},$$

for which we have $\theta = \ln p/(1-p)$ and $\phi = 1$. Since $1-p = (1+e^\theta)^{-1}$ we see that $\kappa(\theta) = n \ln(1-p) = -n \ln(1+e^\theta)$ and $dF_\phi(x) = \binom{n}{x}$.

The probability function for the *Poisson distribution* can be written

$$e^{-m} \frac{m^x}{x!} = e^{x \ln(m)-m} \frac{1}{x!},$$

for which we have that $\phi = 1$, $\theta = \ln(m)$, $\kappa(\theta) = e^\theta$ and $dF_\phi(x) = 1/x!$.

The probability function for the *non-central hypergeometric distribution* can be written (cf. Appendix 5.A.1)

$$F(\psi)^{-1} \binom{n_1}{x} \binom{n_2}{r-x} \psi^x,$$

for which we have that $\theta = \ln \psi$, $\phi = 1$, $dF_\phi(x) = \binom{n_1}{x}\binom{n_2}{r-x}$ and $\kappa(\theta) = \ln F(e^\theta)$.

The density function for the *Gaussian distribution* can be written

$$\frac{1}{\sigma} \varphi\left(\frac{x-m}{\sigma}\right) = \frac{1}{\sigma\sqrt{2\pi}} e^{(mx-m^2/2)/\sigma^2} e^{-x^2/2\sigma^2},$$

so $\theta = m$, $\phi = \sigma^2$, $\kappa(\theta) = \theta^2/2$ and $dF_\phi(x) = (2\pi\phi)^{-1/2} e^{-x^2/2\phi} dx$.

The density function for the *gamma distribution* can be written

$$\frac{a^p}{\Gamma(p)} x^{p-1} e^{-ax} = e^{(-\frac{a}{p}x + \ln \frac{a}{p})/p^{-1}} \frac{p^p x^{p-1}}{\Gamma(p)},$$

so we take $\phi = 1/p$, $\theta = -a/p$, $\kappa(\theta) = -\ln(-\theta)$ and $F_\phi(x) = p^p x^{p-1}/\Gamma(p)$.

In order to estimate θ from data we can use likelihood theory, and for this we note that the part of the log-likelihood that contains information about the parameter θ is given by $(x\theta - \kappa(\theta))/\phi$, where x is the observation of X. (Information about ϕ may be present in the ignored part, so this discussion does not apply to that parameter.) This means that the maximum likelihood estimate of θ is given by the solution to the equation

$$E_\theta(X) = x.$$

In words: the maximum likelihood estimate of θ is the parameter value for which the expected value of X equals the observed value. If we have a sample of size n, it should equal the average of the observations. If we therefore denote the difference between the two sides by $U(\theta) = x - E_\theta(X)$, we have that θ is the solution of the estimating equation $U(\theta) = 0$, for which we have that the variance of $U(\theta)$ is the same as the variance of X. This observation, together with the CLT, gives us a method to compute an approximate (two-sided) confidence function for θ, namely

$$C(\theta) = \chi_s((x - E_\theta(X))^t V_\theta(X)^{-1}(x - E_\theta(X))),$$

where s is the number of components of θ. Note the fundamental difference between the parameters θ and ϕ. In a regression problem our primary focus will be on θ (or some function of its components), whereas ϕ is a measure of dispersion which will not influence the estimation of θ. Its inclusion is, however, of the greatest importance when we wish to make confidence statements about θ, and for that purpose we need to find an estimate of ϕ.

Sometimes only one part of θ is of interest, with the rest being nuisance parameters. When that is the case, one way to obtain knowledge about the interesting components is by finding a conditional distribution which is independent of the nuisance parameters (another is profiling). This was exemplified in Section 9.4, when we introduced the conditional logistic regression models, but can be done in some generality in the exponential family. To be more specific, assume that we have a stochastic variable X with a distribution in the exponential family for which the dispersion parameter is one and which is decomposed into $X = (X_1, X_2)$ (both components can be vectors), with a corresponding decomposition of the canonical parameter $\theta = (\theta_1, \theta_2)$. Then the conditional distribution of X_1 given X_2 also belongs to the exponential family. For an outline of the computations we start with the probability density for X, which is $e^{x_1\theta_1 + x_2\theta_2 - \kappa(\theta_1, \theta_2)} f(x_1, x_2)$. The marginal probability density for X_2 is then given by

$$e^{x_2\theta_2 - \kappa(\theta_1, \theta_2)} \int e^{x_1\theta_1} f(x_1, x_2) dx_1 = e^{x_2\theta_2 - \kappa(\theta_1, \theta_2)} g(x_2, \theta_1).$$

It follows that the density for X_1 given that $X_2 = x_2$, which is the ratio of these two densities, can be written as

$$e^{x_1\theta_1 - \ln(g(x_2, \theta_1))} f(x_1, x_2).$$

This density does not contain the parameter θ_2, so if we want to make inference about θ_1 we can use the confidence function based on the conditional distribution $X_1|X_2 = x_2$. A few examples are listed in Box 9.6, which points out how event data that appear according to Poisson processes are turned into a multinomial distribution if we do the analysis conditional on the total count.

There is one important type of calculation remaining on distributions in the exponential family. It is about allowing the natural parameter to be subject-specific, to vary in the population to account for heterogeneity (e.g., an omitted covariate). We know that this leads to a new distribution, the determination of which involves the computation of an integral. For members of the exponential family there are special complementary parameter distributions for which this computation is easily carried out: for the distribution defined in equation (9.5) we define the (family of) *conjugate distributions* by

$$dQ(\theta) = c(\gamma, \chi)^{-1} e^{\chi\theta - \gamma\kappa(\theta)} d\theta \tag{9.6}$$

Box 9.6 Some important conditional distributions from the exponential family

---

The probability function for two independent Poisson processes with rates λ_i, and observed for times T_i, is

$$\frac{e^{x_1\theta_1+x_2\theta_2-(e^{\theta_1}+e^{\theta_2})}}{x_1!x_2!},$$

where $\theta_i = \ln(\lambda_i T_i)$, $i = 1, 2$. The distribution for $x_+ = x_1 + x_2$ is Poisson with the natural parameter given by $\theta = \ln(e^{\theta_1} + e^{\theta_2})$. Division gives the conditional distribution

$$e^{x_1\theta_1+(x_+-x_1)\theta_2-x_+\theta'}\frac{x_+!}{x_1!x_2!} = \binom{x_+}{x_1}p^{x_1}(1-p)^{x_+-x_1},$$

where $p = T_1/(T_1 + \chi T_2)$ is a function of $\chi = \lambda_2/\lambda_1$ only.

The bivariate distribution of two independent $\text{Bin}(n_1, p_1)$ and $\text{Bin}(n_2, p_2)$ distributions can be written as

$$\frac{e^{x_1\theta_1+x_2\theta_2}}{(1+e^{\theta_1})^{n_1}(1+e^{\theta_2})^{n_2}}b(x_1, x_2) = a(\theta_1, \theta_2)b(x_1, x_2)e^{x_1\psi+(x_1+x_2)\theta_2},$$

where θ_i is the log-odds, $\psi = \theta_1 - \theta_2$ is the log-OR, $a(\theta_1, \theta_2)$ the inverse of the denominator, and $b(x_1, x_2) = \binom{n_1}{x_1}\binom{n_2}{x_2}$. The probability that $x_1 + x_2 = r$ is given by

$$\sum_k a(\theta_1, \theta_2)b(k, r - k)e^{k\psi+r\theta_2} = a(\theta_1, \theta_2)e^{r\theta_2}\sum_k b(k, r - k)e^{k\psi},$$

so the conditional distribution for x_1 given that $x_1 + x_2 = r$ is given by

$$b(x_1, r - x_1)e^{x_1\psi}/F(\psi, r), \quad \text{where } F(\psi, r) = \sum_k b(k, r - k)e^{k\psi},$$

which is the non-central hypergeometric distribution, defined in Box 4.4.

Related to the first example above is the case where we observe k independent Poisson distributions $x_i \in \text{Po}(m_i)$, $i = 1, \ldots, k$. The joint probability function is then

$$p(x) = \prod_{i=1}^{k} e^{-m_i}\frac{m_i^{x_i}}{x_i!} = e^{-\sum_i(x_i\ln m_i - m_i)}\frac{1}{\prod_i x_i!}.$$

The probability function for $x_+ = \sum_i x_i$ is given by $e^{-m_+}m_+^{x_+}/x_+!$, so the distribution of x conditional on $x_+ = n$ will be

$$\frac{n!}{\prod_i x_i!}\prod_i p_i^{x_i} = \binom{n}{x_1, \ldots, x_k}p_1^{x_1}\cdots p_k^{x_k}, \quad p_i = \frac{m_i}{m_+},$$

which is the probability function for a multinomial distribution.

Box 9.7 Some mixed distributions from the exponential family

The CDF for the *Poisson distribution* is given by $dF_\theta(x) = e^{x\theta - e^\theta}/x!$, so according to formula (9.6) the conjugate distribution is

$$dQ(\theta) = e^{p\theta - ae^\theta} \frac{a^p}{\Gamma(p)} d\theta.$$

This means that it is the distribution for e^X when X is a gamma distribution (i.e., the log-gamma distribution), and we identify that $c(a, p) = \Gamma(p)/a^p$. It follows that the mixed distribution is

$$\frac{\Gamma(p+x)/(a+1)^{p+x}}{\Gamma(p)/a^p} \frac{1}{x!} = \binom{p+x-1}{x} \left(\frac{a}{a+1}\right)^p \left(\frac{1}{a+1}\right)^x,$$

which is the *negative binomial distribution*.

Consider the *Bernoulli distribution* (which is the binomial with $n = 1$) for which we have $dF_\theta(x) = e^{x\theta - \ln(1+e^\theta)}$, where θ is the log-odds. Its conjugate distribution is

$$dQ(\theta) = c(\gamma, \chi)^{-1} e^{\chi\theta - \gamma \ln(1+e^\theta)} d\theta = c(\gamma, \chi)^{-1} p^\chi (1-p)^{\gamma - \chi} dp,$$

where $p = e^\theta/(1 + e^\theta)$. The coefficient is identified by comparison with the beta distribution as $c(\gamma, \chi) = B(\chi + 1, \gamma - \chi + 1)$ ($B(a, b)$ is defined in Appendix 6.A.1), and it follows that the mixed distribution for a sample of n is

$$\frac{c(\gamma + n, \chi + x)}{c(\gamma, \chi)} \binom{n}{x} = \frac{B(a+x, b+n-x)}{B(a, b)} \binom{n}{x}.$$

For the *Gaussian distribution* with mean θ we have $dF_{\theta,\phi}(x) = e^{(x\theta - \theta^2/2)/\phi} dF_\phi(x)$, as we have seen, so the conjugate distribution takes the form

$$dQ(\theta) = c(\gamma, \chi)^{-1} e^{\chi\theta - \gamma\theta^2/2} d\theta,$$

which means that it is a Gaussian distribution with variance $1/\gamma$ and mean χ/γ. It follows that $c(\gamma, \chi)^{-1} = \frac{\gamma}{\sqrt{2\pi}} e^{-\chi^2/2\gamma}$, and the mixed probability density is therefore

$$\frac{\sqrt{\gamma} e^{-\chi^2/2\gamma}}{\sqrt{\gamma + 1/\phi} e^{-(\chi+x)^2/2(\gamma+1/\phi)}} dF_\phi(x) = \frac{1}{\sqrt{2\pi(\phi + 1/\gamma)}} e^{-K(x)}$$

for a quadratic form $K(x)$ with coefficients that are functions of the parameters χ, γ and ϕ. This means that it is a Gaussian distribution with variance $\phi + 1/\gamma$, and we know that its mean is the mean of $Q(\theta)$. It follows that the mixed distribution is $N(\chi/\gamma, \phi + 1/\gamma)$.

---

for the appropriate coefficient $c(\gamma, \chi)$. A short calculation then shows that the population averaged density is

$$\int dF_\phi(x, \theta) dQ(\theta) = \frac{c(\gamma + 1/\phi, \chi + x/\phi)}{c(\gamma, \chi)} dF_\phi(x).$$

If we instead have n observations $x_1, \ldots, x_n$, the distribution is the same if we replace ϕ by ϕ/n and x by the arithmetic average $\bar{x}$ of the observations (we also change $F_\phi(x)$, but it still does not contain any parameter other than ϕ). Box 9.7 contains three important examples. Of particular importance is the last case, that the mixture of two Gaussian distributions is another Gaussian distribution, which again shows that in this case it does not really matter whether or not the individual mean response is heterogeneous in the population. We can still analyze the model under the assumption of fixed effects; the heterogeneity will only show up in the residual variance.

The conjugate distributions for distributions from the exponential family are also useful to Bayesian statisticians. In fact, if we take $dQ(\theta)$ as the *a priori* distribution, the *a posteriori* distribution is given by

$$dF(\theta|x) = c(\gamma + 1/\phi, \chi + x/\phi)^{-1} e^{(\chi+1/\phi)\theta - (\gamma+1/\phi)\kappa(\theta)},$$

which is another member of the same family of distributions as the *a priori* distribution. To use such distributions is therefore a way out of the general complexity of Bayesian statistics, and explains why we used beta distributions when we discussed the distribution of a binomial parameter in Section 4.6.2.

9.8   Generalized linear models

A general discussion on regression analysis in the exponential family will include not only standard Gaussian regression and logistic regression, but also such matters as Poisson regression, about which we will have more to say later. Denote the outcome variable by Y, and the covariate vector by X, with corresponding lower case letters denoting observations. In a regression model we specify a function $f(\beta, x)$ such that $E(Y|X = x) = f(\beta, x)$, and the purpose of the regression analysis is to estimate the coefficients β.

Let us first look at the unconditional problem, where we have the mean $E(Y)$ expressed as a function $\mu(\beta)$. We know from the general theory for the exponential family how to estimate the natural parameter. To estimate β requires that we identify the relation between the natural parameter and the mean of the distribution. The estimating equation for β can be shown to be

$$\mu'(\beta)V_\beta(Y)^{-1}(y - \mu(\beta)) = 0. \tag{9.7}$$

(In order to derive this, we express the natural parameter θ as a function $\Theta(\beta)$ of β, which is done using the equation $\kappa'(\theta) = \mu(\beta)$. Insert $\Theta(\beta)$ into the expression for the log-likelihood to obtain $y\Theta(\beta) - \kappa(\Theta(\beta))$ and differentiate. From this we derive the estimating equation $\Theta'(\beta)(y - E_\beta(Y)) = 0$, where $E_\beta(Y)$ is shorthand for $E_{\Theta(\beta)}(Y)$. The final observation is that $\mu'(\beta) = \kappa''(\theta)\Theta'(\beta) = V_\beta(Y)\Theta'(\beta)$.)

Once we have equation (9.7), we can apply this to the regression problem with n observations (x_i, y_i) of (covariate, outcome) pairs. Notation-wise this means replacing $\mu(\beta)$ with $f(\beta, x)$, and we obtain the equation for the maximum likelihood estimate for the regression coefficients β as

$$\frac{1}{n} \sum_{i=1}^{n} \sigma(\beta, x_i)^{-2} f'(\beta, x_i)^t (y_i - f(\beta, x_i)) = 0,$$

where $\sigma^2(\beta, x) = V(Y|X = x)$ is the conditional variance of Y, provided the model is correct. We recognize here equation (9.1), which means that the maximum likelihood estimate is the same as the GLS estimate for distributions in the exponential family.

A regression model for distributions in the exponential family which is such that the regression function $f(\beta, x)$ takes the form $f(\beta, x) = h(x\beta)$ is called a generalized linear model, and the function $h(u)$ is called the response function for the model. This is often expressed in terms of the inverse $g(u)$ of $h(u)$ instead, a function which is called the *link function*. Depending on the nature of the problem, different link functions apply to different problems, as was discussed for the binomial distribution in Section 9.4. Of special interest are those link functions for which we have that $g(\mu)$ actually defines the natural parameter θ. The logistic regression model is such an example, since it is the generalized linear model for binomial distributions with the link function $g(p) = \ln(p/(1 - p))$. Such links are called *natural links* and they have the property that the GLS equation simplifies to

$$\sum_{i=1}^{n} x_i^t(y_i - h(x_i\beta)) = 0.$$

This is what the estimating equation looked like for the logistic regression model.

9.9   Comments and further reading

See Lehmann (1990) and Cox (1990) for general discussions about modeling in statistics, the former with some historical comments. The list of reasons for covariate modeling given in Section 9.2 was adapted from Koch et al. (1982). For an account of the history of least squares and Gauss's justification for changing his method of proof, see Hald (1998, Chapter 13). Chapter 6 in the same book gives some historical remarks on the general problem of fitting data, including the problems with absolute residuals.

For a further discussion on the problem of estimating, and interpreting, effects in a heterogeneous world, and how the meanings of parameters change as we change the model, see Neuhaus et al. (1991). This problem is important, but not for Gaussian data, which we have seen can accommodate misspecification by increasing the dispersion parameter (Ford et al., 1995).

For details on the theory and practical use of GLMs in biostatistics and other fields, the original 'bible' is McCullagh and Nelder (1989), whereas the book by Fahrmeir and Tutz (2001) is a modern treatise covering a wider area. See Morgan (1992) for more on binomial regression, including an example of the multivariate probit model mentioned in Section 3.8. For a general discussion on the roles of conditional tests in statistical inference, see Reid (1995).

References

Cox, D.R. (1990) Role of models in statistical analysis. *Statistical Science*, **5**(2), 169–174.

Fahrmeir, L. and Tutz, G. (2001) *Multivariate Statistical Modelling Based on Generalized Linear Models* Springer Series in Statistics second edn. New York: Springer.

Ford, I., Norrie, J. and Ahmadi, S. (1995) Model inconsistency, illustrated by the Cox proportional hazards model. *Statistics in Medicine*, **14**(8), 735–746.

Hald, A. (1998) *A History of Mathematical Statistics from 1750 to 1930* Wiley Series in Probability and Statistics. New York: John Wiley & Sons, Inc.

Koch, G.G., Mara, I.A., Davis, G.W. and Gillings, D.B. (1982) A review of some statistical methods for covariance analysis of categorical data. *Biometrics*, **38**(3), 563–595.

Lehmann, E.L. (1990) Model specification: The views of Fisher and Neyman, and later developments. *Statistical Science*, **5**(2), 160–168.

McCullagh, P. and Nelder, J.A. (1989) *Generalized Linear Models* Monographs on Statistics & Applied Probability second edn. London: Chapman & Hall.

Morgan, B.J.T. (1992) *Analysis of Quantal Response Data* vol. 46 of *Monographs on Statistics and Applied Probability*. London: Chapman & Hall.

Neuhaus, J.M., Kalbfleisch, J.D. and Hauck, W.W. (1991) A comparison of cluster-specific and population-averaged approaches for analyzing correlated binary data. *International Statistical Review*, **59**(1), 25–35.

Reid, N. (1995) The roles of conditioning in inference. *Statistical Sciences*, **10**(2), 138–199.

10

Analysis of dose response

10.1 Introduction

Finding the appropriate dose for a new drug is important. If the developer makes a bad job of it, there is always a risk that the marketed dose comes out so high that it produces serious side-effects in a few cases. This may lead to the withdrawal of a drug which might be equally effective at a lower dose without these side-effects. Alternatively, if they select for further development a dose which is too low, the drug may fail due to lack of efficacy, and a good drug will not reach the patients who need it.

Although understanding dose–response relationships is important in drug development, the reason why we discuss it here is that it serves as an illustration to covariate modeling; we will use it as a role model for how we can model the mean of an outcome variable in terms of an explanatory variable, in this case the dose given. For this purpose we first discuss how to construct models for dose–response relationships, which is a question of how we expect the outcome to change when we change the dose. Such considerations, which start with an idealized and simplified view of the biology behind the data, provide the basis for the construction of a parametric family of regression functions, which will be fitted to the data. This model should be so constructed that its parameters can be rearranged into biologically meaningful parameters. In the context of dose response, examples of such parameters are the ED_{50}, the dose that provides 50% of maximal effect, and concepts like the relative dose potency and therapeutic ratio.

The next step is the actual data analysis. There are basically two different ways to do this. We can either try to describe the dose response as it looks on a population level, or we can try to do it on an individual level. In the first case we look at how a change in the dose of the drug affects the whole population, on average. In the second approach we try to describe individual dose–response relationships, assumed to have the same functional form for all individuals, but with subject-specific values of some of the parameters describing this relationship. These parameters have a distribution in the population, and the setup leads to a class of statistical models called mixed-effects model. It is a discussion which is related to the discussion in the previous chapter about heterogeneity and omitted covariates.

Understanding Biostatistics, First Edition. Anders Källén.
© 2011 John Wiley & Sons, Ltd. Published 2011 by John Wiley & Sons, Ltd.

10.2 Dose–response relationship

In drug development, whenever a chemical compound is found which shows some efficacy, it is important to find out the appropriate dose(s) that should be investigated further. This is also important in cases where the response is not necessarily a positive effect, but a harmful side-effect which puts limits on what doses can be given to patients. In our discussion on dose–response relationships we will use the words 'response' and 'effect' interchangeably. Strictly speaking, there is a difference: the response is what we measure, the effect is the change in response from when no drug is given. In most cases it does not matter which word we use, but we should be aware of the difference.

Whatever choice of outcome variable we have made, and whatever the reason for the investigation, assessing the dose–response relationship is a regression problem, in which some effect E is a function of the dose D of the compound given. In the overwhelming majority of cases this function has the following key properties:

1. It starts at zero – this is really the definition of the word 'effect', which is the additional response compared to when there is no drug.

2. It is increasing as a function of dose. This is partly a sign convention; sometimes what you measure is an outcome variable for which you want the measured values to decrease, in which case the effect is the difference from no drug to drug, instead of the opposite.

3. There is a maximum effect level that can be attained using the drug, given that we have given sufficient amounts of it (which may be a theoretical assumption when side-effects limit the administration of such doses).

The second criterion here is controversial: some statisticians want to avoid assuming a monotonic dose response, because it sometimes looks as if higher doses have smaller effects than lower doses. We will stick with the assumption and comment further on this at the end of this section.

These properties imply that we can write the dose–effect relationship as

$$E = E_{\max} F(D),$$

where the function $F(x)$ has the properties of a distribution function on $x > 0$. In this context the percentiles of $F(x)$ have special names: the p quantile is denoted ED_{100p}, for the *estimated dose that gives* $100p\%$ *of maximal effect*. For example, the median ($p = 0.5$) of the distribution $F(x)$, which gives 50% of the maximum effect, is ED_{50}. The graph of a typical dose–response function is shown in Figure 10.1.

In many situations it is useful to think of the dose–response relationship in terms of a CDF of a stochastic variable. For example, we may think of a painkiller drug as a drug that blocks pain mediating neurons. More neurons can be blocked with more drug in the body. We can think of $F(D)$ as the proportion of neurons blocked when the subject is given the dose D. What we measure is the subjective pain realized by the patient. Denote the no-drug pain level by P_0 and assume that very large doses can reduce it to zero (all neurons are blocked). The pain that is experienced when a subject is given the dose D is then described by the function $G(D) = P_0(1 - F(D))$, but to be consistent with assumption 2 above (an increasing dose–response relationship) we measure the effect as $E(D) = P_0 - G(D) = P_0 F(D)$.

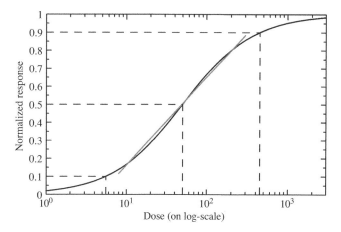

Figure 10.1   A typical dose–response curve and the log-linear approximation with ED_{50} as well as ED_{10} and ED_{90} indicated.

Another example is a β_2-agonist taken by a patient with asthma when he has difficulty breathing. The action of the drug is to relax muscles in the lung, so we can think of $F(D)$ as the proportion of contracted cells that have been relaxed when dose D is given. The lung function is measured by a variable such as FEV_1. If E_0 is the value of FEV_1 when all the relevant muscles are contracted, the FEV_1 measurement (which is a response, not an effect), as a function of dose, can be written as a function $R(D) = E_0 + E_{max}F(D)$. Here $E_0 + E_{max}$ would mean the level of FEV_1 that is obtained when all the contracted muscles are relaxed. The actual effect of the drug is given by the function $E_{max}F(D)$.

Both these examples illustrate an important aspect of dose–response relationships – it does not need to be the same on different occasions, not even for the same patient. In the last example, the extent to which the muscles have been contracted may differ between occasions, and values of E_0 and E_{max} may differ between occasions. It may be that $E_0 + E_{max}$ is relatively fixed for the individual, but E_0 varies as the pre-treatment proportion of contracted muscle cells vary. If we assume that $E_0 + E_{max}$ is constant, this means that the room for improvement depends on the baseline conditions, and these are in turn are intimately connected with the study design, in particular the choice of inclusion criteria for patients to be entered into the study, and on what background medication the study will be carried out.

There is one further aspect of dose–response functions illustrated in Figure 10.1 – they are often almost log-linear over a range of doses. This means that

$$E(D) \approx \alpha + \beta \ln D$$

in some dose range, expected to be symmetric around ED_{50} on the logarithmic scale. This is consistent with the medical tradition of increasing doses by doubling when there is insufficient effect. The parameter β measures how sensitive the response is to changes in the dose. For example, doubling a dose means an increase in response of size $\beta \ln(2) \approx 0.69\beta$.

The dose range over which the log-linear approximation of the dose–response curve is reasonably accurate is the range over which the interesting doses can be found. Smaller doses have only minute effects, and increasing the dose above the upper limit of this interval does not give any measurable further increase in effect. This dose range may be defined by ED_{10}

and ED_{90}, and when we study doses in this dose range, the log-linear model should suffice for all practical purposes.

The consequence of this observation about log-linearity is that we can write the general dose–response function as

$$E(D) = E_{max} \Psi(a + b \ln D),$$

where a and b are parameters, and $\Psi(x)$ is some CDF such that $\Psi(0) = 1/2$, and which is almost linear on some interval around $x = 0$. The condition $\Psi(0) = 1/2$ is only a convenient normalization, and implies that $\ln ED_{50} = -a/b$. The constant b measures the sensitivity to the drug, but is not the same as the β above. (To relate the two, note that the slope of the tangent at the ED_{50} point in Figure 10.1 is given by $bE_{max}\Psi'(0)$, which is therefore an approximation to β.)

The most popular choices for the distribution function $\Psi(x)$ are the two related CDFs discussed in Box 9.4, the standard Gaussian distribution $\Psi(x) = \Phi(x)$, and the logistic distribution $\Psi(x) = (1 + e^{-x})^{-1}$. In the logistic case we get the explicit and much used expression

$$\Psi(a + b \ln D) = \frac{D^b}{ED_{50}^b + D^b}.$$

This model is often referred to as the Emax model, and the coefficient b is called Hill's coefficient, named for the English statistician A. V. Hill.

Example 10.1 In a typical experiment in toxicology, increasing doses/concentrations of a chemical are given to groups of mice, and the number of deaths is counted. In one experiment, in which each dose group consisted of 10 mice, the following data were obtained:

dose:	5	25	100	800
deaths:	1	3	8	10

For each dose level D there is a probability $p(D)$ of death for a mouse given that dose. This is true for all doses; what is particular with the doses we actually study is that for them we can estimate these probabilities. However, we want to use this information to interpolate between doses in order to be able to estimate ED_{50}. This is the dose that kills 50% of the mice and is called LD_{50}, where LD stands for lethal dose. To estimate LD_{50} we need a model for the dose–response relationship.

If we model the probability of death as a function of dose, by using the function $p(D) = \Psi(a + b \ln D)$ with the logistic CDF $\Psi(x)$, we have that

$$\ln \frac{p(D)}{1 - p(D)} = a + b \ln D,$$

where the left-hand side is the log-odds for a mouse death at dose level D. This is a logistic regression model with the log-dose as covariate. Analysis of the data gives us the parameter estimates $a = -4.95$ and $b = 1.36$. From this we can estimate the lethal dose for 50% of the animals, which is given by the formula $LD_{50} = \exp(-a/b)$, as 37.6 with 95% confidence limits (18.4, 76.9). Here we have used Fieller intervals to obtain the confidence limits for the ratio a/b and then exponentiated the result. The predicted dose–response curve is shown in Figure 10.2 as the black curve, and LD_{50} is indicated by a dashed (black) line.

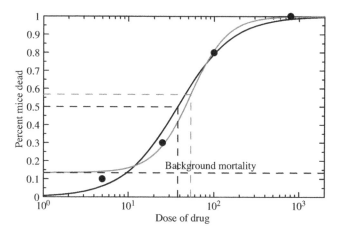

Figure 10.2   Two models for a toxicology experiment. The original model is in black, the modified one that takes background mortality into account is in gray.

However, the fact that we had a death in the lowest dose group raises the question whether there was some background mortality that needs to be taken into account when we model the dose response. The experiment was therefore repeated on another batch of 20 mice, except that these did not receive the chemical. It turned out that 3 out of these 20 animals died. It was therefore decided that a background mortality needed to be incorporated into the model, which we can do by writing it as

$$p(D) = \pi + (1 - \pi)\Psi(a + b \ln D).$$

Here π is the probability that an animal dies during the experiment for reasons independent of the chemical, and the function $\Psi(a + b \ln D)$ now describes the excess risk, which is what we want to relate LD_{50} to. The estimation of these parameters is easily done by use of the GLS method discussed in Section 9.3. The resulting parameter estimates are $a = -8.16$, $b = 2.05$ and $\pi = 0.13$, and from this we obtain the estimate 53.6 with 95% confidence limits (25.6, 111.5) for LD_{50}. Figure 10.2 shows the resulting dose–response curve in gray. Again the LD_{50} value is indicated by dashed (gray) lines. The corresponding level is greater than 0.5, which illustrates that in this case LD_{50} does not correspond to an observed count of 50% dead mice, but to an observed count of 57% dead mice, since $0.13 + 0.50(1 - 0.13) = 0.57$, accounting for natural mortality.

The assumption that the dose–response curve is an increasing function of the dose may not always manifest itself in data, because other factors may start to take effect at higher doses. It may be that the ability to perform the outcome measurement is impeded by some drug-related side-effect; if the drug causes headache the patient may not fully appreciate the improvement in the symptoms of the disease without the headache, despite the fact that the headache has nothing to do with the disease symptoms. If we need to model such complicated aspects of the readouts we may want to subtract, from the strictly increasing function discussed above, another function which starts to take effect at higher doses and describe these other effects, thereby obtaining a dose–response curve that is not increasing over the whole dose range. This leads to more parameters to fit, and we should be very confident that this is the appropriate model to use before applying it. That a mean was observed to be lower for a higher dose, as compared to a lower dose, is not sufficient evidence that this is also true for the underlying true means.

10.3   Relative dose potency and therapeutic ratio

Before we take a closer look at the estimation problems related to dose–response relationships, we consider what we wish to describe with such an analysis. One investigation of interest is to compare the dose–response relationship for two different drugs on the same outcome variable(s), another to compare the dose–response relationships for two different outcome variables for the same drug. We may even want to compare a number of different outcome variables for two different drugs. Similarly to what we did when we compared distribution functions in terms of distributional parameters, we need to find simple ways to summarize and communicate the result.

Figure 10.3 illustrates the main issues in dose–response modeling, at least in drug development. We have two drug treatments, denoted A and B, and for each of these the dose–response relationship for two different outcome variables. One represents a positive effect, the other a side-effect. Inspecting the graph, we see that for the (positive) effect we have $ED_{50} = 10$ for A and $ED_{50} = 20$ for B. In comparative terms we can say that the relative potency, $\rho = ED_{50}(B)/ED_{50}(A)$, is 2 for A versus B. (A is more potent since its ED_{50} is smaller.) The number $\ln \rho$ is the horizontal distance we need to move the dose–response curve of B to the left, in order for it to pass through the ED_{50} for A. It is important to plot the response curve versus dose on a log scale for this in order to assess the extent to which such a description conveys the important features of the situation at hand. In Figure 10.3 the dose sensitivities (the slopes of the curves) of the two treatments differ slightly, so this shift will not put the entire dose–response curve for B on top of that of A. If the outcome has the same dose sensitivity to the two drugs, this is what would happen, and in such a case the difference

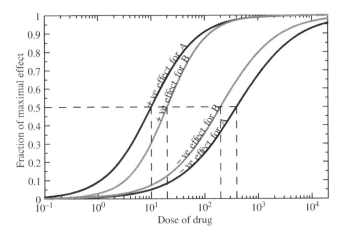

Figure 10.3   Geometric description of the relative dose potency and the therapeutic ratio for two drugs. Drug A is shown in black, drug B in gray. The positive effect is represented by the pair of curves on the left, the side-effect by the pair on the right.

between the two drugs would be completely described by ρ – drug A is twice as potent as drug B. With different dose sensitivities the picture is more complicated, but as a first approximation this description may still be a reasonable summary. This is the same situation as when we compare the means or medians for two distributions that are not true horizontal shifts of each other.

Next we consider the side-effect, for which the difference in sensitivity is smaller. Again it is reasonable to express the difference between the two treatments as the relative potency, and we find that $\rho = 0.5$, since the ED_{50} is 400 for A and 200 for B. If we look within treatment, we have that the ratio of the ED_{50} for the effect to the side-effect is $400/10 = 40$ for A; it is 40 times more potent on the effect side than on the side-effect side. We call this ratio the *therapeutic index*. For B we calculate this number to be 10. The ratio of these therapeutic indices constitutes the *therapeutic ratio* for A versus B, found to be 4 in our case. This means that, in some sense, A is four times better than B at separating positive from negative effects. The therapeutic ratio can also be written as the ratio ρ_+/ρ_-, where ρ_+ (ρ_-) is the relative potency for the two drugs on the positive effect (side-effect). We can therefore compute the relative dose potency for the two drugs for the effect and side-effect separately, and form this ratio to obtain the same parameter (in this case $2/0.5 = 4$). As a summary measure this is most useful if the dose–response curves are parallel shifts of each other for the positive effect and for the negative effect separately, but the two outcomes need not have the same dose sensitivity within drug.

10.4   Subject-specific and population averaged dose response

So far we have discussed the shape of a single dose–response curve. When we study dose–response data from many subjects, we encounter a particular problem, namely that dose–response curves may differ between individuals, and between occasions for individuals. This means that we should expect heterogeneity in the response curve, which is a problem equivalent to that of the omitted covariates in Section 9.6; what is the dose–response function here was the response function there. The present discussion is therefore a further illustration of what was discussed back then.

The problem is illustrated in Figure 10.4 in a very simplified setting. The assumptions underlying this graph are as follows. We study an outcome variable for which the minimum and maximum levels are well defined, so we can normalize these to zero and one. We also assume that the sensitivity to the drug is the same for all individuals; what differs is the dose required to achieve a certain effect. More specifically, we assume that the dose–response curve for an individual is given by the simple formula

$$E(D) = \frac{D}{\theta + D},$$

where the parameter $\theta = ED_{50}$ varies in the population according to some distribution with CDF $P(\theta)$, which we take to be lognormal for our discussion. We have a sample of 20 such dose–response curves, shown in Figure 10.4 as the dashed curves. Each of these is (on the logarithmic dose scale) a translated version of the one defined by $\theta = 1$.

The solid curve in Figure 10.4 represents something different. It is the average of all responses at given doses, that is, the population mean values of the outcome variable at a given dose. This curve has a slope which differs significantly from that of the individual

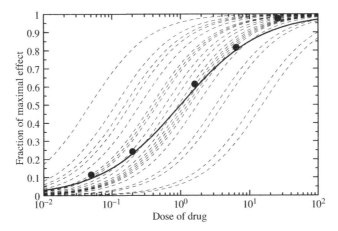

Figure 10.4 The difference between subject-specific curves and the population average curve. The dashed curves shows a sample of dose–response functions for different individuals, whereas the thick curve is the population average. The dots represent mean values for selected doses from the individual curves plotted with a measurement error added – see text.

curves. Mathematically the solid curve is obtained as the average over the population of the individual dose–response curves, and is given by the formula

$$E(D) = \int_{-\infty}^{\infty} \frac{D}{\theta + D} dP(\theta).$$

(We referred to this graph in Section 9.6, where the dashed curves were logistic functions for different values of the missing covariate Z, and the solid curve is the average response function for the population.)

In order to describe the dose response in a population we therefore need to decide what it is we wish to describe:

1. The individual dose–response functions are referred to as *subject-specific* functions. These are usually described by a family of functions $f(\theta, D)$ which depend on some parameters θ and on the dose D. What varies in the population is θ, according to a CDF $P(\theta)$, and when we wish to describe the subject-specific dose response we need to identify the functions $f(\theta, D)$ and $P(\theta)$.

2. The *population averaged* function, on the other hand, is a single function which, for each dose, describes the mean effect in a population when that dose is given to all subjects. In this case there is only one regression function we need to identify.

The regression function may well depend on covariate information; we may, for example, wish to identify different dose–response functions for males and females.

For our example, the subject-specific description is that the dose–response functions take the form $E(D) = D/(e^{\xi} + D)$, where ξ has a Gaussian distribution with mean zero and variance $\sigma^2 = 2$. The population averaged dose response, on the other hand, turns out to be well approximated by the function

$$E(D) = \frac{D^{3/4}}{1 + D^{3/4}}.$$

(This result was obtained from a nonlinear regression analysis, but if we replace the logistic function with a Gaussian CDF we can repeat the argument in Example 9.6 to find $b = 0.769$, which therefore should be a good approximation.)

Which of these two descriptions of the dose response is the most appropriate depends on how we want to use the information. If we want to use it to choose one or more doses to develop further in a drug development program, for ultimate use in the clinic, the population averaged approach seems to be the most appropriate. If, on the other hand, we want a personalized medicine approach and want to introduce a method to decide on the individual dose for each specific patient, we may want to estimate that subject's dose–response curve and how his parameters depend on measurable covariate information.

One further complication is that when we actually measure the outcome, these measurements will probably include observation errors. The observed measurements will therefore not be found on the subject-specific curve, but scattered around, as defined by some random noise. This is illustrated in Figure 10.4 by the point marks, which represent the observed mean values for a few doses. These are obtained from the 20 dose–response curves displayed in Figure 10.4, but where to each observation we have added a measurement error with a Gaussian distribution with zero mean and standard deviation 0.1. Note that we sampled 20 subjects and measured their response on five different doses, as opposed to having sampled 100 different subjects, on each of which we have measured the response to one single dose.

The issue illustrated here is not confined to dose response, nor to experimental studies for that matter. In epidemiology it is important to make a clear distinction between cross-sectional and longitudinal studies. Take growth as an example. When puberty sets in, there is first a short-lived reduction in growth rate, followed by a growth spurt which slowly flattens out as the child approaches final height. This is clearly seen when we investigate the growth curve for individual children. However, puberty occurs at different ages for different children (and earlier in girls than in boys), which means that if we measure height in a cross-sectional study (sampling individuals more or less at a random age) and plot the mean height versus age from such data, we may find a smoothed curve with no distinct puberty-related effect.

10.5   Estimation of the population averaged dose–response relationship

In this section we will illustrate the discussion in Section 9.5 by showing how to estimate the population averaged dose–response relationship. Suppose we have obtained, possibly from an earlier analysis, estimators $\hat{m}_i$ for the population mean response m_i when the patients have been given dose D_i, $i = 1, \ldots, d$, and that these estimates have a Gaussian distribution such that $\hat{m} \in AsN(m, \Sigma)$. Here $m = (m_1, \ldots, m_d)$ is the vector of mean responses (the points in Figure 10.4) and Σ is the covariance matrix for the corresponding estimator, which may depend not only on m but also on some auxiliary parameter ϕ. In the common case where adjusted means were obtained from a linear model with a Gaussian error distribution, we have

---

Box 10.1 The concept of the minimal effective dose

In the context of dose response, it is natural to try to find the minimal effective dose (MED), but the way this is commonly used in drug development leads to some confusion about its true meaning. It should be a medical concept, telling us what is the lowest dose we can give in order to achieve a clinical benefit of the drug. This dose may vary between patients, and we may need to interpret it to mean the lowest dose which, if given to the population, shows a sufficient effect on a population basis. The problem is to define what this means in quantitative terms, so that we can read off the corresponding dose from the dose–response relationship. One way to do this is to define the MED as the dose that gives the MID value (see Box 5.5), which would make sense provided the MID makes sense. However, the MID is a relatively recent concept, and prior to this it was notoriously hard to agree on how to quantify the properties the looked-for dose should have.

Because of this difficulty, the medical community often turned to statisticians for help, who solved the problem by scanning doses top-down as long as there is a statistically significant effect (compared to placebo). When it ceases to be, the MED had been found on the previous dose level. However, this is not a medically meaningful dose (unless you are very lucky); it is only the lowest dose that we can claim (beyond reasonable doubt), from this study alone, to have an effect. It therefore depends on how many patients we studied, and what significance level we use. Such defined MEDs cannot be expected to be the same from two different studies that vary considerably in size.

This definition of the MED may have some relevance for a regulatory authority, for approval purposes, but it is not a medically sensible concept to use in the clinic. The regulator needs statistical evidence that the dose in question has an effect, which is information obtained from a group of patients. The physician's interest is different.

---

that $\Sigma = \sigma^2 \Lambda$, where Λ is derived from the design matrix for the model. We assume we have estimates not only of m, but also of Σ. The model used to obtain these estimates depends on what design the clinical trial had, and which covariates we chose to adjust the mean estimates for.

Once we have done this preliminary analysis, we want to investigate more closely how the mean of the response variable depends on the dose. For simplicity, we first consider the log-linear dose–response model, in which we assume that the relationship is $E = a + b \ln D$. With $\theta = (a, b)$ this model can be written in matrix notation as

$$m = B\theta, \quad \text{where } B = \begin{pmatrix} 1 & \ln D_1 \\ \vdots & \vdots \\ 1 & \ln D_d \end{pmatrix},$$

and from the discussion in Section 9.5 we have the following estimator for θ:

$$\hat{\theta} = (B^t \Lambda^{-1} B)^{-1} B \Lambda^{-1} \hat{m} \in N(\theta, \sigma^2 (B^t \Lambda^{-1} B)^{-1}).$$

The next example illustrates this, and also one particular use of the dose–response function.

Example 10.2 We have a new drug A in an old drug class; such drugs are sometimes referred to as me-too products. We wish to find out what dose of A gives the same effect as the standard dose of one particular competitor, already on the market, which we denote by C and call the control. We study three doses of A, all of which we believe are on the log-linear part of the dose–response curve, together with the control.

At this point we do not need to be specific about what design we use. It may be a parallel group design, or it may be some crossover design. What matters is that we start out with some model of the data that produces unbiased estimates of the means with an approximate Gaussian distribution. Next we construct a model $m = B\theta$ similar to the one described above, but we now add another component to the vector θ, namely the mean for the control treatment, m_C. This means that we add one row (at the bottom) and one column (to the right) to the matrix B above, such that all additional elements are zero except for a one in the lower right corner. With this modification the formula gives an estimate of this augmented θ, together with the distribution of the corresponding estimator.

Next we wish to use this information to find the dose D_{eq} of drug A which has the same mean effect as the control. Mathematically this means that we want to solve the equation $a + b \ln D = m_C$. Therefore $\ln D$ is given as the ratio $(m_C - a)/b$, and we can obtain confidence limits for this parameter using Fieller intervals, if we first determine the bivariate Gaussian distribution for $\eta = (m_C - a, b)$. For this we introduce the design matrix

$$D = \begin{pmatrix} -1 & 0 & 1 \\ 0 & 1 & 0 \end{pmatrix},$$

which is such that $\eta = D\theta$. A little matrix algebra shows that

$$\hat{\eta} = D\hat{\theta} \in N(D\theta, \sigma^2 D(B^t MB)^{-1} D^t),$$

and we can now compute the confidence interval for the ratio as indicated. The final confidence limits for D_{eq} are obtained by exponentiating. The end result is illustrated in gray in Figure 10.5, where we have plotted the straight line approximation to the middle three mean values in the graph, as well as the mean of the control. The gray dashed lines indicate the point estimate of D_{eq}, with the corresponding confidence interval shown close to the dose axis. The estimate and 95% confidence limits for D_{eq} are 3.11 and (2.28, 4.33), respectively.

The data used for this analysis are the data we generated in the previous section (with measurement error), together with additional data for the control. We have used the middle three mean values shown in Figure 10.4 to fit the straight line to, and then used this dose–response function to estimate a quantity of some importance, namely the dose of the new drug that gives the same mean effect as the control. Because of the nature of the data, this analysis must have been preceded by a crossover study analysis to obtain mean and covariance estimates.

Next we wish to do a similar analysis fitting a complete dose–response curve to all five mean values in Figure 10.4. This analysis, to be discussed in the next example, is also illustrated in Figure 10.5, now as the black curve and black confidence interval.

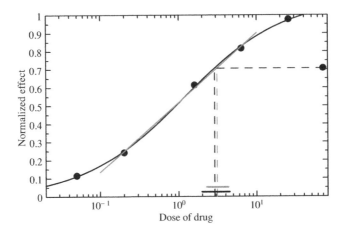

Figure 10.5   Illustration of two methods to estimate the dose that has a mean effect equivalent to that of a control. The corresponding confidence intervals are shown close to the dose axis, with the shorter gray one derived from the log-linear approximation and the black one derived from the sigmoidal curve.

Example 10.3   In order to fit a sigmoidal curve to the response means of a series of doses, we choose the four-parameter family of dose–response functions

$$E(D) = E_0 + \frac{E_{\max} D^b}{ED_{50}^b + D^b}$$

as our model function. This, together with the control mean m_C, provides us with a parameter vector $\theta = (E_0, E_{\max}, ED_{50}, b, m_C)$. Parameter estimation is performed by weighted nonlinear regression to the (adjusted) means, as outlined in Section 9.5. The end result is an estimate of θ, together with an estimate of the corresponding variance matrix. From this we wish to estimate, with confidence limits, the parameter

$$ED_{50} \left(\frac{E_{\max}}{m_B - E_0} - 1 \right)^{-1/b},$$

which is the solution to the equation $E(D) = m_B$, and therefore the D_{eq} from above. The confidence limits for this nonlinear function are obtained in the same way as we obtained confidence limits for functions of binomial parameters in Section 7.7, and which was outlined on page 260 at the end of Section 9.5. Both the estimated dose–response curve and the point estimate with 95% confidence interval for D_{eq} are shown in black in Figure 10.5. Estimates with confidence limits are 2.89 and (1.99, 4.52), respectively, which should be compared with the result of the previous example. The new confidence interval is wider than the previous one.

When we compare the results from these two analyses, we see that we get a considerably shorter interval with the first analysis. This is because the first analysis makes one assumption the second does not: it assumes that the three doses we study are on a straight line. The second analysis is not so sure about that, an uncertainty which is propagated into the corresponding confidence interval.

In the examples above we used one variable to estimate D_{eq}, albeit using two different models. For many diseases the disease status is not accurately measured by only one outcome variable. It is rather easy and direct to extend the methodology above to cover situations where we use a number of outcome variables to simultaneously estimate a common D_{eq}. We start out with a multivariate analysis of a linear model in order to get all the mean values for all the variables in one long vector, together with the variance matrix for this vector. That involves some matrix algebra, but no mysteries. Then we construct the model function for these means in such a way that there is a common D_{eq} defined for all the variables. If we wish, we can allow some of the other parameters to be variable-specific; in particular, it may be reasonable to allow for variable-specific E_0 and E_{max}, whereas variables may be assumed to have a common slope.

10.6 Estimating subject-specific dose responses

The population averaged dose–response function in the previous section corresponds to a randomly sampled patient who has been given a particular dose. We have seen that individual dose–response curves may look quite different from the population averaged curve. If we have data on different doses for individual patients it may be possible to get some information on individual dose–response curves and how these vary in the population. In this section we will consider two methods for estimating subject-specific dose–response curves, methods that are similar to the method we used for the population average approach. They are methods with limitations, and we will revisit the problem of estimating subject-specific curves again in the final chapter of this book, in the context of the more general mixed-effects models.

Whichever estimation method we use, we first need a parametric representation of the individual dose–response function. These should be defined from a common mathematical formula, including subject-specific covariate information and subject-specific parameters, which we denote ξ in this section. Next we want to describe how these subject-specific parameters vary in the population. Conceptually the simplest approach to this is to perform the analysis in a two-step fashion in the same way we did the meta-analysis that accounted for heterogeneity in Section 6.8. In such a two-step approach:

1. the first step is to use data from each individual subject separately, to find the parameter estimate ξ_i^* that provides the best fit of the regression function to that particular subject's data;

2. the second step is to use these individual estimates to estimate a distribution for the true model parameters ξ_i in the population.

The complication is that ξ_i^* is only an estimate of the true parameter ξ_i, but these estimates come with an estimated standard deviation, derived from the analysis of the individual curves and reflecting the precision with which the true parameter is estimated. If the true residual variance for subject i is σ_i^2, this means that $V(\xi_i^*) = V(\xi_i) + \sigma_i^2$, which must be taken into account in the analysis.

Example 10.4 Consider again the individual data described in Section 10.4, and shown in Figure 10.6 (a). Because of the measurement error, some observations are outside the valid range for the true data; there are both negative values and values that are larger than one. The

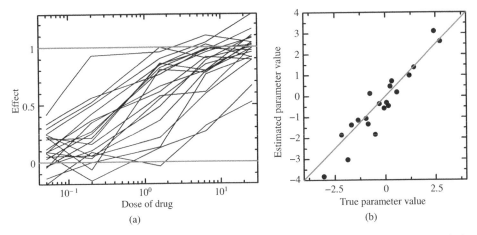

Figure 10.6 (a) shows the dose–response data as piecewise linear curves, (b) the correlation between the (known) true parameter values and the estimated ones. Further details are given in the text.

model assumption is that we have individual regression functions $f(\xi, D) = D/(e^\xi + D)$, with the parameter ξ assumed to follow a $N(\mu, \eta^2)$ distribution (the true value of μ is zero and that of η is $\sqrt{2}$). The measurement error is assumed to have the same standard deviation σ (which we know is 0.1) across doses. Summing up, we have a model with three parameters to estimate, constituting the vector $\theta = (\mu, \eta, \sigma)$.

If we apply the two-stage method to our data, we first determine 20 subject-specific ξ_i by making 20 separate nonlinear regression analyses, one for each subject. The 20 dose–response curves that we obtain are not exactly the same curves as shown in Figure 10.4, since they use estimated parameters instead of true parameters, but they look rather similar. Since we have simulated data, we know what the true parameter values are, and Figure 10.6 (b) shows the relation between the true and the estimated ξ for our 20 individuals. There is a good correlation, but some estimates do differ substantially from their true values. (Also note that the estimated parameters span a wider range than the true ones.)

When we wish to describe the population distribution of ξ we need to take into account the precision of the individual estimates. As already mentioned, we have encountered this problem in Section 6.8, and doing an analogous analysis on the present data, we get the result that the Gaussian distribution describing ξ has a mean value estimated to be $\mu^* = 0.44$ and a standard deviation estimated to be $\eta^* = 1.00$. We can also use this knowledge about the parameter distribution in the population and update our estimates of the individual parameters. Such updates are called empirical Bayes estimates, and are obtained by computing the *a posteriori* distribution for the particular individual, using the Gaussian distribution $N(\mu^*, \eta^{*2})$ as *a priori* distribution, and then taking the value that maximizes this distribution as the updated estimate of ξ for that individual.

We should note (as we did in Section 6.8) that the estimation method in the second step in this case is not a least squares estimation, but a maximum likelihood estimation. This is because of the way σ^2 enters the variance matrix (it is not a proportionality factor).

The drawback with the two-stage approach is that it requires relatively rich data for each individual. There are, however, alternative one-step approaches that can also be carried out with more sparse data. We will discuss here one particular one-step estimation method which is very close in spirit to the population averaged method. It is not a method that is much used in this context (such methods will be discussed in Section 13.5), but it represents a relatively small conceptual leap from what we have done so far. The basic assumption is that the distribution of ξ is described by a CDF $P(\xi)$, which depends on some parameters; in our case this is the CDF of the $N(\mu, \eta^2)$ distribution. Under this assumption the population average dose–response curve $m(D)$ is given by

$$m(D) = \int f(\xi, D)dP(\xi),$$

where $f(\xi, D)$ is the regression function. Note that $m(D)$ depends on parameters, both those in the regression function and those in the distribution $P(\xi)$, which we have suppressed from the notation. Next we perform a weighted non-linear regression analysis, in which we fit observed data (our estimates of the means) to this function. To do the actual fitting we need to estimate the variance matrix for the data, which contains correlations, because some observations are connected by being from the same individual (have the same value of ξ). The next example gives some more mathematical details.

Example 10.5   To carry this out for the data above, we need to numerically evaluate integrals, for which there are a few methods available. We use a classical method based on Hermite polynomials with 100 nodes. The population mean function $E(Y_i) = m(D_i, \theta)$ for subjects given the dose D_i is then given by the integral

$$\int_{-\infty}^{\infty} \frac{D_i}{e^{\mu+\xi} + D_i} \eta^{-1}\varphi(\xi/\eta)d\xi = \pi^{-1/2} \int_{-\infty}^{\infty} \frac{1}{e^{\mu+\sqrt{2}x\eta-\ln(D_i)} + 1} e^{-x^2}dx,$$

which does not depend on the observational error variance σ^2. The different Y_i are correlated, since data are observed at all five dose levels for each individual. In fact, the variance matrix for the vector of observations for an individual is given by

$$V(D, \theta) = \pi^{-1/2} \int_{-\infty}^{\infty} \Delta(\theta, D)^t \Delta(\theta, D)e^{-x^2}dx + \sigma^2 I,$$

where $\Delta(\theta, D) = f(\mu + \sqrt{2}x\eta, D) - m(D, \theta)$. In this notation the arithmetic means are such that $\bar{y} \in N(m(D, \theta), V(D, \theta)/20)$, and parameter estimates are obtained by minimizing the negative log-likelihood

$$20(\bar{y} - m(D, \theta))^t V(D, \theta)^{-1}(\bar{y} - m(D, \theta)) + \ln \det(V(D, \theta)).$$

The solution to this minimization problem is $\mu = -0.18$, $\eta = 1.20$ and $\sigma = 0.100$, and a graphical comparison of the estimated mean curve compared to mean data as well as to the true mean curve (available since we have simulated data) is given in Figure 10.7. The description of the result is that the subject-specific dose–response curves look like $f(\xi, D) = D/(\xi + D)$, where the $\xi = ED_{50}$ follow a lognormal distribution in the population, the median of which is given by $e^{-0.18} = 0.84$, with a coefficient of variation of $100\sqrt{e^{1.20^2} - 1} = 180\%$.

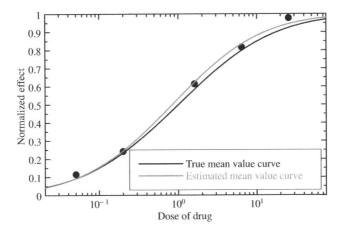

Figure 10.7 The agreement between the true dose–response curve (black) and the one obtained from the one-step estimation method described in the text (gray).

We see that the second, technically more involved, method produced the better estimates of the two in this case. However, as already pointed out, none of these methods is really the preferred method to use. The preferred method is based on constructing the log-likelihood for each individual and then constructing a global log-likelihood by summing these. Each individual likelihood will involve an integral of the kind encountered above. Finally, we maximize the likelihood. To get numerically feasible estimation methods, certain approximate methods are commonly used, a topic we will discuss briefly in Section 13.5.

10.7   Comments and further reading

As in many other areas of biostatistics, there are some controversies about how to do proper dose-finding for a drug. These controversies were not the purpose of this chapter, but the interested reader may find an overview with references in Senn (2007,Chapter 20). Our purpose is to use dose-finding as a role model for modeling how a particular outcome may depend on a continuous covariate. For a classical discussion on dose–effect relationships, see Holford and Sheiner (1981). The discussion in this chapter is closely related to the discussion in Källén and Larsson (1999). The extension of how to estimate the relative dose potency using more than one outcome variable, mentioned on page 284, is discussed with examples in Källén (2004).

References

Holford, N.H. and Sheiner, L.B. (1981) Understanding the dose-effect relationship: clinical application of pharmacokinetic models. *Clinical Pharmacokinetics*, **6**(6), 429–453.

Källén, A. (2004) Multivariate estimation of the relative dose potency. *Statistics in Medicine*, **14**, 2187–2193.

Källén, A. and Larsson, P. (1999) Dose response studies: how do we make them conclusive?. *Statistics in Medicine*, **18**, 629–641.

Senn, S. (2007) *Statistical Issues in Drug Development*. Chichester: John Wiley & Sons, Ltd.

11

Hazards and censored data

11.1 Introduction

In this and the next chapter we will address a different kind of data of considerable biomedical importance. This is time-to-event data, with survival data as a role model. Such data are special for a few reasons. One is that we often only have partial information for some of the individuals; we may only know that the time to the event is longer than the time we have observed the individual. Another reason is the very nature of the outcome variable, which is time. Whereas the data we have analyzed so far are static – the observations just appear – time-to-event data are dynamic in the sense that, in gathering them, we repeatedly ask whether the event has occurred or not. Therefore the mathematics involved here is different from what we have discussed so far, and the questions we ask are also different to some extent: they are often conditional questions such as 'given that someone is 75 years of age now, what is the probability that he will celebrate his 90th birthday?'

In Chapter 6, which was about complete data, we indicated how we can estimate the CDF in the presence of censoring mechanisms, using the Kaplan–Meier form of the e-CDF. Here we will conduct a more detailed investigation into the properties of this estimator. It is a good estimate of the CDF, or at least parts of it, because it is derived from something with a more direct biological interpretation, the intensity, or hazard, function, which was introduced in Section 2.7.

We also need to understand why censoring occurs. We often measure a particular kind of survival, for example time to death in a specific cancer. A patient under study may, however, die from another cause, such as a myocardial infarction. The presence of such competing events poses some important questions, not only about what the Kaplan–Meier estimator really estimates, but also about what the overall effect on the population would be if we took action to modify one particular competing risk. If we eliminate all cancer deaths, people will still die from other causes, so what impact would such a scientific breakthrough have on how long we live? One of the earliest attempts to address such a question was made by the Dutch-Swiss mathematician Daniel Bernoulli, when he investigated what the elimination of smallpox would mean for overall mortality, in a model we will discuss below.

Understanding Biostatistics, First Edition. Anders Källén.
© 2011 John Wiley & Sons, Ltd. Published 2011 by John Wiley & Sons, Ltd.

The risk, or hazard, for the event is specific to the individual, and one question is how these risks relate to the overall population risk. Heterogeneity in individual risks often distorts the picture and dilutes the effects. The intuitive reason for this is that events that occur early occur in frailer individuals, whereas events that occur later occur in less frail individuals, and therefore the distribution of this frailty (the word used for heterogeneity in this context) will have an impact on what is seen from a population perspective. But this also means that it becomes very important to try to explain frailty in terms of covariates. Much of this discussion will be taken up in the next chapter; here we take a preliminary look at the relationship between individual risks and the population risk. This also includes a preliminary discussion about the situation when we may have repeated events within subject.

The final part of this chapter is devoted to the mathematics of this kind of data. We need this in the next chapter when we obtain confidence statements about group differences and derive the famous Cox model. For this we need to understand the basic estimating equations and the associated variance estimates. This is an area of mathematics called counting process theory, and we outline some aspects of it at the end of this chapter. It is not necessary to read those parts in order to understand the main points in the next chapter, but they provide some information on the details of variance calculations and large-sample theory for counting processes.

11.2 Censored observations: incomplete knowledge

Data are censored when we do not know their precise value, but only have some bounds on them. An observation t is left-censored if we only know that $t < c$ for some c, it is right-censored if we only know that $t > c$ and it is interval-censored if what we know is that $a < t < b$ for some numbers a and b. Censored observations occur also for data that are not time-to-event data, as the following example illustrates.

Example 11.1 Left-censored data occur when we measure concentrations of drugs in some blood compartment. The assay that is used for such measurements typically has a lower limit below which quantification is not possible (or not considered reliable), the limit of quantification (LOQ) or lower limit of quantification (LLOQ). This limit may depend on the blood volume taken, and may therefore vary from sample to sample.

Figure 11.1 shows an example. The curve shows the CDF for the drug concentration measured one hour after a drug was given. The vertical dashed line shows the LOQ limit, which intersects the CDF at a level slightly greater than 0.3. If we take the data as given, and ignore the $<$ sign, the corresponding CDF would have a jump from zero to 0.3 at $x = 0.5$, and

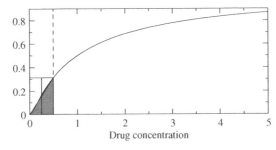

Figure 11.1 Illustration of how a distribution changes when there is a limit of quantification.

coincide with the original CDF for $x > 0.5$. If we computed summary statistics on such data we would get a mean that is larger than the true mean. To see how much, recall (see page 156) that the mean value can be visualized as the area above the CDF curve (up to the level one). The bias introduced when we use truncated data is therefore given by the gray area in the graph. To reduce this bias, it is common practice to impute the value LOQ/2 for data below the LOQ. In our case such imputation would be rather accurate in terms of mean estimation. In fact, the (white) area above the curve in the small rectangle with base (0.25, 0.5) is of about the same size as the gray area under the curve in the interval (0, 0.25), so these areas more or less cancel each other out.

Data below the LOQ are really missing data, the curse of which was addressed in Section 3.7. When we replace such data with LOQ/2 in descriptive statistics, we make an imputation for missing values. However, in a statistical analysis we may not need to use imputed data. If our choice of analysis is to do a rank test, what value we impute does not matter (as long as the same value is imputed for all individuals and this is below the LOQ). If we use a parametric model we can often avoid imputation altogether by using a likelihood method.

Censored data are most common in the time-to-event context. Right-censored data are particularly relevant in that context; we may follow some patients after initiation of a new treatment to see how long they live, but since the trial itself usually is restricted in time, we may not be able to follow them all to their death. Instead we stop the trial at some specific point in time, and for those still alive at that point, we only know that the survival time is longer than the time we have observed them, not the exact value. There are a number of other reasons why follow-up of a subject may cease before the event of interest has occurred. As we will discuss in some detail later, for a proper analysis to be conducted, the reason why the event has not been observed in an individual must be independent of his underlying risk for the event (non-informative censoring), or there will be some issues around the interpretation of the results. There will be bias in the results if there is a systematic withdrawal of either high- or low-risk patients.

Left-censored data may also appear in this context, but not for randomized studies in which the 'clock starts' at the point of randomization. In observational studies, however, when we analyze the natural history of a disease, we might be interested in having birth as the origin of time, and, depending on how subjects present themselves in the study, we may have left-censored data; we may know that an event has happened prior to the first investigation, but not when.

Interval-censored data also occur in some clinical trials. We may want to determine the time to a particular event, but the determination of whether or not this event has occurred can only be made at visits to the clinic, where the appropriate measurement can be obtained. When to schedule visits to the clinic is defined in the protocol, and we may for a particular individual only know that the event had not occurred at visit 3 after 23 days of treatment, but had occurred at visit 4 after 56 days of treatment. This means that we know it occurred somewhere in the time interval (23, 56) days, but not on which day.

11.3 Hazard models from a population perspective

To describe data for events that occur with a particular intensity, we first need to define what we mean by an intensity, or hazard (these two words will be used interchangeably;

intuitively 'intensity' is more neutral, whereas 'hazard' is associated with risks). We take a deterministic view and assume we have a very large population for which a function $F(t)$ describes the fraction of individuals who have experienced the event up to time t (the CDF for the corresponding stochastic variable). The (possibly time-dependent) *intensity function* $\lambda(t)$ for this event is defined by the differential equation (see also Section 2.7)

$$F'(t) = \lambda(t)F^c(t).$$

In words, the rate at which new events occur in the population is proportional to the number of individuals who have not yet experienced it, that is, the population still at risk. We assume that an individual only can experience the event once. If we use the fact that $F^c(0) = 1$, high-school calculus shows that $F^c(t) = e^{-\Lambda(t)}$, where $\Lambda(t) = \int_0^t \lambda(s)ds$ is the cumulative intensity (or hazard) function. The cumulative intensity $\Lambda(t)$ relates to the intensity $\lambda(t)$ in the same way as the CDF relates to the probability density.

We can write the differential equation above in the notation of Box 4.8 as

$$dF(t) = F^c(t-)d\Lambda(t), \tag{11.1}$$

noting that the proportion at risk at time t is actually the left-hand limit $F^c(t-)$. At a jump point of $F(t)$ the relation, expressed as probabilities, reads

$$P(T = t) = P(T \geq t)P(T = t|T \geq t),$$

which means that the hazard $d\Lambda(t) = P(T = t|T \geq t)$ is the conditional probability that we have a jump at time t, given that it has not yet occurred. This observation helps us also to understand the meaning of the hazard $d\Lambda(t)$ at points where $F(t)$ is continuous, as the *instantaneous* conditional probability that the event occurs now, given that it has not yet occurred. Equation (11.1) therefore means that the jump size at time t is the product of the number at risk and the hazard at that time.

From equation (11.1) we learn how to obtain $F(t)$ from $\Lambda(t)$. For smooth functions this is the differential equation mentioned above; in the general case we integrate the expression to get an integral equation for $F(t)$. We will discuss this integral equation and its connection to the Kaplan–Meier estimate of the CDF further in Section 11.8.

More important at this point is that we have the relation $d\Lambda(t) = dF(t)/F^c(t-)$. If we multiply the numerator and denominator on the right-hand side by the sample size, this means that $d\Lambda(t) = dN(t)/Y(t)$, where $N(t)$ is the expected number of events that have occurred at time t and $Y(t)$ the expected number at risk at the same time t. This immediately leads us to the *Nelson–Aalen estimator* $\Lambda_n(t)$ of the cumulative hazard from the relation

$$d\Lambda_n(t) = \frac{dN_n(t)}{Y_n(t)},$$

where $N_n(t)$ is the (cumulative) number of observed events at time t and $Y_n(t)$ the number observed to be at risk for experiencing an event at the same time. Since $d\Lambda(t)$ is an instantaneous probability, we can compute it from only those we observe, as long as we adjust $N_n(t)$ to mean observed events only. This estimates the true hazard as long as any censoring mechanisms operating are not related to the event we observe (more on this in the next section).

Box 11.1 The Gompertz distribution

A survival distribution much used to describe life expectancy in insurance contexts is a distribution which was introduced in 1825 by Benjamin Gompertz in an article with the title 'On the Nature of the Function Expressive of the Law of Human Mortality'. The distribution is motivated by assuming

> the average exhaustion of a man's power to avoid death to be such that at the end of equal infinitely small intervals of time he lost equal portions of his remaining power to oppose destruction which he had at the commencement of these intervals.

This assumption leads to the differential equation

$$\frac{d}{dt}\lambda(t)^{-1} = -b\lambda(t)^{-1}$$

(where t is age) which is equivalent to the simpler equation $\lambda'(t) = b\lambda(t)$, and implies that $\lambda(t) = \lambda e^{bt}$. The corresponding distribution is a generalization of the exponential distribution (which is the case $b = 0$); when $b \neq 0$ the survival function is given by

$$F^c(t) = e^{-\lambda(e^{bt}-1)/b}.$$

There is a connection between this distribution and the logistic distribution in Box 9.4, via the generalized logistic function defined by

$$F'(t) = F(t)(1 - F(t)^{\nu})/\nu,$$

with ν is a further parameter. If we let $\nu \to 0$, this equation becomes $F'(t) = -\ln(F(t))F(t)$, the solution of which is the Gompertz function.

In 1874 W.M. Makeham expanded this to a distribution with one additional parameter, deriving what is known as the Gompertz–Makeham law of mortality, which states that the death rate is the sum of an age-independent component (the Makeham term) and an age-dependent component (the Gompertz function). This appeared to be particularly suitable as an approximation to many empirical life tables and therefore soon became an important tool in actuarial theory.

The parametric distributions $F(t)$ that appear for time-to-event data are typically not the same as those that appear in the analysis of other types of measurement data. The Gaussian distribution is usually not relevant here, being too symmetric for most applications, but the lognormal distribution is a candidate, at least in some situations. However, it has a long right-hand tail, and for this type of data, in particular mortality data, tails are often to the left instead, which calls for a different set of distributions.

It is natural to derive the relevant distribution from the behavior of the intensity function. The simplest case is when there is a constant intensity λ, so that $\Lambda(t) = \lambda t$ and $F(t) = 1 - e^{-\lambda t}$. In such a case T has an *exponential distribution* with parameter λ, denoted $\mathrm{Exp}(\lambda)$ (the same as $\Gamma(1, 1/\lambda)$). Like the lognormal distribution, it has a long tail to the right. Its mean value is $1/\lambda$ and the graph of the function $\Lambda(t)$ is a simple straight line through the origin with slope λ. It has the important and interesting property that it has no memory. This means

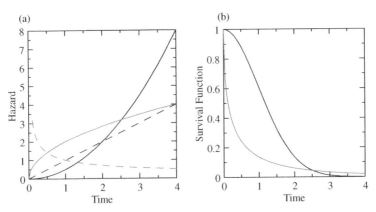

Figure 11.2 Illustration of the Weibull distribution. Dashed curves in (a) are graphs of the hazard functions t and $1/\sqrt{t}$, corresponding to Weibull distributions with exponents $\gamma = 2 > 1$ (black) and $\gamma = 0.5 < 1$ (gray), respectively. Solid curves in (a) are the corresponding cumulative hazards, and the corresponding survival functions are shown in (b).

that if the event has not occurred at time s, the time that remains until it occurs has the same distribution as it had from the start; the distribution is insensitive to when we start our clock. It is the only continuous distribution with this property. (In fact, since $P(T > t + s | T > t) = F^c(t + s)/F^c(t)$, the 'no-memory property' is equivalent to the functional equation $F^c(t + s) = F^c(t)F^c(s)$, for which the only (admissible) solution is $F^c(t) = e^{-\lambda t}$.) This property and the constancy of the intensity say the same thing: events occur completely at random. So, when events occur completely at random, the time-to-event distribution is the exponential distribution.

There are different directions in which we can generalize the exponential distribution for time-to-event data. One such direction is the gamma distribution, which occurs when there is a series of p events that build up the final event, and the times between individual sub-events are independent with a common exponential distribution. Another generalization is the Weibull distribution (Figure 11.2), which we write $\text{Wei}(\lambda, \gamma)$. Compared to the exponential distribution it has an additional exponent γ on the time variable, so that

$$F^c(t) = e^{-\lambda t^\gamma}, \quad \Lambda(t) = \lambda t^\gamma.$$

The corresponding hazard $\lambda(t) = \lambda \gamma t^{\gamma-1}$ is increasing with time if $\gamma > 1$ and decreasing with time if $\gamma < 1$. The Weibull family of distributions constitutes a flexible family for data when one wants to understand how 'non-exponential' a particular distribution is. Furthermore,

$$P(\ln T > t) = F^c(e^t) = e^{-\lambda e^{\gamma t}},$$

and this distribution of $\ln T$ is called the smallest extreme value distribution. We will see later that some important models for time-to-event data (AFT models) are actually regression models for $\ln T$, which explains why this observation is important.

We close this section with a description of the likelihood for a parametric model for time-to-event data in the presence of right-censoring. We want to express the likelihood in the

---

Box 11.2 The likelihood for parametric hazard models with right-censored data

---

In order to compute the likelihood for a parametric model for continuous survival data in the presence of right-censoring, let $F^c(t, \theta) = e^{-\Lambda(t,\theta)}$ where $\Lambda(t, \theta)$ is the cumulative hazard for $\lambda(t, \theta)$. For complete data the likelihood is given by

$$L(\theta) = dF(t_1, \theta) \ldots dF(t_n, \theta) = \lambda(t_1, \theta) \ldots \lambda(t_n, \theta) F^c(t_1, \theta) \ldots F^c(t_n, \theta),$$

where t_i is the observed event time for subject i. If we also have another m observations about which we only know that they are larger than τ, for a fixed number τ, then we multiply this likelihood by $F^c(\tau, \theta)$ a total of m times. The resulting likelihood can be written as

$$L(\theta) = \prod_i \lambda(t_i, \theta)^{\delta_i} F^c(t_i, \theta),$$

where $\delta_i = 1$ if this is an event ($t_i \leq \tau$) and $\delta_i = 0$ if the observation is censored and we have that $t_i = \tau$ for censored observations.

We now introduce a right-censoring process which is independent of the survival time and with CDF $C(t)$ and intensity function $\mu(t)$ not containing θ. What we observe is the minimum of the time to event and the time to censoring, so what we observe has a CDF $Q(t, \theta)$ which is such that $Q^c(t, \theta) = F^c(t, \theta) C^c(t)$. The likelihood $L(\theta)$ now becomes

$$\prod_i \lambda(t_i, \theta)^{\delta_i} \mu(t_i)^{1-\delta_i} Q^c(t_i, \theta) = \prod_i \lambda(t_i, \theta)^{\delta_i} F^c(t_i, \theta) \prod_i \mu(t_i)^{1-\delta_i} C^c(t_i).$$

The relevant (θ-free) part of the log-likelihood can therefore be written as

$$\ln L(\theta) = \sum_i (\ln \lambda(t_i, \theta) \Delta N_i(t) - \Lambda(t_i, \theta)),$$

where we have written δ_i as $\Delta N_i(t)$ instead. In integral notation this is equation (11.2).

---

intensity function instead of distribution function, in order to smoothly handle censoring. Let $\lambda(t, \theta)$ be the intensity function. As is outlined in Box 11.2, we have that

$$\ln L(\theta) = \int_0^\infty \ln \lambda(t, \theta) \, dN_n(t) - \int_0^\infty Y_n(t) \lambda(t, \theta) dt, \tag{11.2}$$

where $N_n(t)$ and $Y_n(t)$ are as before. To find the maximum of this function, we differentiate. The equation that determines the estimate of θ from data is then given by

$$\int_0^\infty \frac{\partial_\theta \lambda(t, \theta)}{\lambda(t, \theta)} (dN_n(t) - Y_n(t) d\Lambda(t, \theta)) = 0. \tag{11.3}$$

This equation will, in various forms, prove to be fundamental in the next chapter.

11.4   The impact of competing risks

There is a special consideration for right-censored data which is not relevant to complete data, namely an understanding of the censoring mechanism. We have already pointed out that the Kaplan–Meier estimator estimates the CDF. Its use may be uncontroversial if we are studying time to death and a data point is censored for an individual only when he has yet to die. If, however, we want to study mortality in a particular disease, it is a complicating factor that patients may die from other causes. It may be that understanding time to death in the particular disease is what is important to the investigator, because that is what a particular treatment targets, but from the patient's perspective the situation may be different. He may want to learn how much longer he is expected to live if he takes the treatment, especially if there is a price for taking it, such as side-effects. It is also possible that there are some negative effects of a treatment, which leads to an increased mortality in other diseases, such as heart attacks. The cause of death may be less important to the patient.

Suppose we are studying a particular event, such as death in a specific disease, in the presence of competing events which, when they occur in an individual, do not allow us to study the event we are interested in. What we can always estimate is the function $G(t)$, which is the probability that an event has occurred by the time t (it has occurred in the interval $(0, t]$). It is simply the proportion of individuals for whom this event has occurred, and is called the *cumulative incidence function* (CIF) for the event. It is not a distribution function, since in the presence of competing risks we expect to have that $G(\infty) < 1$. But it has all the other properties of a CDF, and is therefore called a sub-distribution function. If we denote by T the time we can observe, which is when the first of the competing events occurs, the following relation holds:

$$dG(t) = P(T_1 = t | T \geq t)P(T \geq t),$$

where T_1 denotes the time to the event we are interested in. If $F(t)$ is the CDF for T, this relation can be written as

$$dG(t) = F^c(t-)d\Lambda^*(t)$$

where $d\Lambda^*(t)$ is the event intensity among those that have not experienced any event yet. This is not the same as the hazard $d\Lambda(t)$ defined earlier, which is associated only with the event. The hazard $d\Lambda^*(t)$, on the other hand, is very much defined by what competing risks operate, because it measures intensity on patients who so far have survived these. When we eliminate one particular risk, it is not necessary that the hazards for the remaining ones are unchanged. If we take away a risk factor for a particular cause of death, we may well affect other causes as well: tobacco smoking not only induces lung cancer but also has effects on cardiovascular death risks, as well as on other causes of death. If we give a drug that reduces blood clotting we may reduce the risk of cerebral infarction, but increase the risk of cerebral hemorrhage, leaving the overall risk of dying from a stroke unchanged.

We can estimate $d\Lambda^*(t)$ by the Nelson–Aalen ratio of number of observed events divided by number at risk for this. The question then is when this estimates $d\Lambda(t)$, so that we can deduce the CDF for the time variable we are interested in. This question assumes that there is a well-defined stochastic variable T_1, free of the environment, that we can discuss. This is not necessarily the case.

Example 11.2   Suppose we are studying a chronic but fluctuating disease for which there naturally occur periods of symptom worsening, so-called exacerbations. Our treatment is expected to reduce the intensity of the occurrence of such exacerbations, and we have a placebo control. As part of the protocol the patient is allowed to discontinue participation in the study whenever he wishes, and in doing so he should state his reasons for discontinuation. One such reason may be a deterioration of the disease under study. Even though such a withdrawal is not an exacerbation *per se*, not fulfilling the precise medical criteria for this, it may well be closely related to it. It may, for example, precede it. If we study the time to the first exacerbation and censor these withdrawals, we probably will not do a meaningful analysis. It may be better to redefine the event under study as either exacerbation or discontinuation due to deterioration of the disease.

If the variable T_1 exists, we have that $d\Lambda(t) = P(T_1 = t | T_1 \geq t)$. We can only measure $d\Lambda^*(t)$, which works conditionally on $T \geq t$, so if we want to substitute $d\Lambda^*(t)$ for $d\Lambda(t)$, we need the cause-specific hazard to stay the same if we remove subjects experiencing competing events. Essentially that means that our event and the competing events must act independently of each other.

This, like the ITT analysis we discussed in Section 3.7, is something that is often difficult for medical researches to accept: 'why can't I analyze non-fatal and fatal myocardial infarctions separately?' The answer is that even though it may make medical sense to do this, there is simply not enough data to do a valid statistical analysis. Instead we need to make a fundamental assumption, that competing risks act independently of the event of interest, so it is the shortcomings of statistics (or, rather, the amount of information) that hinder the fulfillment of this wish.

If we can somehow assume independence, we can estimate $\Lambda(t)$ from data, considering those individuals with a competing event to be censored, and thereby obtain the Kaplan–Meier estimate of the CDF for T_1. It may also be that we want to eliminate only some censoring reasons. Suppose we are studying the events 'death from myocardial infarction' and 'death from other causes', but also that patients in the study may be lost to follow-up. In order to be able to estimate the CIF for death from myocardial infarction in the presence of other causes of death, we want to eliminate the problem posed by those who were lost to follow-up and to study termination. This is done by use of the formula

$$G(t) = \int_0^t F^c(s-)d\Lambda(s),$$

where $d\Lambda(t)$ is cause-specific and is estimated from the Nelson–Aalen estimator, and $F^c(t)$ is estimated by the Kaplan–Meier estimate for total survival. In this way we get an estimate of the CIF, eliminating the study-specific censoring mechanisms.

One lesson from this discussion is that the Kaplan–Meier estimate relates to an abstract time-to-event variable, free of any competing events. To describe the results of an analysis in terms of the Kaplan–Meier estimate may therefore be somewhat artificial in situations where there are competing risks that cannot be eliminated. In such a situation it may be more relevant to compute something like the function $G_1(t)/G_2^c(t)$. This represents the conditional probability for the event prior to time t, provided that the none of the competing risks have yet occurred. This is how the Kaplan–Meier estimate is sometimes (erroneously) interpreted. The Kaplan–Meier estimate is an estimate of the CDF for the event in the absence of other, competing, events. The suggested function provides the probability of obtaining an event of

---

Box 11.3 Bernoulli's model as differential equations

Bernoulli's analysis did not discuss vaccination against smallpox, since Edward Jenner's work on inoculation with cowpox was still 30 years in the future. Instead it was concerned with variolation, in which infectious material is inoculated into the skin of the susceptible in order to induce a mild infection. This was not an altogether harmless process; children could die from it, and it could trigger small epidemics. At the time physicians argued about whether the benefits of inoculation outweighed the risks, and the objective of Bernoulli's paper was to provide some numbers to facilitate decision making. As such it is one of the earliest attempts at evidence-based medicine.

In his analysis Bernoulli used the then recently developed methods of calculus. Introducing the two unknowns $x(a)$, the number of susceptibles at age a, and $n(a)$, the total number surviving to age a, he wrote down the two differential equations

$$x'(a) = -(\lambda + \mu(a))x(a), \quad n'(a) = -p\lambda x(a) - \mu(a)n(a),$$

subject to the initial conditions $x(0) = n(0) = n$. Here λ is the force of infection, $\mu(a)$ the age-dependent death rate, and p is the probability of dying from smallpox. From this he derived the following differential for the prevalence $f(a) = x(a)/n(a)$ of susceptibles:

$$f'(a) = \lambda f(a)(pf(a) - 1), \quad f(0) = 1,$$

which we solve to get $f(a) = 1/(p + (1 - p)e^{\lambda a})$. Bernoulli could then estimate the number of deaths due to smallpox, and deduce that in an environment free of smallpox the fraction surviving to age a should be given by $e^{\lambda a}/(p + (1 - p)e^{\lambda a})$. He described the result of his investigation by adding new columns to Halley's tables, containing 'would-be' data.

---

the type under study, given that a competing event has not occurred. It will depend on the environment in which it operates.

The competing risk discussion is important when we try to assess what the overall effect would be if we took action to modify one of the competing risks. What does the elimination of one particular cause of death do to overall mortality? This may well be the relevant health economic question for a new drug treatment, whereas the relevant biological question may address what happens in an environment free of competing risks. The former type of question was actually where the mathematical theory of competing risks started, and its history preceded mainstream statistics when, in the mid eighteenth century, Daniel Bernoulli of the (among mathematicians) famous Bernoulli family, wanted to understand what influence smallpox had on overall mortality. The specific question he addressed was the following. Published life tables reflect the mortality of the population for which they were calculated, taking into account all causes of death, including smallpox. How would these life tables change if, because of mandatory vaccination, deaths from smallpox were entirely eliminated?

We will address this question in a way similar to Bernoulli's approach, but investigating a probability model, where Bernoulli used differential equations. An outline of his approach is given in Box 11.3. His calculations were based on a life table compiled in 1693 by the English astronomer Edmond Halley (of Halley's comet) from the records of the then German city of Breslau (now Wrocław in Poland) on the age at death of each individual in the town.

For a person who is alive and still susceptible to smallpox (i.e., has not yet been infected), at age a, there are two competing risks:

1. He may contract smallpox, from which he dies with probability $p(a)$. Let $S(a)$ denote the distribution for the age at which the subject contracts smallpox in an 'environment free of other causes of death'.

2. He may die for reasons unrelated to smallpox. Let $H(a)$ denote the lifetime CDF in a 'smallpox-free environment'.

Bernoulli's objective was to estimate the survival function $H^c(a)$. His argument was essentially as follows. Someone who is alive at age a either is alive and susceptible to smallpox, which occurs with probability $H^c(a)S^c(a)$, or has had smallpox, survived it and not yet died from other causes. The probability of having contracted smallpox and survived to age a is given by $\int_0^a (1 - p(\alpha))dS(\alpha)$, from which we deduce that the overall probability of survival to age a is given by

$$l(a) = H^c(a) \left(S^c(a) + \int_0^a (1 - p(\alpha))dS(\alpha) \right).$$

Note the independence assumptions here. The left-hand side is (approximately) provided by Halley's table, so we obtain the expression

$$H^c(a) = \frac{l(a)}{S^c(a) + \int_0^a (1 - p(\alpha))dS(\alpha)}.$$

If we assume, almost certainly incorrectly, that the probability of dying from smallpox is age-independent, $p(a) = p$, we can write this as $H^c(a) = l(a)/((1 - p) + pS^c(a))$. Bernoulli also assumed that there is an age-independent risk λ of contracting smallpox, so that $S^c(a) = e^{-\lambda a}$, and therefore arrived at the expression

$$H^c(a) = \frac{l(a)}{1 - p + pe^{-\lambda a}}.$$

To obtain parameter values he assumed that during one year smallpox attacks one in eight individuals, so that $\lambda = 1/8$, and that it causes death in one in eight who are infected, so that $p = 1/8$ as well. Under these assumptions the eradication of smallpox would be expected to increase the median age at death by 12.4 years from about 11.4 to 23.9 years, which is illustrated in Figure 11.3. Under the model we can also compute the mean life expectancy from Halley's table, which predicts an increase of 3.1 years, from about 26.6 to 29.6 years.

As already noted, there are some strong assumptions in these calculations, an important one being the independence between death from smallpox and death from other causes. But this is a modeling exercise, and when we finally inspect the result and compare with our predictions, we must bear in mind that this assumption may explain some of the discrepancy. Likewise, when we discuss the modeling predictions we must assess how sensitive they are to the assumptions, not only the independence assumption but also other assumptions made. This makes modeling an endeavor with special problems, especially if you cannot test the model before you need to make your decision. Climate modeling today is a case in point.

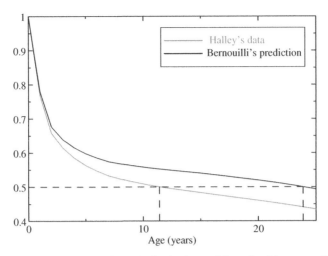

Figure 11.3   Life table data based on Halley's data with and without smallpox. The gray curve shows Halley's tabulated data, the black curve Bernoulli's prediction of how it would change if smallpox were eradicated.

11.5   Heterogeneity in survival analysis

It is tempting to interpret the hazard, or the survival function, computed on population data as individual risks. This is often not appropriate, basically because biological systems do not consist of cloned, identical individuals living in identical environments; instead there is biological variation. In fact, one of the purposes of a statistical analysis is to explain part of this variation in terms of covariates. We have already seen that omitted covariates have different effects in different statistical models. For some statistical models unobservable variation is simply random noise with no effect on the size of the signal. This is true for methods such as the t-test for means and ordinary linear regression. It is not true for time-to-event data because, in a heterogeneous world, high-risk subjects will experience the event early and with time we will gradually study more and more low-risk subjects. The hazard we see will therefore decline with time because of a selection procedure, which may give us a false impression that the individual hazards disappear with time. Although this may be true on a population level, it may not apply on the individual level. Since the life expectancy is dependent on how frail, or fragile, someone is, heterogeneity in this context is referred to as frailty.

To discuss this frailty problem we make a simple basic assumption about the nature of the heterogeneity, which may or may not be realistic. At least it will help us understand the problem and its consequences. The model in question, referred to as the frailty model, assumes that individuals have proportional hazards, so that for each individual there is a value θ, such that the hazard for that individual is given by $\theta d\Lambda(t)$, where $d\Lambda(t)$ is some reference hazard, common to all individuals. The θ tries to capture the individual frailty, which includes factors such as disease severity, genetic makeup or aspects of the (constant) environment. We refer to the θs as frailties and assume that their distribution in the population is described by a CDF $P(\theta)$. In order to be able to uniquely specify the reference hazard and the frailty distribution, we assume that the latter has a known mean value. Let $F(t)$ denote the CDF corresponding to $\Lambda(t)$, and let $F(t, \theta)$ denote the CDF with cumulative hazard

---

Box 11.4 What can we deduce about frailty based on the outcome?

This box contains some mathematical details about the discussion on frailty the text.

If we randomly sample an individual from the population, the bivariate distribution of the time-to-event variable T and the frailty θ has a density given by $dF(t, \theta)dP(\theta)$. The proportional hazards model means that

$$dF(t, \theta)dP(\theta) = [\theta F^c(t-, \theta)dP(\theta)]d\Lambda(t),$$

a relation that contains a lot of information. First, we can integrate out θ to obtain the marginal density for T:

$$d\Psi(t) = S^0(t)d\Lambda(t), \quad \text{where } S^0(t) = \int F^c(t-, \theta)\theta dP(\theta).$$

Here $F^c(t-, \theta)dP(\theta)$ is the probability (density) that we get an individual with frailty θ and for whom $T \geq t$. It follows that $F^c(t-, \theta)dP(\theta)/\Psi^c(t-)$ is the conditional density for θ, given that $T \geq t$, and therefore that $S^0(t)/\Psi^c(t-)$ is the expected value of θ, given that $T \geq t$. The population hazard is therefore given by

$$d\Lambda_\Psi(t) = \frac{d\Psi(t)}{\Psi^c(t-)} = E(\theta|T \geq t)d\Lambda(t).$$

If we instead divide the bivariate density $dF(t, \theta)dP(\theta)$ by the density $d\Psi(t)$, we get the conditional probability that the frailty is θ, given that we know that the outcome is $T = t$. An alternative expression for this is obtained from above as

$$\frac{dF(t, \theta)}{d\Psi(t)}dP(\theta) = \frac{F^c(t-, \theta)\theta d\Lambda(t)}{S^0(t)d\Lambda(t)}dP(\theta) = \frac{F^c(t-, \theta)\theta dP(\theta)}{S^0(t)}.$$

This is the density of the conditional frailty CDFs $P(\theta|T = t)$ shown in Figure 11.5.

---

function $\theta\Lambda(t)$. Before proceeding, we note that the famous Cox model, which is so often used in the analysis of survival data, has precisely this set-up, except that instead of a frailty distribution it tries to explain the different frailties in terms of measured covariates. More precisely, it assumes that there is a relation of the form $\theta = e^{z\beta}$, where z are the covariates and β are regression coefficients that are to be estimated. This model is discussed in some detail in Section 12.6 and our derivation of it will build on the observations we make in the current section.

Now back to the basic setup. As seen in Box 11.4, we can derive some useful information about the frailty of an individual based on knowledge about when his event occurred. The marginal CDF for the time-to-event variable T is obtained by integrating θ out:

$$\Psi(t) = \int F(t, \theta)dP(\theta).$$

This is the distribution we see in the population, and with some mathematical trickery, outlined in Box 11.4, we can show that the corresponding population hazard is given by the expression

$$d\Lambda_\Psi(t) = E(\theta|T \geq t)d\Lambda(t).$$

Box 11.5 Frailties with a gamma distribution

A popular choice for the frailty distribution $P(\theta)$ is the gamma distribution with mean 1 and variance $1/a$. Its popularity is largely due to a simple Laplace transform:

$$\mathcal{L}(s) = \int_0^\infty e^{-s\theta} dP(\theta) = (1 + s/a)^{-a}.$$

For a continuous variable, the population survival function for a frailty model is given by

$$\Psi^c(t) = \int e^{-\theta\Lambda(t)} dP(\theta) = \mathcal{L}(\Lambda(t)).$$

For the gamma frailty case this implies that $\Psi^c(t) = (1 + \Lambda(t)/a)^{-a}$ and that

$$d\Lambda_\Psi(t) = \frac{d\Lambda(t)}{1 + \Lambda(t)/a}.$$

Since $\Lambda(t)$ is increasing, this means that the hazard ratio $d\Lambda_\Psi(t)/d\Lambda(t)$ is decreasing with time and the steeper $\Lambda(t)$ is (the more common the events) the greater is the role played by frailty.

Consider an epidemiological situation where the baseline hazard is $\Lambda(t)$ for the exposed, but $r\Lambda(t)$ for the unexposed, $r < 1$. Assume that the heterogeneity is described by the same gamma distribution for the two groups. If we denote by $\Lambda_i(t)$ the two population hazards, we find that

$$\frac{d\Lambda_2(t)}{d\Lambda_1(t)} = r\frac{1 + \Lambda(t)/a}{1 + r\Lambda(t)/a}.$$

This instantaneous hazard ratio start out as r, but for very large t it approaches one. The initial relative risk therefore seems to disappear. But this is only a reflection of the fact that, for large t, we only study robust individuals for whom the event rarely occurs. (With a different choice of frailty distribution we may even get a crossover, so that the ratio approaches something greater than one.)

This qualifies the intuitive discussion above, that the population hazard at time t is the mean hazard, evaluated among the survivors. It is also time-dependent when the reference hazard is constant, because what we know about the hazard at a later point in time is based on a selected subgroup of individuals (those who survived that long) and not on all individuals. For more on the important case where the frailty distribution is a gamma distribution, see Box 11.5.

Example 11.3 Figure 11.4 is a graphical illustration of the consequences of the heterogeneity for the cumulative hazard. We have two groups with constant hazards, one twice the other in magnitude (dashed lines). These serve as (constant) baseline hazards in frailty models with gamma-distributed frailties, with the corresponding population hazards shown as the solid curves, matched by level of grayness. We see that the diverging baseline hazards get replaced by population hazards that become parallel with time. The asymptotic vertical shift is easy to

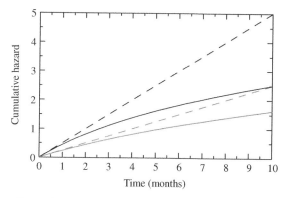

Figure 11.4   The effect of population heterogeneity on survival and cumulative hazard functions. The assumption is a gamma-distributed frailty distribution with variance one.

compute as

$$a \ln(1 + (\lambda t)/a) - a \ln(1 + (\lambda t/2)/a) = a \ln \left(1 + \frac{\lambda t}{2a + \lambda t} \right) \approx a \ln 2.$$

In summary, in a world of frailty differences between people, the individual hazard may look quite different from what we find when we study a random sample from the population. The proportional hazards model can be true on an individual level, but the marginal Kaplan–Meier estimate may look quite different.

For the sake of completeness (and for other reasons which will become apparent at the end of next chapter) we also want an expression for what we expect the frailty to be for a patient for whom we have observed an event at time t. Box 11.4 also outlines how we can derive the corresponding conditional frailty CDF $P(\theta|T = t)$, and these distributions are illustrated in Figure 11.5, with similar assumptions as in the example above (constant baseline hazard and gamma frailty). We see that $P(\theta|T = t)$ puts weight on smaller θ as t increases, which is precisely what we should expect. This is generalized to the following statement about what

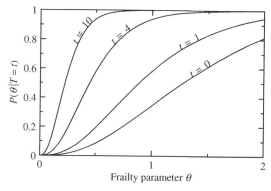

Figure 11.5   Illustration of the conditional frailty CDF $P(\theta|T = t)$ for some choices of t.

we expect of the value of a covariate Z, given that we know that the subject had an event at time t:

$$E(Z|T = t) = S^0(t)^{-1} \int E(Z|\theta)F^c(t-, \theta)\theta dP(\theta), \qquad (11.4)$$

where $E(Z|\theta)$ denotes the mean of Z for those subjects that have frailty θ. (For notation, see Box 11.4.) This observation, together with the rest of the material discussed in this section, will be important when we discuss the famous Cox model in the next chapter.

11.6 Recurrent events and frailty

In this section we digress a little from the main discussion on events that can occur only once in an individual, and add a few comments about the case where there may be recurrent events for individuals, bearing in mind that how prone individuals are to experience the event may differ. Examples of recurrent events in biostatistics include the study of exacerbations of chronic conditions such as asthma and other inflammatory diseases, epileptic seizures and the recurrence of tumors. In such situations we may want to assess the event incidence by using all observed events. The simplest way to do this is to reduce the data to the total number of events for each individual, not concerning ourselves with when they occurred.

When we study individuals who can only experience a particular event once, for which the (continuous) hazard is $d\Lambda(t)$, the process $N(t)$ that counts the number of events can only take on two values, zero (with probability $e^{-\Lambda(t)}$) and one (with probability $1 - e^{-\Lambda(t)}$). If there are recurrent events within subjects, and if we assume that all events are described by the same hazard, the process that counts the events is a Poisson process (see Box 2.10), so that $N(t) \in \text{Po}(\Lambda(t))$. Even if the exact conditions required are not necessarily fulfilled, this may still be a reasonable description of the total count within a subject. However, there are almost certainly different hazard functions for different individuals. As before, the frailty model means that we assume that for each individual there is a θ such that $N(t) \in \text{Po}(\theta\Lambda(t))$, and that θ has CDF $P(\theta)$ in the population. In order to compute the marginal distribution of the number of events for individuals we fix the time point t and write $\mu = \Lambda(t)$. For a randomly sampled individual the probability that precisely k events have occurred at time t is then given by

$$p(k) = \int_0^\infty e^{-\theta\mu} \frac{(\theta\mu)^k}{k!} dP(\theta).$$

If we again consider the convenient case where the frailty distribution is gamma with mean one and variance $1/a$, we can compute this integral to obtain the probability function

$$p(k) = \frac{\Gamma(k+a)}{k!\Gamma(a)} \frac{(\mu/a)^k}{(\mu/a + 1)^{k+a}}.$$

This is (see Box 11.6) the probability function for a negative binomial distribution with mean μ and variance $\mu + a^{-1}\mu^2$. The convention here is to call $\phi = a^{-1}$ the overdispersion parameter, since it inflates the variance (dispersion) compared to that of the Poisson distribution. When $\phi \to 0$ the negative binomial distribution becomes the Poisson distribution. To see the effect of larger ϕ on individual probabilities, consider Figure 11.6 in which we compare the CDF

Box 11.6 The negative binomial distribution

To explain why the negative binomial distribution has this name, a small mathematical tour is useful. The binomial theorem,

$$(1 + x)^n = \sum_k \binom{n}{k} x^k,$$

holds true also when n is not a positive integer. For a positive integer n it shows that with $p = x/(1 + x)$ we have that

$$\sum_{k=0}^{n} \binom{n}{k} p^k (1 - p)^{n-k} = 1,$$

which defines the binomial distribution. Similarly, for $n > 0$, but not necessarily an integer,

$$(1 - x)^{-n} = \sum_{k=0}^{\infty} (-1)^k \binom{-n}{k} x^k = \sum_{k=0}^{\infty} \binom{n+k-1}{k} x^k$$

which leads us to the negative binomial distribution with probability function $(p = 1 - x)$

$$\binom{n+k-1}{k} p^n (1 - p)^k.$$

Since $p^n (1 - p)^k$ is the probability of k failures and n successes, if p is the probability of success in identical and independent trials, this will be the distribution for the number of failures until we have had n successes. The case $n = 1$ is the geometric distribution, also called the binomial waiting time distribution, since it describes the number of trials until the first success.

The negative binomial is often parameterized in μ, where $p = 1/(1 + \mu)$. The probability of k events is then given by

$$\binom{n+k-1}{k} \frac{\mu^n}{(\mu + 1)^{n+k}},$$

and with this parameterization it can be shown that the mean is $n\mu$ and the variance $n\mu(1 + \mu)$.

---

for a $Po(\mu)$ distribution with that of a few negative binomials with the same mean, in order to see the effect of the overdispersion. We see how the CDFs flatten out as we increase ϕ, which is what is to be expected: by increasing ϕ we introduce both more individuals with no events and more individuals with many events.

One of the effects of using the negative binomial instead of the Poisson distribution is therefore to increase the probability of no events. It could also be the case that a proportion of the population cannot experience the event at all; they could be cured of the disease or immune to it. It is not that they have a low intensity for the event, it simply does not happen to them.

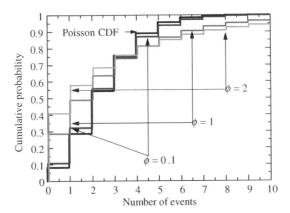

Figure 11.6   A comparison of the CDF for the Po(2.5) distribution (black) to that of the negative binomial distribution with the same mean and a few choices of the dispersion parameter ϕ.

The negative binomial does not fully capture this feature of two subpopulations (neither does the Poisson distribution). For this we need a model with a CDF of the form $p + (1 - p)F(x)$, where p is the fraction in the subpopulation for which the event cannot occur, and $F(x)$ is the CDF for the number of events for those that can experience it. In the case where $F(x)$ is a Poisson distribution, this is called a zero-inflated Poisson model.

In this discussion on recurrent events we have reduced data to a single within-individual measurement, the total number of events per individual. This may not be what we really want to do, but a more complete use of data requires more complicated methods. We will address this problem in some detail in Section 13.4, where we will provide an example of the application of this theory, and also carry the analysis further. The reason why we have chosen to postpone giving an example comparing the Poisson and the negative binomial models will then be clear – we want to introduce the so-called robust covariance estimate first.

11.7   The principles behind the analysis of censored data

In order to analyze time-to-event data statistically we need to know more about their mathematical/statistical properties. We made a remark about the likelihood theory for parametric models earlier; now we are interested in an analysis without parametric assumptions. The actual analysis will be more extensively discussed in the next chapter when we not only compare two groups, but also consider linear models for a proportional hazards model (the Cox model). To carry this out, we need to understand not only how to obtain estimating equations, but also how to compute the corresponding variances. We therefore need to consider time-to-event data in their proper mathematical context, which is different from that of other data. As mentioned in the introduction, this is because we do not really observe an event time, instead we repeatedly observe an individual and note, at each time point, whether the event has occurred or not. The collection of such variables constitutes a stochastic process and (if only one event can occur for each individual) this process can only transition from the state 0 to the state 1, which it does at the event time. Adding such processes for all individuals gives us a stochastic process which counts the total number of events that have occurred at each time point. This puts this type of data within the framework of counting process theory.

For complete data we use the e-CDF $F_n(t)$ to estimate the CDF $F(t)$. In order to understand how good our estimate is, we investigate the difference $F_n(t) - F(t)$. For time-to-event data the fundamental quantity is not the CDF, but the hazard $d\Lambda(t)$. We estimate it by the Nelson–Aalen estimate, but in order to investigate how good such an estimate is, we look at the difference $F_n(t) - \int_0^t F_n^c(t-)d\Lambda(t)$. The expected value of this is zero, as can be seen from an integration of the fundamental relation in equation (11.1). This becomes more convenient if we multiply by the total sample size, so that this difference can be written as

$$\xi(t) = N_n(t) - \int_0^t Y_n(s)d\Lambda(s).$$

As before, $N_n(t)$ is the total number of events that have occurred at time t and $Y_n(t)$ is the number of subjects at risk of an event at time t. The process $\{\xi(t); t \geq 0\}$ is simply the difference between what we observe and what we predict from an assumption on the intensity $d\Lambda(t)$. The process $Y_n(t)$ is a *predictable process*, which means that if we know its value just prior to time t, what we expect next is precisely what we see now.

We can rewrite $\xi(t)$ as an integral,

$$\xi(t) = \int_0^t (dN_n(s) - Y_n(s)d\Lambda(s)),$$

which means that it is a cumulative sum of components $dN_n(t) - Y_n(t)d\Lambda(t)$. Here $dN_n(t)$ is the rate of new events at time t, which we can think of as having a binomial distribution $\text{Bin}(Y_n(t), d\Lambda(t))$. Its expected value is therefore $Y_n(t)d\Lambda(t)$ and its variance is $Y_n(t)d\Lambda(t)(1 - d\Lambda(t))$. For the variance, the last factor only comes into play if there is a jump in $\Lambda(t)$, which means we can write it as $Y_n(t)(1 - \Delta\Lambda(t))d\Lambda(t)$. These observations imply two important things:

- The process $\xi(t)$ is a *martingale*. Such processes are generalizations of cumulative (partial) sums of small stochastic variables with mean zero, and are such that, conditionally on what we know at time s, the prediction (conditional mean) for $\xi(t)$ at a time $t > s$ is the present observation, $\xi(s)$. In particular, the global mean is zero, since $\xi(0) = 0$.

- The process

$$\langle\xi\rangle(t) = \int_0^t Y_n(s)(1 - \Delta\Lambda(s))d\Lambda(s)$$

takes on the role of a variance. Since it is a stochastic process, it is not the true variance, but there is a CLT for martingales (described in Appendix 4.A and further explored in Appendix 11.A) which asserts that it can be used for this very purpose.

There is one important observation more to be made. Test statistics are usually weighted averages of a predictable process $\{H(t); t \geq 0\}$, weighted by the martingale $\{\xi(t); t \geq 0\}$ (weighted averages of the differences between what is observed and what is predicted):

$$\xi_H(t) = \int_0^t H(s)d\xi(s).$$

Such a weighted process is also a martingale (use the fact that the conditional expectation of $d\xi_H(t) = H(t)d\xi(t)$, given the state just prior to time t, can be computed by pulling $H(t)$ out of the expectation), and its variance process is given by

$$\langle \xi_H \rangle(t) = \int_0^t H(s)^2 d\langle \xi \rangle(s) = \int_0^t H(s)^2 Y_n(s)(1 - \Delta\Lambda(s))d\Lambda(s).$$

There now follow two important applications of this. The first explains how the mathematics works, when we derive information about right-censored processes.

Example 11.4   A censoring process $C(t)$ is a predictable process which is one at time t if the individual is observed, and zero otherwise. If $N(t)$ is the number (zero or one) of events that have occurred for an individual at time t, the number of *observed* events for that individual is given by the integral $\int_0^t C(s)dN(s)$. Similarly, the number (zero or one) that are observed to be at risk at time t will be $C(t)Y(t)$. We then see that the process $\{\xi(t); t \geq 0\}$ above, which was previously discussed in the absence of any censoring mechanism, has the properties stated above also when $N_n(t)$ counts only the observed events and $Y_n(t)$ denotes the number observed to be at risk.

Important special cases of the censoring mechanism are obtained as follows. A stochastic time variable Z such that $C(t) = I(Z \geq t)$ is a censoring process is referred to as a censoring time and corresponds to a right-censoring mechanism. More generally, we can have two independent continuous stochastic variables, U, V such that $0 \leq V \leq U$ and define $C(t) = I(V < t \leq U)$. This is called Aalen filtering and includes both right-censoring and left-censoring as special cases.

The next example shows how we can obtain knowledge about $\Lambda(t)$ from the Nelson–Aalen estimator $\Lambda_n(t)$. What we need is the variance of $\Lambda_n(t) - \Lambda(t)$.

Example 11.5   For the Nelson–Aalen estimator we have that

$$\Lambda_n(t) - \Lambda(t) = \int_0^t \frac{1}{Y_n(s)}(dN_n(s) - Y_n(s)d\Lambda(s)) = \int_0^t \frac{d\xi(s)}{Y_n(s)},$$

and since $H(t) = Y_n(t)^{-1}$ is a predictable process, we have that the variance process for $\Lambda_n(t)$ is given by

$$\int_0^t \frac{1}{Y_n(s)^2} Y_n(s)(1 - \Delta\Lambda(s))d\Lambda(s) = \int_0^t \frac{1}{Y_n(s)}(1 - \Delta\Lambda(s))d\Lambda(s).$$

If we know that there are no jumps in $\Lambda(t)$, the immediate estimate of this is obtained by replacing $d\Lambda(t)$ with $d\Lambda_n(t)$. In the general case, recalling the discussion on page 158 on estimation of the binomial variance, we obtain the tie-corrected variance estimate

$$\int_0^t \frac{Y_n(s) - \Delta N_n(s)}{Y_n(s)^2(Y_n(s) - 1)} dN_n(s).$$

Note that when $\Delta N_n(t) = 1$ the integrand reduces to the simpler expression $dN_n(t)/Y_n(t)^2$. Using this as a variance estimate, we can derive confidence intervals for $\Lambda(t)$.

The last observation in this example justifies a quick discussion about ties for time-to-event data. Even though there might be situations where the design of a study (or outcome variable) is such that we could have pure jumps in the hazard, this is very rare. With continuous time we therefore have that $\Delta\Lambda(t) = 0$. However, we only measure time to a certain precision, such as days. This means that two events that occurred at different time points may be recorded as having occurred at the same time point. Had we measured to the precision of hours, many of these ties would probably have been broken, and with more detailed timings even more so. The proper way to handle ties that occur this way is to modify the Nelson–Aalen estimator to account for what is really happening. In fact, if we have no censored observations at a time t but d events, and two events cannot occur simultaneously, the estimate of $d\Lambda(t)$ should be

$$d\Lambda_n(t) = \frac{1}{Y_n(t)} + \frac{1}{Y_n(t) - 1} + \ldots + \frac{1}{Y_n(t) - d + 1},$$

because first one occurs, then the next, etc. If there were censored events at the same time, this would not be true, but still better than the crude estimate. If there is a mixture of events and censorings, $d\Lambda_n(t)$ should be the average of all possible ways in which what we observe can occur, which makes this a combinatorial problem.

11.8 The Kaplan–Meier estimator of the CDF

We conclude this chapter with a more mathematical derivation of the Kaplan–Meier estimator for the CDF from the Nelson–Aalen estimate of the hazard rate. Our main objective is to find its variance in the presence of right-censored data, which allows us to investigate the CDF $F(t)$, using the methods discussed in Chapter 6.

For a continuous CDF we have that $F^c(t) = e^{-\Lambda(t)}$, and it is therefore tempting to use the function $e^{-\Lambda_n(t)}$ as an estimator for $F^c(t)$. This has indeed been suggested, and is called the Breslow estimator of the CDF. However, it is not the natural estimator of $F(t)$ based on $\Lambda_n(t)$, a role that is taken by the Kaplan–Meier estimator. Before we look closer into why this is, we compare these two survival function estimators in a numerical example.

Example 11.6 Suppose that the logarithm of the sputum count in Section 6.2 instead describes the survival time (in years) of 20 subjects after they have had a particular cancer diagnosis. There are no censored data. Figure 11.7(a) shows the Nelson–Aalen estimate (solid gray curve) of the true cumulative hazard (dashed curve) as well as pointwise confidence limits (solid black curves) for the (true) cumulative hazard. The jumps in $\Lambda_n(t)$ become larger as t increases, for the obvious reason that the size of a jump is inversely proportional the number at risk, of which there are fewer later than early on.

Figure 11.7(b) shows the true $F^c(t)$ (dashed line) as well as the two estimates of the CDF discussed above, the Kaplan–Meier and Breslow respectively. We see that although they are similar, they differ in details. In particular, we see that the Breslow estimator lies above the Kaplan–Meier estimator everywhere. This is always true because of the inequality $e^{-x} \geq 1 - x$, which holds for all x, since it implies that

$$F_n^c(t) = \prod_{s \leq t}(1 - d\Lambda_n(s)) \leq \prod_{s \leq t} e^{-d\Lambda_n(s)} = e^{-\Lambda_n(t)}.$$

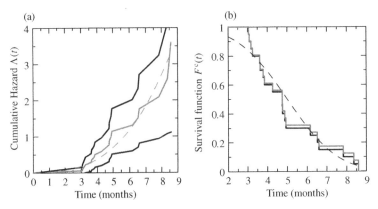

Figure 11.7   (a) shows the cumulative hazard with confidence limits (linearly interpolated), with the true function as the dashed curve. (b) shows the true survival function (dashed) together with the Kaplan–Meier estimate (black) and Breslow estimate (gray) thereof (estimates as step functions).

In order to see where the Kaplan–Meier estimator comes from mathematically, we rewrite the equation (11.1) as an integral equation in the survival function:

$$F^c(t) = 1 + \int_0^t F^c(s-)d(-\Lambda(s)). \qquad (11.5)$$

Its solution can be written as the 'product integral'

$$F^c(t) = \prod_{s \le t}(1 - d\Lambda(s)),$$

a result due to the Swedish mathematician Ivar Fredholm in the early twentieth century (see also Box 11.7). To obtain the Kaplan–Meier estimate of $F(t)$ from this, we simply replace $d\Lambda(t)$ with the Nelson–Aalen estimate $d\Lambda_n(t)$ to get the expression

$$F_n^c(t) = \prod_{s \le t}(1 - d\Lambda_n(s)).$$

In order to describe what knowledge we have obtained about $F(t)$ from the Kaplan–Meier estimator $F_n(t)$, we mostly rely on large-sample theory, and for this we need to compute its variance. In the absence of censored data we know that $F_n(t)$ is the standard e-CDF, for which the variance is $F(t)F^c(t)/n$. The presence of censored data is expected to increase this variance (compared with if the censored data were actually observed), because censoring means an increase in uncertainty. We seek an expression for the variance $V(F_n(t))$ in $F^c(t)$ and $d\Lambda(t)$, and the trick for this is the observation that

$$\frac{F(t)}{F^c(t)} = \frac{1}{F^c(t)} - 1 = \int_0^t \frac{dF(s)}{F^c(s)F^c(s-)} = \int_0^t \frac{d\Lambda(s)}{F^c(s-) - \Delta F(s)}.$$

Box 11.7 The product-limit estimator

For a given function $A(t)$, let the function $B(t)$ satisfy the integral equation

$$B(t) = 1 + \int_0^t B(s-)dA(s).$$

We can write $B(t+h)$ as $B(t) + \int_t^{t+h} B(s-)dA(s) \approx B(t)(1 + \Delta A(t))$, which motivates why, if $A(t)$ is a step function, the solution of the integral equation is given by

$$B(t) = \prod_{s \in [0,t]} (1 + \Delta A(s)) = \mathcal{P}_0^t(A).$$

This defines the *product-integral* $\mathcal{P}_0^t(A)$ for step functions, and by imitating how we define the Stieltjes integral, this can be extended to many functions $A(t)$ (those of bounded variation). If $A(t)$ is continuous we have that $B(t) = e^{A(t)}$. (Note that $A(t)$ and $B(t)$ can be matrix functions, which is useful in the study of Markov processes.) A consequence of this is that the relation between $\Lambda(t)$ and $F(t)$ defined by equation 11.5 is that $F^c(t) = \mathcal{P}_0^t(-\Lambda)$, and that the Kaplan–Meier estimator is defined by $F_n^c(t) = \mathcal{P}_0^t(-\Lambda_n)$.

The key result for product integration is *Duhamel's formula*,

$$\mathcal{P}_0^t(A) - \mathcal{P}_0^t(B) = \int_0^t \mathcal{P}_0^{s-}(A)(dA(s) - dB(s))\mathcal{P}_{s+}^t(B).$$

Applied to $A = \Lambda_n$ and $B = \Lambda$, it shows that

$$F_n^c(t) - F^c(t) = \int_0^t F_n^c(s-)(d\Lambda_n(s) - d\Lambda(s))F^c(t)/F^c(s),$$

which can be rewritten as

$$\frac{F_n^c(t) - F^c(t)}{F^c(t)} = \int_0^t \frac{F_n^c(s-)}{F^c(s)}(d\Lambda_n(s) - d\Lambda(s)) = \int_0^t \frac{F_n^c(s-)}{F^c(s)Y_n(s)}d\xi_n(s). \qquad (11.6)$$

The compensator of this martingale is

$$\int_0^t \frac{F_n^c(s-)^2(1 - \Delta\Lambda(s))d\Lambda(s)}{F^c(s)^2Y_n(s)} \approx \int_0^t \frac{dN_n(s)}{Y_n(s)(Y_n(s) - \Delta N_n(s))},$$

where $\approx$ means 'can be estimated by'.

Estimating this integral in the obvious way, we find that approximately

$$V(F_n(t)) = F^c(t)^2\sigma_n^2(t), \qquad \sigma_n^2(t) = \int_0^t \frac{dN_n(s)}{Y_n(s)(Y_n(s) - \Delta N(s))}.$$

Martingale theory confirms this (see Box 11.7). In the notation most often used, with d_j as the number of events at time t_j, and r_j as the number at risk at the same time, we have that

$$\sigma_n^2(t) = \sum_{t_j \leq t} \frac{d_j}{r_j(r_j - d_j)}. \qquad (11.7)$$

From this we deduce the classical *Greenwood formula* approximation to the variance for $F_n(t)$, which is $V(F_n(t)) \approx F_n^c(t)^2 \sigma_n^2(t)$. When there are no censored data we have that $r_{j+1} = r_j - d_j$, and this variance expression becomes $F_n(t)(1 - F_n(t))/n$.

When we make inference about $F(t)$ from the Kaplan–Meier estimator $F_n(t)$, we can use large-sample theory, which is outlined in Appendix 11.A, to deduce that

$$\frac{(F_n^c(t) - F^c(t))^2}{F^c(t)^2 \sigma_n^2(t)} \in As\chi^2(1).$$

Once we have this, we can do the same thing as we did when we described $F(t)$ from complete data, using the standard e-CDF. From this observation we can obtain confidence regions for $F(t)$ as well as confidence intervals for percentiles, using the methods discussed in Chapter 6.

11.9   Comments and further reading

A short introduction to what makes survival data special is given by Hougaard (1999), which expands on some of the points made above. When it comes to textbooks on the analysis of this kind of data, there is a choice available for almost any taste – theoretical, example-driven, for dummies, etc. – and new books appear more or less every year. Some (historically) important ones can be found among the references below.

The problem of competing risks is a controversial subject (Hougaard, 2000, Chapter 12). The basic problem with analyzing a particular cause of death in the presence of competing causes is that there is not sufficient information available to settle the problem, which is why we need to make additional assumptions. This creates a philosophical dilemma regarding what can legitimately be done: can we think about cause-specific survivals at all, using models that are such that we can identify these (like an independence assumption, or using so-called copulas) or should we restrict ourselves to what we actually can observe? Examples of advocates for these two positions are Zheng and Klein (1994) and Prentice et al. (1978), respectively.

The paper that started it all, by Daniel Bernoulli, has been republished (Bernoulli, 2004) in a review which starts with the following quote from the author in 1760: 'I simply wish that, in a matter which so closely concerns the well-being of the human race, no decision shall be made without all the knowledge which a little analysis and calculation can provide.' The data he used (Halley, 1693) can, at the time of writing, be found on the internet at http://www.pierre-marteau.com/editions/1693-mortality.html. We should note that Bernoulli added a 'guesstimate' to this table for the number of births.

A more detailed discussion on the smallpox model, its history and some generalizations can be found in Dietz and Heesterbeek (2002), including a discussion on a dispute Bernoulli had with the French mathematician d'Alembert. The latter advocated a simpler competing risk model, in which you either die from smallpox or not. The model he suggested was the general approach to competing risks we use today, whereas Bernoulli's model is restricted to infection-type diseases. Modeling competing risks is a first step toward the wider subject of multi-state models, to which Andersen and Keiding (2002) provide an introduction, and about which most reasonably advanced textbooks have something to say.

An introduction to the frailty problem in survival analysis is given by Aalen (1994), whereas a more extensive discussion is found in the book by Hougaard (2000), who suggests using a three-parameter family of distributions to describe frailty, as an alternative to the gamma distribution.

References

Aalen, O. (1975) *Statistical inference for a family of counting processes*, PhD thesis University of California, Berkeley.

Aalen, O. (1978) Nonparametric inference for a family of counting processes. *Annals of Statistics*, **6**(4), 701–726.

Aalen, O. (1994) Effects of frailty in survival analysis. *Statistical Methods in Medical Research*, **3**(3), 227–243.

Andersen, P., Borgan, Ø., Gill, R. and Keiding, N. (1993) *Statistical Models Based on Counting Processes*. New York: Springer.

Andersen, P.K. and Keiding, N. (2002) Multi-state models for event history analysis. *Statistical Methods in Medical Research*, **11**(2), 91–116.

Bernoulli, D. (2004) An attempt at a new analysis of the mortality caused by smallpox and of the advantages of inoculation to prevent it. *Reviews of Medical Virology*, **14**(5), 275–288.

Cox, D.R. and Oakes, D. (1984) *Analysis of Survival Data* Monographs on Statistics and Applied Probability. London: Chapman & Hall.

Dietz, K. and Heesterbeek, J. (2002) Daniel Bernouilli's epidemiological model revisited. *Mathematical Biosciences*, **180**, 1–21.

Fleming, T.R. and Harrington, D.P. (1991) *Counting Processes and Survival Analysis*. New York: John Wiley & Sons, Inc.

Halley, E. (1693) An estimate of the degrees of the mortality of mankind, drawn from curious tables of the births and funerals at the city of Breslau, with an attempt to ascertain the price of annuities upon lives. *Philosophical Transactions of the Royal Society*, **17**, 569–610.

Hougaard, P. (1999) Fundamentals of survival data. *Biometrics*, **55**(1), 13–22.

Hougaard, P. (2000) *Analysis of Multivariate Survival Data*, Statistics for Biology and Health. New York: Springer.

Kalbfleisch, J.D. and Prentice, R.L. (2002) *The Statistical Analysis of Failure Time Data*. Hoboken, NJ: John Wiley & Sons, Inc.

Prentice, R.L., Kalbfleisch, J.D., Peterson, A.V., Flournoy, N., Farewell, V.T. and Breslow, N.E. (1978) The analysis of failure time data in the presence of competing risks. *Biometrics*, **34**, 541–554.

Zheng, M. and Klein, J.P. (1994) A self-consistent estimator of marginal survival functions based on dependent competing risk data and the assumed copula. *Communications in Statistics – Theory and Methods*, **23**, 2299–2311.

11.A   Appendix: On the large-sample approximations of counting processes

The natural mathematical framework for time-to-event data is the theory of counting processes, as noted by the Norwegian statistician Odd Aalen (1975; see also 1978). For a comprehensive account of the theory, see Andersen et al. (1993) or Fleming and Harrington (1991), each of which contains references to earlier work. Other important books on the subject are Cox and Oakes (1984) and Kalbfleisch and Prentice (2002). In this appendix we will give a heuristic outline of some large-sample theory for counting processes with applications in practical statistics. Martingales and predictable processes in continuous time were introduced in Appendix 4.A, where the variance process was given its proper name, the compensator. Recall that if $\{\xi(t); t \geq 0\}$ is a martingale and $\{H(t); t \geq 0\}$ is a predictable process, then the process defined by the integrals $\xi_H(t) = \int_0^t H(s)d\xi(s)$ is also a martingale. This integral is a stochastic integral, which is a non-trivial thing to define since it involves two limiting processes, as indicated in Box 11.8.

In order to review large-sample theory for counting processes, let (as before) $N_n(t)$ be the total number of observed events and $Y_n(t)$ the number at risk. If $\Lambda_n(t)$ is the Nelson–Aalen estimator of $\Lambda(t)$, the process $\{z_n(t) = \sqrt{n}(\Lambda_n(t) - \Lambda(t)); t \geq 0\}$ has compensator

$$\langle z_n \rangle(t) = n \int_0^t \frac{1}{Y_n(s)}(1 - \Delta\Lambda(s))d\Lambda(s),$$

which converges to the function $\tau_0(t) = \int_0^t Y_C(s)^{-1}(1 - \Delta\Lambda(s))d\Lambda(s)$ as n increases. Here $Y_C(t)$ is the limit of $Y(t)/n$. Using the CLT for martingales (see Appendix 4.A), we can deduce that

$$\{\sqrt{n}(\Lambda_n(t) - \Lambda(t)); \ t \geq 0\} \to \{w(\tau_0(t)); \ t \geq 0\} \quad \text{in distribution,}$$

where $\{w(t); t \geq 0\}$ denotes the Wiener process. To generalize this, consider a predictable process $H_n(t)$ such that $H_n(t) \to h(t)$ in probability, where $h(t)$ is a function. We can then define the stochastic integral

$$I_n(t) = \sqrt{n} \int_0^t H_n(s)d(\Lambda_n(s) - \Lambda(s)) = \int_0^t H_n(s)dz_n(s),$$

whose compensator is

$$\langle I_n \rangle(t) = \int_0^t H_n(s)^2 d\langle z_n \rangle(s) \to \int_0^t h(s)^2 d\tau_0(s).$$

From the discussion above it follows that $\{I_n(t); t \geq 0\}$ has the same asymptotic distribution as the stochastic process $\{w(\int_0^t h(s)^2 d\tau_0(s)); t \geq 0\}$. If we combine this observation with equation (11.6) in Box 11.7 we obtain the limit theorem for the Kaplan–Meier estimator, which says that

$$\frac{\sqrt{n}(F(t) - F_n(t))}{F^c(t)} = \int_0^t H_n(s)dz_n(s), \quad H_n(t) = \frac{F_n^c(t-)}{F^c(t)}.$$

Box 11.8 Stochastic integrals

In order to define the integral $I(g) = \int_0^\infty g(t)dx(t)$, where $\{x(t); t \geq 0\}$ is a stochastic process with $E(x(t)) = 0$ for all t, we start by defining it for a piecewise constant function such that $g(x) = g_k$ when $t_k < t \leq t_{k+1}$, as

$$I(g)(\omega) = \sum_{k=1}^n g_k \Delta_k x(\omega), \quad \Delta_k x(\omega) = x(t_{k+1}, \omega) - x(t_k, \omega).$$

If the increments $\Delta_k x$ over disjunct intervals are uncorrelated, it follows that

$$E(I(g)^2) = \sum_k g_k^2 E((\Delta_k x)^2) = \sum_k g_k^2 \Delta_k V = \int_0^\infty g(t)^2 dV(t),$$

where $V(t) = E(x(t)^2)$. We can then show that this is also true for a more general class of functions $g(t)$, including continuous functions. The problem starts when we want to allow $g(t)$ to be a stochastic process, that is, a function that depend on ω as well. For a piecewise constant process (fixed jump points) we would have

$$I(g)(\omega) = \sum_{k=1}^n g_k(\omega) \Delta_k x(\omega),$$

but there is no general limiting process that allows us to define this in more general terms unless we make some additional assumptions. One set of assumptions under which the construction goes through is if the process $\{x(t); t \geq 0\}$ is a martingale, and the process $\{g(t); t \geq 0\}$ is a predictable process.

Since $H_n(t)$ converges to $h(t) = F^c(t-)/F^c(t)$ in probability, it follows from this that

$$\{\sqrt{n}(F_n(t) - F(t)); t \geq 0\} \to \{F^c(t)w(\tau(t)); t \geq 0\} \quad \text{in distribution,}$$

where

$$\tau(t) = \int_0^t \left(\frac{F^c(s-)}{F^c(s)}\right)^2 \frac{1 - \Delta\Lambda(s)}{Y_C(s)} d\Lambda(s) = \int_0^t \frac{dF(s)}{Y_C(s)F^c(s)}.$$

For complete data we have that $Y_C(t) = F^c(t-)$ and therefore that $\tau(t) = F(t)/F^c(t)$. The limiting distribution is then $F^c(t)w(F(t)/F^c(t))$, and if we compute the covariance structure for this process, we see that it is $F(s) \wedge F(t) - F(s)F(t)$. This means that the limiting distribution is that of $w^0(F(t))$, where $w^0(t)$ is the Brownian bridge, as was found in Appendix 6.A.3.

12

From the log-rank test to the Cox proportional hazards model

12.1 Introduction

We will now address for censored time-to-event data what we outlined for complete data in Chapter 8: how to compare two groups. The problem is really the same, we can compare distributions by comparing the value of their e-CDFs at specific points, or by investigating percentiles, but instead of using the conventional e-CDF, we use the Kaplan–Meier estimator of the CDF with its associated variance. Because mean values usually are of minor interest in this context (we often have only partial knowledge about the CDF for large times) there is no t-test for this situation. There are, however, non-parametric tests, and as in our previous discussion on complete data we will focus on the Wilcoxon test. In doing this we will find that the Mantel–Haenszel test is buried inside most of these extensions; a variation of it, called the log-rank test, provides the building blocks. This test is actually the important test in this context, overshadowing the Wilcoxon test.

The most important models for time-to-event data are different from those for most other data. They are models for hazards, the two key examples being the AFT model and the proportional hazards model. The first of these is often analyzed within the framework of parametric models, using particular distributions, such as the Weibull distribution. This distribution is also useful for proportional hazards models, but in that context the use of parametric methods is completely overshadowed by non-parametric methods.

Working under the assumption of a homogeneous world, it is a direct extension to go from a two-group non-parametric test to a semi-parametric proportional hazards regression model. In the case of the log-rank test this is better known as the Cox proportional hazards model, or Cox model for short, one of the brightest stars in the firmament of biostatistics; nowadays it is so important in cancer research, that when you do not use it for the analysis of survival data, you may need to provide explicit excuses. The Cox model tries to explain the frailties in terms of specified covariates, and the log-rank test is simply the case where we only have the

Understanding Biostatistics, First Edition. Anders Källén.
© 2011 John Wiley & Sons, Ltd. Published 2011 by John Wiley & Sons, Ltd.

group indicator variable available. The Cox model is a nonlinear model, and somewhat similar to the logistic regression model for binary data. Like it, there are consequences of omitting important and predictive covariates in the model, in that effects get diluted. We will end with an example of this, which also gives us an opportunity to compare a regression model to a stratified analysis in this setting.

12.2   Comparing hazards between two groups

In this section we will discuss some of the more immediate ways to compare two groups with respect to a time-to-event variable with censored data. In order to illustrate the different methods we will use the data described in the next example.

Example 12.1   In order to investigate whether a certain drug increases the risk of a particular cancer, an experiment was carried out on 150 female rats from 50 litters. One pup from each litter was chosen for drug treatment, together with two control animals. The rats were followed for the occurrence of a tumor for 2 years, after which they were sacrificed; the maximum observed time is therefore 104 weeks.

The overall result can be described as follows. Of the 50 drug-exposed rats, 21 died from the cancer, whereas of the 100 controls, 19 died from the cancer. The probability of cancer death in the drug-treated group is therefore estimated to be 0.42, whereas it is about half that for the controls, 0.19. Using these numbers only, we apply Fieller's method as described in Section 5.4.2 to obtain a risk ratio of 2.21 with 95% confidence interval (1.31, 3.98).

This analysis is an end-point analysis of the occurrence of the event cancer death in the presence of other causes of death. In fact, more rats died from other causes than from the particular cancer under study. This is illustrated in Figure 12.1, which shows the CIFs (see Section 11.4), both for all-cause mortality and for cancer deaths. The (right) end-point of the black curves corresponds to the result in Example 12.1.

To assess the effect of the drug, we want to analyze the cancer mortality in an environment free of competing causes of death. This is what we (try to) do with the Kaplan–Meier estimator,

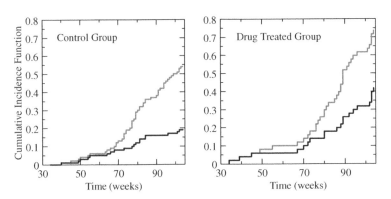

Figure 12.1   The cumulative incidence functions for the two groups. The gray curves show the all-cause mortality, and the black curves show the mortality for a particular cancer. The difference is the mortality for other causes.

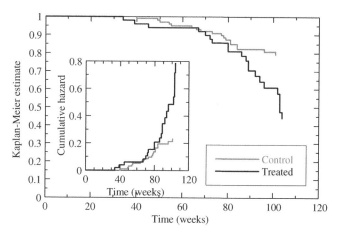

Figure 12.2   The larger graph shows the Kaplan–Meier estimates of the survival distributions for the drug-exposed rats (black) and their controls (gray). The inset graph shows the corresponding Nelson–Aalen estimates of the cumulative hazards.

which are shown in Figure 12.2 for the two groups. From this graph the probabilities of cancer death within 2 years are estimated to be 0.56 and 0.22 in the two groups, respectively. These numbers differ from, and are higher than, those in Example 12.1. They also give a different risk ratio, namely 2.49. With the same kind of analysis as in Example 12.1, but using Greenwood's variance estimate instead, the corresponding 95% confidence interval is (1.52, 4.45). The difference between this analysis and the previous one is that the new estimates address the mortality ratio when the only cause of death for the rats is the particular cancer, under the assumption that competing risks of death act independently of the risk of interest. The inset graph in Figure 12.2 displays the (Nelson–Aalen) estimates of the cumulative hazards. We see that nothing happens for about 30 weeks, whereafter the cumulative hazard increases sharply, most pronounced for the drug-treated group. From now on most of our discussion will be concerned with comparing two CDFs, which means censoring relevant competing risks. We will, however, make the occasional remark on the competing risk case as well.

With these analyses, both of which can be made more precise using profiling techniques that take all variability into account, we have small p-values associated with the hypothesis that the true relative risk is one.

An alternative way to compare $F(t)$ and $G(t)$ would be to compare some specified percentiles. To do this we can apply the methods described in Section 8.3, again modified so that they use the Greenwood variance estimate for the Kaplan–Meier estimators. All such comparisons of two survival functions are comparisons of a single aspect only, and we want to find methods which use the full Kaplan–Meier estimates when we compare groups. It is to this that we now turn.

12.3   Nonparametric tests for hazards

When we design tests to compare time-to-event data for two groups, we want these so constructed that they compare the hazards, not the CDFs. On a high level this is immaterial, because of the relationship between the hazard and the CDF, but because of the nature of time-to-event data it is natural to model what holds true instantaneously. This also allows us to handle censored data smoothly.

There are a number of tests available for this situation, many of them from the 1960s and 1970s. The first to attain widespread use was Gehan's generalization of the Wilcoxon test, which was constructed by extending the Mann–Whitney score (defined on page 219) in such a way that it is set to zero when it is not known which of the two variables is the largest. At about the same time Mantel used a Mantel–Haenszel type of argument to propose a test that nowadays is known as the log-rank test. Further tests have been proposed by others, but most are variations on a theme.

The test construction process starts with equation (11.1), which shows how we build $F(t)$ from knowledge about the hazard $d\Lambda(t)$ (which we denote $d\Lambda_F(t)$ when we wish to emphasize its relation to the CDF $F(t)$) and the proportion at risk $F^c(t-)$. The basic idea for test construction is that, under a specific assumption about the relation between the two distributions, we can express $d\Lambda(t)$ in the CDFs $F(t)$ and $G(t)$, thereby providing a test statistic which should be close to zero if the model is correct.

To be more specific, we weight the differences $dF(t) - F^c(t-)d\Lambda(t)$ with a particular weight function $w(t)$, which means that we define

$$\Delta = \int_0^\infty w(t)(dF(t) - F^c(t-)d\Lambda(t)). \tag{12.1}$$

This is by definition zero when $\Lambda(t) = \Lambda_F(t)$, which is the important observation to bring forward. The choice of weight function $w(t)$ is subject to some constraints. First of all, we want to use weights constructed from the CDFs of the problem. Second, we want $w(t)$ to be estimated by predictable processes, so the function $w(t)$ should be continuous from the left. With these constraints an immediate choice would be to take

$$w(t) = a(\Psi^c(t-)) \tag{12.2}$$

for some function $a(u)$. This is not a necessary choice; we could use some other function of $F^c(t-)$ and $G^c(t-)$. The particular choice $a(u) = u^\rho$ defines, varying $\rho \in [0, 1]$, the Fleming–Harrington family (of tests). The two border cases $\rho = 0$ and $\rho = 1$ in this family are of particular interest. As can be seen in Box 12.1, the choice $\rho = 1$ gives us the Wilcoxon test, while the choice $\rho = 0$ is what defines the fundamental log-rank test.

In this section our discussion will address the null hypothesis of no group difference. In other words, we assume that $G(t) = F(t)$, which implies that the hazard for the first group is the same as the hazard for the combined sample: $d\Lambda(t) = d\Lambda_\Psi(t)$. Under this assumption we wish to find an estimator of the parameter Δ in the presence of right-censored data. The obvious choice is to estimate $d\Lambda(t)$ with the Nelson–Aalen estimator of the combined sample, which gives us the stochastic variable

$$\hat{\Delta} = \int_0^\infty \hat{w}(t)\left(dN_n(t) - Y_n(t)\frac{dN_+(t)}{Y_+(t)}\right). \tag{12.3}$$

Here the single subscript n refers to data from the first group (with n subjects) and the subscript $+$ means that we sum over both groups. We have ignored a proportionality factor. The expected value of $\hat{\Delta}$ is zero under the null hypothesis of no group difference, so we can use this test statistic to test the null hypothesis. For this purpose we need to derive an

Box 12.1 The limits of the Fleming–Harrington family

The parameter Δ defined by equation (12.1) is of particular interest in the following two cases.

Case $\rho = 1$. If we take $a(u) = u$ and change the order of integration in a double integral, we find that

$$\Delta = \int_0^\infty \Psi^c(t-)dF(t) - \int_t^\infty dF(s)d\Psi(t) = \int_0^\infty (\Psi^c(t-) - \int_0^t d\Psi(s))dF(t)$$

$$= \int_0^\infty (1 - (\Psi(t-) + \Psi(t)))dF(t)$$

$$= 2\left(\frac{1}{2} - \int_0^\infty \frac{G(t-) + G(t)}{2}dF(t)\right).$$

The condition $\Delta = 0$ is therefore equivalent to equation 8.6, which defines the Wilcoxon test.

Case $\rho = 0$. If we take $a(u) = 1$ we find that

$$\Delta = 1 - \int_0^\infty F^c(t-)d\Lambda_\Psi(t) = \int_0^\infty (1 - \Lambda_\Psi(t))dF(t),$$

which leads to the log-rank test. The name is justified because for continuous distributions the right-hand side can be written as $1 + \int_0^\infty \ln(\Psi^c(t))dF(t)$.

Like the Wilcoxon test, the log-rank test defines a rank test for complete data; it can be written as

$$\int_0^\infty \left(1 - \int_{-\infty}^t \frac{d\Psi_{mn}(s)}{\Psi_{mn}^c(s-)}\right) dF_n(t) = 1 - \frac{1}{n}\sum_{i=1}^n a(R_{mn}(x_i)),$$

where (assuming no ties) we have

$$a(k) = \sum_{i=1}^k \frac{1}{n + m + 1 - i}.$$

These scores are the expected value of the kth-order statistic in a sample of size $n + m$ from a Exp(1) distribution and were originally introduced by Leonard Savage in order to test the null hypothesis of equal distributions against the alternative that $G(x) \geq F(x)$ with strict inequality at least one point; he proved that it was the best test for the one-parameter model (Lehmann alternative) $G(x) = F(x)^\theta, \theta > 1$. Note that if there are no ties (and no censoring) then the Nelson–Aalen estimator for the cumulative hazard is $\Lambda_{nm}(t_k) = a(k)$.

estimate of its variance, and then appeal to the CLT. The choice of $\hat{w}(t)$ is not unique even if we have decided on the weight function, because we can estimate $\Psi^c(t-)$ in different ways. One choice is to use the Kaplan–Meier estimate for $\Psi(t)$, lagged one time step to ensure predictability. Alternatively, we can estimate it by $Y_+(t)/(n + m)$. These different tests have slightly different interpretations. If we use the Kaplan–Meier estimate we take weights from an environment that is free of other risks, whereas if we use the second version we take weights that depend on the competitive risks present when the data were collected. For the Wilcoxon test, if we choose the Kaplan–Meier estimator when we estimate $\Psi^c(t-)$, we get either the Prentice version of the Wilcoxon test for censored data, or a variant due to Peto and Peto which depends on details we ignore. If we instead estimate it from the 'at-risk' function, we derive Gehan's version of the test, for which the test statistic is

$$\int_0^\infty \frac{Y_+(t)}{n+m}\left(dN_n(t) - Y_n(t)\frac{dN_+(t)}{Y_+(t)}\right).$$

In the sequel, when we refer to the Wilcoxon test for censored data, we mean this version.

The (partial) log-rank test (up to time t) can be written as

$$N_n(t) - \int_0^t \hat{p}(s)dN_+(s), \quad \hat{p}(t) = \frac{Y_n(t)}{Y_+(t)}.$$

The entity $\hat{p}(t)$, which is a predictable process, is an estimate of the conditional probability that an event which we know occurs at time t, occurs in the first group. The test is therefore simply the difference between the number of events that have occurred in the first group and our prediction of what should happen, conditional on the situation just before each event. With this interpretation we see that the variance of the log-rank test can be derived from the variance of the binomial distribution as

$$\int_0^t \hat{p}(s)(1 - \hat{p}(s))dN_+(s).$$

We can use this to compute the p-value for the test of the null hypothesis, at least if there are no ties. In the presence of ties we need to split these and, referring to the observation on page 158, we have the adjusted formula:

$$\int_0^t \hat{p}(s)(1 - \hat{p}(s))\frac{Y_+(s) - \Delta N_+(s)}{Y_+(s) - 1}dN_+(s).$$

The only reason we mention this correction is that it helps us to understand how Nathan Mantel arrived at the log-rank test. For this we write the integral explicitly as a sum. The integral in equation (12.3), with $\hat{w}(t) = 1$, is a sum over event times t_j, namely

$$\sum_j \left(d_{1j} - \frac{n_{1j}d_j}{n_j}\right),$$

where n_{ij} is the number at risk in group i at time t_j, d_{ij} the corresponding number of events in the respective groups and n_j, d_j the total number at risk and number of events, respectively. In this notation the variance above is given by

$$\sum_j \frac{n_{1j}n_{2j}d_j(n_j - d_j)}{n_j^2(n_j - 1)}.$$

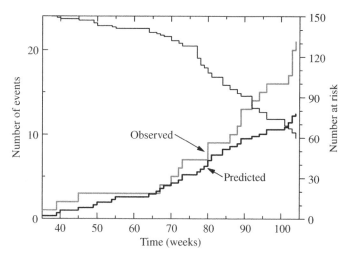

Figure 12.3   The observed and predicted number of deaths in the drug-treated rat group, as well as the number 'at risk' (the decreasing curve in the upper half of the graph) at each time point.

This means that the test statistic is formally the same as the Mantel–Haenszel test in Section 5.5. The strata are the 2×2 tables we obtain at the event times, with the analysis within each table done conditionally on margins.

Example 12.2   Figure 12.3 shows the key information for the log-rank test and the Wilcoxon test for the rat data in Example 12.1. The three curves represent the number of cancer deaths, $N_n(t)$, in the drug-treated group (Observed), the predicted number of cancer deaths in that group based on the combined sample, assuming no difference between the groups (Predicted), and the total number at risk, $Y_+(t)$ (the decreasing step function with y-axis to the right). One summary of the log-rank test is as follows:

Group	N	Observed	Expected
Control	100	19	27.55
Drug-treated	50	21	12.45

This test compares $N_n(\infty) = N_n(104)$ to the corresponding predicted value, and gives us the p-value 0.0034. The Wilcoxon test is different: it weights the differences between observed and predicted at each time point, with the weights given by the number of subjects at risk at the same time. In Figure 12.3 we see that this down-weights the parts of the data where the difference is the largest, so it should come as no surprise that the p-value in this case is larger than that for the log-rank test, namely 0.026.

So far we have discussed the time to cancer death as an isolated phenomenon, when all other causes of death are eliminated. If we wish to understand the effect of the drug in the presence of these competing risks, we need to take a different approach. The analysis should now focus on the CIF $G(t)$ (see Section 11.4) for cancer death, as we did in Example 12.1. In general this function may also be prone to censoring; subjects may be lost to follow-up or subject to other censoring mechanisms that are not considered to be real competing risks. This means that we want to apply the survival analysis methods to $G(t)$, despite the fact that it is not a proper distribution function. The methodology only uses the fact that we can write $G(t) = e^{-\Lambda(t)}$, with a corresponding expression as a product-limit operator at jump points. We can therefore perform any of the tests discussed above, based on this $\Lambda(t)$, in order to derive a comparison of the event rate in the presence of competing risks. It means that $\Lambda(t)$ is defined through the relation $d\Lambda(t) = dG(t)/G^c(t-)$, and we arrive at what is called Gray's test. If there are no other censoring mechanisms present, this test can be computed by redefining the stochastic variable, so that observations that are censored due to a competing risk are replaced by infinitely large values. We then analyze this modified variable with a log-rank or Wilcoxon test.

Example 12.3 If we perform the log-rank test on the modified variable for the toxicological rat data in Example 12.1, we get the p-value 0.00445. This is larger than the p-value we found when we compared survival with other causes of death removed, but still statistically significant at the conventional 5% level. The conclusion is that the drug also has an effect on cancer survival in the presence of competing causes of death.

When we apply Gray's test we must be careful not to conclude that an intervention is beneficial for one event type, when it instead increases the incidence of a competing event. To draw conclusions in a competitive environment is more complicated than in the non-competing world targeted by the Kaplan–Meier approach, which helps to explain the popularity of the latter approach.

12.4   Parameter estimation in hazard models

The methodology we used in the previous section has an immediate extension to an estimation method for the appropriate hazard model. Among the models previously discussed, the shift model in equation (8.1) is not really relevant in this situation, in contrast to the model in equation (8.2) which can be expressed in terms of hazards as

$$\Lambda_G(t) = \Lambda_F(t/\theta). \tag{12.4}$$

This model is called the accelerated failure time (AFT) model, and we will discuss it further in Section 12.5. Since the right-hand side is the cumulative hazard for a process with time parameter θT, we call θ an acceleration factor. It describes how much faster the biological clock runs in the second group compared to the first. The most popular model for hazards, however, is arguably the proportional hazards model

$$\Lambda_G(t) = \theta \Lambda_F(t), \tag{12.5}$$

which we encountered in our discussion on frailty in Section 11.5. Here we will focus on this model and discuss how to estimate the model parameter θ. (We may note in passing that, because of the relation between the log-rank test and the Mantel–Haenszel test, this test is really a test of proportional odds in each 2×2 table of events

$$\frac{d\Lambda_G(t)}{1 - d\Lambda_G(t)} = \theta \frac{d\Lambda_F(t)}{1 - d\Lambda_F(t)}.$$

The denominators are one here, because the terms subtracted are zero at a continuity point, which would not be the case if we have truly discrete distributions, instead of continuous ones.)

The problem here is how to estimate $d\Lambda(t)$ from the combined sample, when we assume that the proportional hazards model holds. The way to do this was given in Section 11.5, if we note that the frailty distribution here is the $\theta \text{Bin}(1, 1 - r)$ distribution, which takes values 0 and θ with probability r and $1 - r$, respectively. This means that

$$d\Psi(t, \theta) = (rF^c(t-) + (1 - r)\theta G^c(t-))d\Lambda(t),$$

and we can estimate $d\Lambda(t)$ from the combined sample by

$$\frac{dN_+(t)}{S^0(t, \theta)}, \quad \text{where } S^0(t, \theta) = Y_n(t) + \theta Y_m(t).$$

(Intuitively, if we use the combined sample and there is a twofold increased hazard for group 2, each individual in that group counts as two when we compute the probability, which is why we multiply $Y_m(t)$ by θ in the denominator.) The log-rank test corresponds to the observation that the mean of $N_n(\infty)$ is equal to the mean of

$$\int_0^\infty \hat{p}(t, \theta)dN_+(s), \quad \text{where } \hat{p}(t, \theta) = \frac{Y_n(t)}{S^0(t, \theta)}.$$

If we choose a weight process $\hat{w}(t)$ and apply the discussion above, we arrive at the estimating equation $U(\theta) = 0$ for θ, where

$$U(\theta) = \int_0^\infty \hat{w}(t)(dN_n(t) - \hat{p}(t, \theta)dN_+(t)).$$

Different choices of statistical tests (which means weight function $\hat{w}(t)$) produce different estimates for θ. In order to apply a test to real data and obtain confidence information about θ, we need to have an estimate of the variance of $U(\theta)$. Such an estimate is

$$\hat{V}(\theta) = \int_0^\infty \hat{w}(t)^2 \hat{p}(\theta, t)(1 - \hat{p}(\theta, t))dN_+(t),$$

provided there are no ties. (The true variance depends on the exact censoring mechanism and is therefore seldom possible to compute.) An approximative (two-sided) confidence function for θ is now given by

$$C(\theta) = \chi_1(U(\theta)^2/\hat{V}(\theta)).$$

Example 12.4 In Example 12.2 we applied both the log-rank test and a Wilcoxon test to the rat data; now we wish to estimate the corresponding parameter for the proportional hazards model. The two confidence functions for these tests are shown in Figure 12.4. They are similar in shape, with the one for the Wilcoxon test lying to the left of that for the log-rank test, and provide the following hazard ratio estimates:

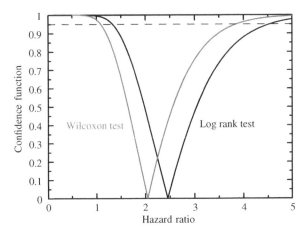

Figure 12.4 Two-sided confidence functions for the hazard ratio parameter using the log-rank and Wilcoxon test statistics.

Test	Estimate	95% CI
Log-rank	2.46	(1.33, 4.53)
Wilcoxon	2.05	(1.09, 3.86)

The fact that the estimate of θ is smaller with the Wilcoxon test is consistent with the observation in Example 12.2 that this test down-weights the parts where the group difference happens to be largest.

So far we have discussed non-parametric group comparison methods under the proportional hazards model. What about parametric analysis? For such an analysis we prefer to use a family of distributions that is closed under the proportional hazards model. This means that if we take one member from such a family and define a new distribution by equation (12.4), this new distribution will belong to the same family. An important example is the Weibull family discussed in Section 11.3, which includes the exponential distributions as a special subfamily. We will use the maximum likelihood method for estimation, and therefore pick up the discussion where we left off in Section 11.3, allowing for individual hazards $\lambda_i(t, \psi)$ including some unknown parameter vector ψ. In this notation, the estimating equation is written as

$$\sum_i \int_0^\infty \frac{\partial_\psi \lambda_i(t, \psi)}{\lambda_i(t, \psi)} (dN_i(t) - Y_i(t)d\Lambda_i(t, \psi)) = 0.$$

For a family closed under the proportional hazards model there is a baseline hazard density $\lambda_0(t, \alpha)$, defined by some parameters α. The complete model is then that $\psi = (\alpha, \theta)$ and $\lambda_i(t, \psi) = \lambda_0(t, \alpha)$ for subjects in the first group, and $\lambda_i(t, \psi) = \theta\lambda_0(t, \alpha)$ in the second group. This means that the estimating equation for α is

$$\int_0^\infty \frac{\partial_\alpha \lambda_0(t, \alpha)}{\lambda_0(t, \alpha)} (dN_+(t) - S^0(t, \theta)d\Lambda_0(t, \alpha)) = 0,$$

Box 12.2 Power analysis of the log-rank test

The log-rank test works conditionally on information about when each event happens, which means that for power calculations we need to make some simplifying assumptions. One such assumption is that the ratio $R = Y_n(t)/Y_m(t) = r/(1 - r)$ does not depend on t. This should be approximately true if either the fraction of individuals with events is small, or θ is close to one. This approximation means that $\hat{p}(\theta, t) = R/(R + \theta)$ at all time points, and that the test statistic is $Z(\theta) = U(\theta)/\sqrt{V(\theta)}$, where $V(\theta) = DR\theta/(\theta + R)$ and D is the total number of events observed. To test the hypothesis $\theta = 1$ we use

$$Z(1) = Z(\theta)\sqrt{\frac{V(\theta)}{V(1)}} + \frac{e(\theta) - e(1)}{\sqrt{V(1)}}$$

where

$$e(\theta) = \int_0^\infty \hat{p}(\theta, t) dN_+(t) = \frac{DR}{R + \theta}.$$

For most relevant θ we have that $V(\theta) \approx V(1)$, and with this approximation we see that

$$Z(1) = Z(\theta) - DV(1)^{-1/2} \frac{R(\theta - 1)}{(\theta + R)(1 + R)} = Z(\theta) - \sqrt{DR} \frac{\theta - 1}{\theta + R}.$$

From this we can compute the power function for a one-sided test:

$$\beta(\theta) = P_\theta(Z(1) \leq -z_\alpha) = \Phi\left(-z_\alpha + \frac{\sqrt{DR}}{\theta + R}(\theta - 1)\right).$$

A further approximation is $(\theta - 1)/(\theta + R) \approx (\ln \theta)/(1 + R)$, which gives the power function

$$\beta(\theta) = \Phi(-z_\alpha + \sqrt{Dr(1 - r)} \ln \theta).$$

In order to find out what θ we are looking for, it may be helpful to note that under the proportional hazards model we have that

$$\theta = \frac{\ln(G^c(\infty))}{\ln(F^c(\infty))},$$

so we reengineer θ from our perception of the percentage of individuals that should not experience the event during the study for each group.

One remaining question is how many patients we need to study in order to get the number of events this calculation assumes. Many clinical trials on survival have an accrual period a, during which patients enter the study, and a follow-up period, f, from the end of the accrual period until the end of the study. In order to assess the number of patients needed, previous information on the survival distribution for one of the treatments from a similar protocol is needed (or some qualified guess). The proportion of patients who will survive is the average of this survival function in the interval $(f, a + f)$, provided the patients enter the trial at a constant rate.

with the same $S^0(t, \theta)$ as above, whereas the estimating equation for the proportionality constant θ is simpler:

$$\theta^{-1} \int_0^\infty (dN_m(t) - \theta Y_m(t)d\Lambda_0(t, \alpha)) = 0.$$

This is essentially the same equation as we have for the log-rank test, except that it is written for the other group. If our parametric model is flexible enough, so that $\Lambda_0(t, \alpha)$ is a reasonable approximation of the true hazard for some α, we therefore do not expect much difference in the result between the parametric model and the log-rank test. (Another observation is that the estimating equation for θ indicates that it may be worthwhile to parameterize in $\beta = \ln \theta$ instead of in θ. This discussion will be picked up again in Chapter 13.) If we know the baseline hazard we can solve this equation and get the estimate

$$\hat{\theta} = \frac{N_m(\infty)}{\int_0^\infty Y_m(t)d\Lambda_0(t, \alpha)},$$

which is the ratio of the number of events we see in the second group, compared to what we expect, based on the known hazard. In other words, it is the standardized mortality ratio.

Example 12.5   We have seen that the Weibull family supports a proportional hazards model, and that the exponential family is a subfamily of it. We therefore compare the two groups of rats in Example 12.1, using these distributions:

Baseline distribution	Estimate	95% CI	p-value
Exponential	2.30	(1.23, 4.27)	0.0086
Weibull	2.47	(1.32, 4.62)	0.0049

Comparing with the result in Example 12.4, we find that for the Weibull distribution we get a result which is very similar to what we obtained with the log-rank test, whereas the result for the exponential distribution is somewhat different. The reason why these two analyses differ can be inferred from Figure 12.2, where we see that if we were to approximate the baseline hazard by a power function (as for the Weibull distribution), the exponent must be greater than one. The shape forced by the exponential distribution, a straight line through the origin, is therefore a poor fit to it.

12.5 The accelerated failure time model

Analysis of time-to-event data under the AFT model in equation (12.4) uses the observation that $\ln T$ fulfills the shift model with shift $\ln \theta$. We can therefore apply the Wilcoxon test to the logged data when there is no censoring. If we have censored data, we can potentially still estimate $\ln \theta$ by the Hodges–Lehmann estimate, if we use the Kaplan–Meier form of the e-CDF and can estimate sufficiently large parts of the distributions. An alternative is to construct an estimation equation $U(\theta) = 0$ as follows. Consider the rat data. Given θ, multiply the observed times in the drug-treated group by it. Under the assumption that the two groups have the same survival distribution for this modified variable, compute the test function for one

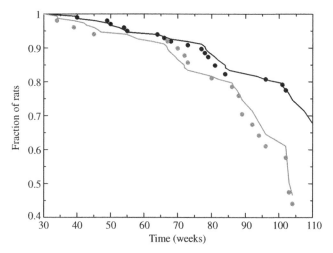

Figure 12.5 Non-parametric AFT models fitted to the rat data. The symbols represent the Kaplan–Meier estimates for the two groups (black is control, gray drug-treated), and the curves are obtained from a Kaplan–Meier estimate of a common survival function, using model adjusted times.

of the tests discussed in Section 12.3. If we choose the Wilcoxon test, which seems natural, and solve the corresponding estimating equation, we find that 'true' time for drug-treated rats is really the clock time multiplied by $\theta = 1.17$ with 95% confidence interval $(1.01, 1.55)$. This means that time (to death) runs 17% faster for drug-treated rats than for the controls. To see how this model fits the data, consider Figure 12.5, in which we have estimated the survival functions for the two groups from the combined Kaplan–Meier estimate, with survival times adjusted by the acceleration factor.

Alternatively, we may use some parametric form for the distributions involved. This is the most common approach to the AFT model, using distribution families that are closed under this model. Again one example is the family of Weibull distributions, since $F^c(t/\theta) = e^{-\lambda\theta^{-\gamma}t^\gamma}$ defines the CDF for a Wei$(\lambda\theta^{-\gamma}, \gamma)$ distribution. The Weibull distribution is therefore closed under both of the most important models for time-to-event data – the proportional hazards model and the AFT model – and if we fit data to this distribution we can choose between a proportional hazards model and an AFT model interpretation. We may note that if θ_{PH} is the proportionality constant in the proportional hazards model, and θ_{AFT} the constant in the AFT model, we have the relation $\theta_{AFT}^\gamma = \theta_{PH}$.

Another important distribution which defines a family closed under the AFT model is the log-logistic distribution, where $\ln T$ follows a logistic distribution. Its survival function is given by

$$F^c(t) = P(\ln T > \ln t) = \frac{1}{1 + \lambda e^{\gamma \ln t}} = \frac{1}{1 + \lambda t^\gamma}.$$

That this family is closed under the AFT model follows from the observation that $F^c(t/\theta) = 1/(1 + \lambda\theta^{-\gamma}t^\gamma)$, so we do the same replacement as for the Weibull distribution.

A special case of the family of Weibull distributions is the family of exponential distributions, and we have already noted that this family can be generalized in another direction as well, to the gamma distribution. This too defines a family closed under the AFT model. In fact, any family for which the CDF is a function not of t but of λt for some parameter λ will be closed under the AFT model.

Example 12.6 The following table shows three different parametric AFT analyses of the rat data in Example 12.1:

Distribution	Estimate	95% CI	p-value
Weibull	1.27	(1.06, 1.51)	0.0083
log-logistic	1.26	(1.04, 1.52)	0.018
gamma	1.28	(1.05, 1.56)	0.016

We see here a consistent message from the different models: time to death runs about 25% faster for drug-treated rats than for controls. This conclusion is independent of which family of distributions we analyze, but the estimate is larger than that found in the non-parametric analysis above.

Figure 12.6 show the survival functions for these models. The individual models are not labeled, because the choice of model does not make much of difference in this case. The data points are the Kaplan–Meier estimates from Figure 12.2, and we see that none of the models provides a very good fit to them.

As already noted, the fit to the data for the Weibull distribution is the same whether we consider an AFT model or a proportional hazards model. It is the same function; it is only a matter of how we parameterize it. In fact, the estimate of γ in the proportional hazards model previously analyzed was 3.79, which means that $\theta_{\text{AFT}} = \theta_{\text{PH}}^{1/\gamma} = 2.47^{1/3.79} = 1.27$, in agreement with the analysis above.

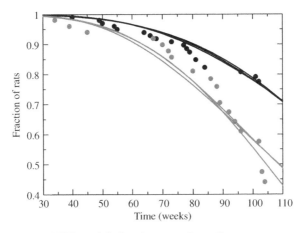

Figure 12.6 Parametric AFT models fitted to rats data. Curves represent estimated group CDFs for the Weibull, log-logistic and gamma distributions, points represent empirical Kaplan–Meier estimates of the CDFs (from Figure 12.2).

Box 12.3 How to estimate the parameter in a failure time model

Parameter estimation in the nonparametric proportional hazards (PH), accelerated failure time (AFT) and accelerated hazards (AH) models is very similar. In all cases the parameter θ is estimated by the equation $U(\theta) = 0$, where

$$U(\theta) = \int_0^\infty a(Y_+(t))(dN_n(t) - \hat{p}(\theta, t)dN_+(t, \theta)), \quad \hat{p}(\theta, t) = \frac{Y_n(t)}{S^0(t, \theta)}.$$

What differ are the functions $S^0(t, \theta)$ and $N_+(t, \theta)$:

PH: $S^0(t, \theta) = Y_n(t) + \theta Y_m(t)$ and $N_+(t, \theta) = N_n(t) + N_m(t)$;
AFT: $S^0(t, \theta) = Y_n(t) + Y_m(t/\theta)$ and $N_+(t, \theta) = N_n(t) + N_m(t/\theta)$;
AH: $S^0(t, \theta) = Y_n(t) + Y_m(t/\theta)/\theta$ and $N_+(t, \theta) = N_n(t) + N_m(t/\theta)$.

The expressions $Y_m(t/\theta)$ and $N_m(t/\theta)$ are obtained from an analysis of the variable θT for group 2.

For all models the variance of $U(\theta)$ is given by

$$V(U(\theta)) = \int_0^\infty a(Y_+(t))^2 \hat{p}(\theta, t)(1 - \hat{p}(\theta, t)dN_+(t, \theta)),$$

and we can obtain confidence intervals and p-values by using the confidence function

$$C(\theta) = \Phi\left(\frac{U(\theta)}{\sqrt{V(U(\theta))}}\right),$$

or its two-sided counterpart.

The proportional hazards and accelerated failure time models are not the only models available for survival data. In particular, there is a compromise between the two which should be mentioned. The assumption of the AFT model is a time acceleration of the integrated hazard, so that $\Lambda_G(t) = \Lambda_F(t/\theta)$. This means that the instantaneous hazard is a mixture of proportional hazards and accelerating time, because $d\Lambda_G(t) = d\Lambda_F(t/\theta)/\theta$. This leads naturally to the alternative suggestion that the difference between the hazards is a difference in time scale only, $d\Lambda_G(t) = d\Lambda_F(t/\theta)$, a model which is called the accelerated hazards model. We can write down the estimating equation for this model (see Box 12.3), but the parameter estimate can also be obtained as follows: for a given θ, perform a log-rank (or Wilcoxon) test on time-adjusted data and estimate a proportional hazards constant θ^* for that data. It is then the case that $d\Lambda_G(t) = \theta^* d\Lambda_F(t/\theta)/\theta$, so we seek out the θ for which we have that $\theta^* = \theta$. For our rat data and the log-rank test, this acceleration factor is estimated to be $\theta = 1.30$.

12.6 The Cox proportional hazards model

The log-rank test is a special case of one of the most celebrated models in biostatistics, the Cox proportional hazards model for survival data. In order to understand the relation between them we will first rederive the former. It is the same derivation as before, but in new notation adapted to a more general situation. The starting point is the repeated means formula $E(Z) = E(E(Z|T))$ which is valid for all stochastic variables. In our application T will be the time-to-event variable, and we can introduce censoring into this by a censor process which is independent of T. We then have that $E(C(T)Z) = E(C(T)E(Z|T))$, which we can write as

$$E(C(T)Z) = \int_0^\infty E(Z|T = t)d\Psi(t).$$

The left-hand side here is the expected value of Z among those individuals for whom we observe an event, multiplied by the fraction of these among all. $\Psi(t)$ is the sub-distribution function describing observed events, and if we have a model from which we can deduce the conditional means $E(Z|T = t)$, the right-hand side is what the model predicts about Z in individuals with an event. Replacing the left-hand side with what we observe, and $d\Psi(t)$ with the Nelson–Aalen estimator, this gives us a relation that can be used to fine-tune the model that defines the conditional means. The log-rank test corresponds to the case where Z is one for those in group 1, and zero for those in group 2. The left-hand side is then the number of events, and the proportional hazards model tells us how to compute the conditional means. The relation is therefore exactly what we use to estimate the hazard ratio parameter from data (the fine-tuning referred to above). Note that this is the same interpretation as we had for the estimating equation for the logistic equation, as was discussed in Section 9.4.

However, the derivation above is more general than the log-rank test, and we can make it even more general by replacing Z with a predictable stochastic process. For our purposes we settle for less, and replace Z with $a(Y(T))Z$, where $Y(t)$ is the fraction at risk at time t, $a(u)$ a function, and Z a stochastic variable (actually a vector). Suppose that we have a model which depends on a parameter β, such that we can compute the function $\bar{z}(t, \beta) = E(Z|T = t)$. This gives us the stochastic variable $U(T, \beta) = a(Y(T))(Z - \bar{z}(T, \beta))$ about which we know that $E_\beta(U(T, \beta)) = 0$. If we have a sample of n from the population with observed event times t_i, we can estimate this mean with the average of the observations. This gives us the following estimating equation for β:

$$U_n(\beta) = \frac{1}{n}\sum_i \int_0^\infty a(Y(t))(z_i - \bar{z}(t, \beta))dN_i(t) = 0. \tag{12.6}$$

Since $a(Y(t))$ is a predictable process the variance of $U_n(\beta)$ is estimated by

$$\frac{1}{n^2}\sum_{i=1}^n \int_0^\infty a(Y(t))^2(z_i - \bar{z}(\beta, t))^2 dN_i(t),$$

a fact we need when we want to derive a confidence function for β.

It remains to compute $\bar{z}(t, \beta)$, for which we need a specific model. The log-rank test was derived under the assumption of a proportional hazards model, so we assume it here as well. This model will explain the frailty θ in terms of the covariate vector Z, so that there is a vector of regression coefficients β such that $\theta = e^{Z\beta}$. The choice of the exponential link here is convenient, but not necessary. It simplifies some calculations and it is the assumption of the Cox model, so we stick to it. Equation (11.4) means that this model estimates the conditional mean $\bar{z}(t, \beta)$ by

$$\frac{\hat{S}^1(t, \beta)}{\hat{S}^0(t, \beta)} = \partial_\beta \ln \hat{S}^0(t, \beta),$$

where

$$\hat{S}^1(t, \beta) = \frac{1}{n} \sum_i z_i e^{z_i \beta} Y_i(t), \quad \hat{S}^0(t, \beta) = \frac{1}{n} \sum_i e^{z_i \beta} Y_i(t).$$

If we plug this into equation (12.6) we get our final estimating equation for β. Varying the weight function defines a whole family of proportional hazard estimating equations (and therefore tests), of which the original Cox model used $a(u) = 1$, and we get a Wilcoxon-type test if we choose $a(u) = u$. We will return to these models in Chapter 13, where we will further discuss the confidence function and how it can be used in situations where the assumption of independence between event times is not fulfilled, as is the case when we analyze recurrent events.

Cox originally derived this model in a different way. To see how he did it, we write the estimating equation that determines β (which we have written as an integral above) as a sum:

$$U_n(\beta) = \sum_i (z_i - \partial_\beta \ln \hat{S}^0(t_i, \beta)) = 0.$$

The sum is over observed event times. It follows that $U_n(\beta)$ is the derivative of $\ln PL(\beta)$, where

$$PL(\beta) = \prod_{t_i} \frac{e^{z_i \beta}}{\hat{S}^0(t_i \beta)}.$$

This is called the *Cox partial likelihood*. The factors in this product are the conditional probabilities that an event, observed at time t_i, is from the individual with covariate value z_i, among those who are still at risk. This follows the idea of survival analysis in general, that observed times are analyzed conditionally on the state of the world at that time. Technically this is not the model likelihood, but Cox treated it as if it was, from an analysis point of view.

In order to see how the partial likelihood above relates to the true likelihood, we need to write down the latter explicitly. For this purpose we recall from Section 12.4 that the log-likelihood for this type of data is given by

$$\sum_i \left(\int_0^\infty \ln \lambda_i(t, \beta) dN_i(t) - \int_0^\infty Y_i(t) \lambda_i(t, \beta) dt \right).$$

The Cox model corresponds to the assumption that $\lambda_i(t, \beta) = \lambda_0(t) e^{z_i \beta}$ for some baseline hazard $\lambda_0(t)$. If we insert this expression for $\lambda_i(t, \beta)$ into the log-likelihood we get

$$\int_0^\infty \left(\sum_i (z_i \beta + \ln \lambda_0(t)) dN_i(t) - \int_0^\infty \lambda_0(t) \hat{S}^0(\beta, t) dt \right).$$

If we know β, we can use $dN_+(t)/\hat{S}^0(\beta, t)$ to estimate $\lambda_0(t) dt$, and if we insert this into the expression above, we see that the log-likelihood is

$$\sum_i \int_0^\infty (z_i \beta - \ln \hat{S}^0(\beta, t)) dN_i(t)$$

plus a term that does not involve β. The derivative of this is the $U_n(\beta)$ above, which means that the Cox partial likelihood is essentially a profiled likelihood, where we have profiled out the unknown baseline hazard by estimating it with the Nelson–Aalen estimator.

There is one more important question we need to address. It has to do with what will happen when the Cox model is true, but we have omitted to include one of the covariates in the analysis. This is the same discussion as we had in Section 9.6, but for this type of data/model. Let Z represent the observed covariates and ξ the omitted one. The explicit assumption is that $\theta = e^{Z\beta+\xi} = \eta e^{Z\beta}$, where $\eta = e^\xi$. Estimation in the Cox model is based on the functions $\hat{S}^0(t, \beta)$ and $\hat{S}^1(t, \beta)$ previously defined. In the presence of heterogeneity the expected values of these are given by

$$S^k(t, \beta) = \int z^k \left(\int F^c(t-, \eta e^{z\beta})\eta e^{z\beta} dP(\eta) \right) dF(z) = -\int z^k \mathcal{L}'(e^{z\beta}\Lambda(t))e^{z\beta} dF(z),$$

where we assume that η is independent of Z in the population. ($\mathcal{L}(u)$ is the Laplace transform of the frailty distribution $P(\theta)$.) The true conditional expectation above is therefore

$$E(Z|T = t) = \frac{\int w(t, e^{z\beta})z e^{z\beta} dF(z)}{\int w(t, e^{z\beta})e^{z\beta} dF(z)}, \quad w(t, \theta) = \mathcal{L}'(\theta\Lambda(t)).$$

With no missing covariates, we have that $w(t, \theta) = F^c(t, \theta)$, but if we have missed some important predictor for the event in our model, and we analyze the data using a Cox model, we will underestimate the true effect of the covariate on survival time. The next example investigates this in more detail for the log-rank test. It may be skipped, as one can read the next section without these details. For a heuristic explanation of this material, see Box 12.4.

Example 12.7   Assume that we have continuous distributions, that the length of the study is τ, and that there is no other censoring. The hazard factor θ^* in a log-rank test is the solution to the equation which equates the expected number of events observed in the first group to what is expected using the combined hazard, which is the equation

$$rF(\tau) = \int_0^\tau \frac{rF^c(t)}{rF^c(t) + (1-r)\theta^* G^c(t)} d\Psi(t).$$

Assume there is heterogeneity in the population such that if the frailty for a patient is η without treatment, it becomes $\theta\eta$ with treatment. This means that θ is the individual hazard ratio, assumed constant. In that case the survival functions in the formula above are obtained as Laplace transforms of the frailty distribution, computed for the values of the cumulative hazards. This defines a relation between θ^* and θ. If we change the variable in this integral to $u = \Lambda(t)$, this relation is

$$1 - \mathcal{L}(\Lambda(\tau)) + \int_0^{\Lambda(\tau)} \frac{\mathcal{L}(u)(r\mathcal{L}'(u) + (1-r)\theta\mathcal{L}'(\theta u))}{r\mathcal{L}(u) + (1-r)\theta^*\mathcal{L}(\theta u)} du = 0.$$

This relation is illustrated in Figure 12.7, assuming gamma frailty and that $\theta^* = 0.91$. We see that if there is considerable heterogeneity, expressed as a large variance for the frailty distribution, we may have a true treatment effect that is as large as a reduction of almost 20%.

Box 12.4 Bias due to heterogeneity: a heuristic explanation

The proportional hazards model with frailty for a time-to-event variable T says that, conditionally on the frailty η, $\Lambda(t|\eta) = \eta e^{z\beta}\Lambda_0(t)$. Assume that $\Lambda_0(t) = \lambda t^\gamma$ is from a Weibull distribution. This can also be expressed as an AFT model, which means a linear model in $\ln T$:

$$\ln T = \gamma^{-1}(-\ln\lambda - z\beta + X + Y),$$

where X has the smallest extreme value (SEV) distribution and Y has the distribution of $-\ln\eta$. It is convenient to assume that the frailty has a lognormal distribution with mean one, which means that $Y \in N(\sigma^2/2, \sigma^2)$ and therefore that the variance of $\ln T$ is $(\eta^2 + \sigma^2)/\gamma^2$, where η^2 is the variance of the SEV distribution.

Next suppose that we ignore the frailty, but assume that $\Lambda_0(t) = \mu t^\nu$. This means that

$$\ln T = \nu^{-1}(-\ln\mu - x\beta^* + X),$$

and the variance of $\ln T$ is η^2/ν^2. Solving for ν gives $\nu^2 = \gamma^2\eta^2/(\eta^2 + \sigma^2)$. If we estimate the regression coefficients by least squares, we should have the same estimate in the two situations: $\beta/\gamma = \beta^*/\nu$, which implies that

$$\beta^* = \frac{\beta}{\sqrt{1 + \sigma^2/\eta^2}}.$$

This shows that what we consider the treatment effect moves toward no effect ($\beta^* = 0$) in the presence of frailty. How much depends on the magnitude of heterogeneity.

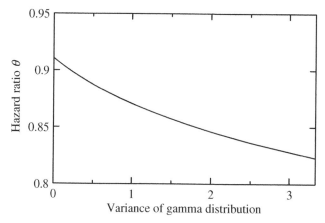

Figure 12.7 The relation between a subject-specific hazard ratio θ and the degree of heterogeneity in the population, when the population hazard ratio (obtained by a log-rank test) is 0.91. The assumptions are a gamma frailty and no censored data except for a finite study length.

What we measure with the log-rank test (θ^*) is not necessarily what matters to the individual patient, who is probably more interested in θ. That parameter is what determines the effect on him, although on a relative scale, so it might not be easily translated to entities such as number of years of added life. But this discussion shows that it is important to find prognostic variables that can explain as much as possible of the heterogeneity when we analyze time-to-event data.

12.7   On omitted covariates and stratification in the log-rank test

In this section we will illustrate the consequences of omitting covariates in the Cox model, using as background a real-life example. For various reasons, the exact details of the study in question will not be described, and they are not important for this discussion anyway. Suffice it to say that it was a placebo-controlled study, where the outcome was survival after start of treatment; it was a two-armed parallel group study which was randomized 2 : 1 between the active drug and placebo. The overall log-rank analysis table comparing the two treatments gives us the key outcome data:

	N	Observed	Expected
Active	1129	634	654.4
Placebo	563	342	321.6

The hazard ratio was estimated to be 0.91 with 95% confidence interval (0.80, 1.04) and with $p = 0.16$ for the hypothesis of no treatment effect. Although there were fewer deaths than expected (assuming no treatment effect) in the active group, there is not enough evidence to claim that the drug has an effect on survival.

However, we cannot rest with this. The effect we are looking for gets attenuated in the presence of heterogeneity, and we wish to explain as much of the heterogeneity as possible, in order to home in on the hazard ratio we are interested in. At our disposal we have six covariates, each of which is dichotomous in nature. The one which on its own is the most predictive of survival is related to the patient's performance status according to a WHO scale. We adjust for this variable in the analysis by carrying out a Cox regression with two factors, treatment and the WHO scale, both dichotomous. Now the estimated hazard ratio is 0.855 with 95% confidence interval (0.75, 0.98), which gives us $p = 0.020$ for the hypothesis of no effect. With this single adjustment we have decreased the hazard ratio so much that we now have sufficient evidence at the conventional (two-sided) 5% significance level that the drug has an effect on survival. There is no reason to believe that we have captured all the heterogeneity, but all we can do with available data is see what the effect is when we include all the covariates (additively) into a Cox regression model. The result is that the treatment hazard ratio is estimated as 0.863 with 95% confidence interval (0.76, 0.99) and $p = 0.029$. Although most of these individually have an effect on survival, including them all seems not to explain any more heterogeneity than is explained already by the first covariate.

There is more to say about this. The original model, the log-rank test for the two treatment groups, corresponds to a model in which the placebo group has hazard $d\Lambda(t)$ and the active group has hazard $\theta d\Lambda(t)$. The model with the WHO covariate is such that the hazard is

WHO scale	Placebo	Active
0 or 1	$d\Lambda(t)$	$\theta d\Lambda(t)$
2 or 3	$rd\Lambda(t)$	$r\theta d\Lambda(t)$

where r is the proportionality factor for the covariate. This is a stronger assumption than assuming that each WHO scale subgroup has its own hazard, which is the table

WHO scale	Placebo	Active
0 or 1	$d\Lambda_0(t)$	$\theta d\Lambda_0(t)$
2 or 3	$d\Lambda_1(t)$	$\theta d\Lambda_1(t)$

The first model corresponds to the assumption $d\Lambda_1(t) = rd\Lambda_0(t)$. The second model can be analyzed by a stratified log-rank test, in which we compute the estimating function for each stratum (i.e., subgroup defined by the WHO scale variable) and from which an estimating equation for θ is obtained by equating a weighted average of these to zero. The convention here is to take equal weights for the strata, and the result of this analysis is summarized in the following table:

WHO scale	Treatment	N	Observed	Expected
0 or 1	Active	387	283	299.3
	Placebo	174	135	118.7
2 or 3	Active	739	351	365.5
	Placebo	389	207	192.5

The corresponding hazard estimate is 0.867 with 95% confidence interval (0.76, 0.99) and $p = 0.034$. We see that the result is very similar to the Cox regression result presented above. However, if we instead stratify on all six covariates the results of the two methods differ more, and in a crucial way: the treatment hazard ratio is estimated to be 0.89 with 95% confidence interval (0.77, 1.02) and $p = 0.087$. Numerically the difference is not large, but the p-value moves over to the other side of the conventional cut-off limit of 5%. It therefore becomes important to understand whether the first analysis was based on faulty assumptions, or whether the explanation for the discrepancy is to be found elsewhere. In exploring this we will highlight an important risk with a stratified analysis.

Figure 12.8 shows the estimates and confidence intervals for the two models (Cox regression in black, stratified test in gray) at different degrees of stratification, in such a way that we start with the WHO scale variable as a single covariate, and then add one new covariate at each step according to how predictive they are for survival on their own. Not much happens, with one important exception, which is when we add the last covariate to the stratified test (a covariate that is not even shown to be predictive on its own). In this test we have stratified on six dichotomous variables, which means that we divide the population into $2^6 = 64$ cells. Four of these are empty, and 23% of the remaining cells have at most 3 patients. In such small

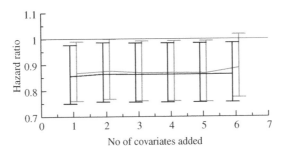

Figure 12.8   Illustration of how confidence intervals for hazard ratio change as we increase the number of variables to stratify on. The gray data are for the stratified test, the black data for the (additive) Cox regression model.

cells it is not unlikely that only one treatment will be represented, a risk that is augmented by the fact that we had a 2 : 1 randomization. In fact, 11 cells have only one treatment and of these, 9 contain only active drug. These cells do not contribute to the stratified test, which means that we effectively loose 27 patients on active drug and 2 on placebo in the analysis. Among these there are in total 14 deaths (one on placebo) that no longer contribute to the analysis. This is a loss of power and explains much of the effect we see when we add the sixth covariate to the analysis. It means that drawing conclusions from this test is not a sensible thing to do, however prespecified the analysis may have been.

The lesson is simple. Do not over-stratify! You must make certain that no cells are too small. This is of course true for all stratified tests. The idea behind stratification is this: we have a heterogeneous population, and in order to apply a test which assumes a homogeneous population we divide the population up into strata, such that within each stratum the population is homogeneous. Thereafter we pool the strata. However, the quest for homogeneity strives toward small strata, and in a small stratum there is a severe risk that the treatment groups are unbalanced. Unbalanced comparisons are less effective than balanced ones, so we loose power as the number of strata increases. On the other hand, if we take fewer strata, these may be heterogeneous with treatment bias as a result. Note that if we stratify when we do not need to, when our population is actually homogeneous, we may have a substantial loss of power, at least if some cells become small. In all, this makes the application of stratified tests problematic if one is forced, as is often the case in the pharmaceutical industry, to prespecify in detail the analysis to be performed, in order to gain credibility (in the eyes of the regulatory agencies).

12.8   Comments and further reading

The rat data we have used in this chapter to illustrate different methods was originally given in Mantel et al. (1977,Table 1), and is reproduced in Hougaard (2000,Table 1.5). The original paper is of independent interest, because it illustrates how the log-rank test is related to the Mantel–Haenszel technique in a very explicit way.

Much of the material in this chapter is covered in major books on the statistical analysis of survival data, some of which were referenced in the last chapter. An overview of how traditional non-parametric tests are expressed and analyzed in a counting process theory context is given by Gill (1983). The accelerated hazards model is less used, but is discussed by Chen and Wang (2000). Gray's analysis in a competitive environment is discussed in Gray (1988).

The power calculation in Box 12.2 is based on the validity of the proportional hazards model. In the design stage we assume a certain (subject-specific) hazard ratio, but when we do the analysis, in order to achieve this assumption, we may need to include a series of predictive variables in the analysis model (Schoenfeld, 1983). In other words, we use the formula for the log-rank test when we compute the number of patients needed, and also if we plan for a more extensive Cox regression model. If we apply the log-rank test and ignore the predictors, the loss of power comes from the fact that the treatment effect is time-dependent and we estimate a parameter which corresponds to a smaller effect than the true one.

The original article by David R. Cox (1972) on the proportional hazards model has had a huge number of citations and its author has received a large number of honors. Our derivation of his model is not the traditional one and is deliberately sketchy; missing details may be found in papers by Sasieni (1993) and Tsiatis (1981). The value of this derivation is that it emphasizes the underlying connection between the model and the problem of explaining heterogeneity. It emphasizes that on an individual level we may well have proportional hazards, even when it does not appear so from the overall population (Kaplan–Meier) perspective. The traditional derivation can be found in most books on survival analysis, many of which contain numerous applications. There are different ways to extend the Cox model (Therneau and Grambsch, 2000) to situations where its basic assumptions are not fulfilled, some of which will be touched upon in the next chapter.

The heuristic idea for the bias (if that is the proper word) in the presence of frailty, or omitted covariates, in the Cox model, described in Box 12.4, is essentially taken from Keiding et al. (1997). A fuller discussion of this bias is given by Henderson and Oman (1999). The amount of bias depends on the frailty distribution, and is actually more pronounced with complete data than if there are censored data. Another discussion about the balancing act between stratification with small cells versus the problem of heterogeneity can be found in Akazawa et al. (1997) with a related discussion in Stavola and Cox (2008) for a Poisson process setting.

References

Akazawa, K., Nakamura, T. and Palesch, Y. (1997) Power of logrank test and Cox regression model in clinical trials with heterogeneous samples. *Statistics in Medicine*, **16**, 583–597.

Chen, Y.Q. and Wang, M.C. (2000) Analysis of accelerated hazards models. *Journal of the American Statistical Association*, **95**(450), 608–618.

Cox, D.R. (1972) Regression models and life-tables (with discussion). *Journal of the Royal Statistical Society, Series B*, **34**, 187–220.

Gill, R.D. (1983) *Censoring and Stochastic Integrals* vol. Mathematical Centre Tracts 124. Amsterdam: Mathematisch Centrum.

Gray, R.J. (1988) A class of *K*-sample tests for comparing the cumulative incidence of competing risks. *Annals of Statistics*, **16**, 1141–1154.

Henderson, R. and Oman, P. (1999) Effect of frailty on marginal regression estimates in survival analysis. *Journal of the Royal Statistical Society, Series B*, **61**(2), 367–379.

Hougaard, P. (2000) *Analysis of Multivariate Survival Data* Statistics for Biology and Health. New York: Springer.

Keiding, N., Andersen, P.K. and Klein, J.P. (1997) The role of frailty models and accelerated failure time models in describing heterogeneity due to omitted covariates. *Statistics in Medicine*, **16**, 215–224.

Mantel, N., Bohidar, N. and Ciminera, J. (1977) Mantel-Haenszel analyses of litter-matched time-to-response data, with modifications for recovery of interlitter information. *Cancer Research*, **37**, 3863–3868.

Sasieni, P. (1993) Some new estimators for Cox regression. *Annals of Statistics*, **21**(4), 1721–1759.

Schoenfeld, D.A. (1983) Sample-size formula for the proportional-hazards regression model. *Biometrics*, **39**, 499–503.

Stavola, B.L.D. and Cox, D.R. (2008) On the consequence of overstratification. *Biometrika*, **95**(4), 992–996.

Sun, J. (2006) *The Statistical Analysis of Interval-Censored Failure Time Data* Statistics for Biology and Health. New York: Springer.

Therneau, T.M. and Grambsch, P.M. (2000) *Modeling Survival Data: Extending the Cox Model* Statistics for Biology and Health. New York: Springer.

Tsiatis, A.A. (1981) A large sample study of Cox's regression model. *Annals of Statistics*, **9**(1), 93–108.

12.A Appendix: Comments on interval-censored data

With interval-censored data there are some adjustments that need to be made to the way we compute things. The Kaplan–Meier e-CDF can be computed only when we know the situation at each time point, so we need to find another way to obtain an e-CDF. Suppose, then, that we have n patients, for each of whom we have an interval $(l_i, r_i]$, such that the event has occurred somewhere within that interval. By going to the limit $r_i - l_i \to 0$ we can include exact observations, and by taking $r_i = +\infty$ we can also include right-censored events. For this discussion we assume that all intervals are proper, finite intervals. Let $t_1 < t_2 < \ldots < t_m$ denote the unique elements from the list of left and right interval limits.

An e-CDF $F_n(t)$ will be a step function with jumps at the t_j of magnitude $\Delta_j = F_n(t_j) - F_n(t_{j-1})$. To determine Δ_j, let I_j^i be the indicator variable which is one if the censor interval for subject i contains the point t_j (i.e., if $(t_{j-1}, t_j] \subset (l_i, r_i]$), otherwise zero. The contribution of subject i to the e-CDF at point $t_j \in (l_i, r_i]$ is then given by $\Delta_j/(F(r_i) - F(l_i))$. But the average over all individuals at that point is the actual jump size, so we have the relation

$$\Delta_j = \frac{1}{n} \sum_{i=1}^{n} \frac{I_j^i \Delta_j}{\sum_k I_k^i \Delta_k}.$$

This defines the jump sizes and therefore what is called the Turnbull e-CDF for interval-censored data. We may note that Δ_j can only be non-zero if t_{j-1} is a left end point of the original data and t_j a right end point, but not necessarily from the same censored interval. (The intervals $(t_{j-1}, t_j]$ are called Turnbull intervals, and identifying them first is helpful for computational reasons.)

Given the Turnbull e-CDF $\Psi_{mn}(t)$, we can define the log-rank test for interval-censored data as follows. Instead of observed event times, use predicted event times, so that

$$N_+(t) = \sum_{i=1}^{n+m} \frac{I^i(t) \Delta \Psi_{mn}(t)}{\sum_k I^i(t_k) \Delta \Psi_{mn}(t_k)},$$

where $I^i(t)$ is an indicator for the interval $(l_i, r_i]$. We can also define the predicted number at risk by

$$Y_+(t) = n \sum_{t_k \geq t} \Delta \Psi_{mn}(t_k).$$

Together with similar estimates for one group alone we derive a log-rank test, or Wilcoxon test, for interval-censored data that is analogous to what they are for right-censored data, except that we use these predicted entities instead of observed ones. The extension to parameter estimation is immediate.

We can alternatively construct a generalized log-rank test for interval-censored data based on the expression $1 + \int_0^\infty \ln(\Psi^c(t)) dF(t)$, which underlies the log-rank test. It is estimated by

$$1 + \frac{1}{n} \sum_i \ln \Psi_{mn}^c(t_i) = \frac{1}{n} \sum_i (1 + \ln \Psi_{mn}^c(t_i)).$$

A primitive function of $1 + \ln x$ is $x \ln x$, which means that a generalization from complete data to interval-censored data can be done by using

$$\frac{1}{n} \sum_i \frac{\Psi_{mn}^c(l_i) \ln(\Psi_{mn}^c(l_i)) - \Psi_{mn}^c(r_i) \ln(\Psi_{mn}^c(r_i))}{\Psi_{mn}^c(l_i) - \Psi_{mn}^c(r_i)}.$$

In the limit this reduces to the previous expression. For the details necessary for practical implementation of this, see Sun (2006), for example.

13

Remarks on some estimation methods

13.1 Introduction

We have a few loose ends to tie up before we conclude this book. They are all concerned with estimation methods and model misspecification and are mathematical in nature. The mathematically less inclined reader may wish to skip these discussions, but is advised to have at least a casual look, since we will address the analysis of some more complex data that may be of general interest. We need to reiterate what was said in the preface – even if this chapter is mathematical in nature, it is an overview, with less attention to details. The chapter is divided into two main parts: the first is a holistic discussion about estimating equations in general and maximum likelihood theory in particular, and the second discusses how to analyze data with repeated measurements within subjects.

We have often estimated parameters from an (estimating) equation $U(\theta) = 0$, and used the distributional properties of the stochastic variable $U(\theta)$ to obtain knowledge (e.g., confidence intervals) about the parameter θ. In the first part of this chapter we will discuss this approach in more generality. This will also help us to gain some understanding of why we sometimes parameterize in θ, sometimes in its logarithm. We will link this to the well-known maximum likelihood theory, which (essentially) is the special, but very important, case where the estimating equation is the derivative of some function. This discussion will introduce a special variance estimate, the robust variance, which is an alternative estimator of the variance also for likelihood estimators. The robust variance is less accurate than the one derived from likelihood theory, but has the virtue of being less sensitive to model misspecification, since it is derived from the estimating equation alone. This gives us an opportunity to use the wrong model and still draw the right conclusion. One illustration of this will be the analysis of recurrent events in a heterogeneous population, a discussion that starts the second part of this chapter.

Understanding Biostatistics, First Edition. Anders Källén.
© 2011 John Wiley & Sons, Ltd. Published 2011 by John Wiley & Sons, Ltd.

This second part continues the discussion on the subject-specific approach to dose response, initiated in Section 10.6, except that this time we will not estimate the model by reducing it to a population averaged model. Here we will take as our starting point the likelihood for the problem. The problem with the likelihood approach for these non-linear, though not for the linear, mixed-effects models is that their likelihood consists of multidimensional integrals that are hard to compute accurately. Methods exist, but they are often too computer-intensive at present for routine use, so we often need to use various approximations. We will have a quick look at some of these approximations and demonstrate why we need to understand how the estimation method works, in order to understand what we are actually estimating – the population averaged or the subject-specific aspect of the problem.

13.2   Estimating equations and the robust variance estimate

The way we have carried out estimation in statistical models has mostly followed a certain pattern. We start with a function $U(x, \theta)$ of the data x (both outcome variables and explanatory variables) and an unknown parameter vector θ, such that the equation $U(x, \theta) = 0$ defines our estimate of θ as a function of x. Important examples of such functions $U(x, \theta)$ include $U(x, \theta) = x_{11}x_{22} - \theta x_{12}x_{21}$ and $U(x, \theta) = x_{11} - E_\theta(x_{11})$ used to estimate the odds ratio in Section 5.3, the function $\int G_m(x, \theta)dF_n(x) - \frac{1}{2}$ associated with the Wilcoxon test, and the different functions discussed in Chapter 12, modeling survival data. Some of these will be revisited below. The function $U(x, \theta)$, together with the sampling scheme, defines a stochastic variable which depends on a parameter, which we denote by $U(\theta)$. The estimating equation is then written as $U(\theta) = 0$. As usual the point estimate of θ obtained from a particular set of data is denoted θ^*, whereas the general solution of the equation (which is a function of the data) is the estimator of θ, is a stochastic variable and is denoted by $\hat{\theta}$.

The basic requirement on $U(\theta)$ is that it is unbiased, which in this case means that it has mean zero:

$$E_\theta(U(\theta)) = \int U(x, \theta)dF_\theta(x) = 0.$$

Here $F_\theta(x)$ is the CDF for the data set, a multidimensional CDF which we do not need to be explicit about. It is important to note, and will be discussed further below, that it is the equation that is unbiased; the estimator $\hat{\theta}$ need not be so. Usually the setup is such that we have a stochastic variable X, representing a unit of data (data from one subject), together with a function $U(x, \theta)$ such that $E_\theta(U(X, \theta)) = 0$, and we have at our disposal a set of n independent observations x_i of X from which θ should be estimated. This means that we estimate θ from the average of the observed values of $U(X, \theta)$,

$$\frac{1}{n} \sum_{i=1}^{n} U(x_i, \theta) = 0.$$

(Note that when θ is a vector with s components, we need the estimating function to consist of s independent components in order for this to be solvable.) This is useful, since it allows us to appeal to large-sample theory when we wish to derive confidence statements about θ. In fact, we often appeal to the CLT to obtain that

$$U(\theta) \in AsN_s(0, V_\theta(U(\theta))).$$

The key implication of this assumption is that

$$U(\theta)^t V_\theta(U(\theta))^{-1} U(\theta) \in As\chi^2(s), \tag{13.1}$$

which we can use to define the (approximative) confidence function

$$C(\theta) = \chi_s(U(\theta)^t V_\theta(U(\theta))^{-1} U(\theta))$$

for θ, from which we can subsequently deduce confidence intervals and p-values. In order to capitalize on this we need to know the variance or, at least, have an estimate of it. In some cases where we only have an estimate of the variance to insert into the expression above, we might wish to modify the distributional statement in equation (13.1) in order to obtain better precision in the confidence statements. If, for example, we analyze a linear model of Gaussian variables we will have an F distribution after insertion of the sample variance as an estimate of the true variance. In many other applications we instead appeal to large-sample theory and keep the $\chi^2(s)$ distribution, accepting that the results are approximate.

This discussion is the general version of approaches we have used for specific problems. Now is the place to compare this with what is standard practice, which is to make one further approximation in equation (13.1), in order to obtain an approximation to the variance of the estimator $\hat{\theta}$ of θ. Recall from the definition of $\hat{\theta}$ that we have that $U(\hat{\theta}) = 0$, which means that when $\hat{\theta}$ is close to the true value θ, we have approximately that

$$0 = U(\hat{\theta}) \approx U(\theta) + U'(\theta)(\hat{\theta} - \theta).$$

This gives us a way (with the appropriate assumptions, which we ignore together with all technical details) to find an approximation to the distribution of $\hat{\theta}$. For this we make one further approximation and replace $-U'(\theta)$ with its expected value,

$$I(\theta) = -E_\theta(U'(\theta)),$$

which is called the *information matrix* (at least in the context of likelihood theory). This gives us the approximation $U(\theta) \approx I(\theta)(\hat{\theta} - \theta)$, using which we can rewrite equation (13.1) as

$$(\hat{\theta} - \theta)^t I(\theta)^t V_\theta(U(\theta))^{-1} I(\theta)(\hat{\theta} - \theta) \in As\chi^2(s).$$

We therefore have that the variance of $\hat{\theta}$ is given, at least approximately, by the matrix

$$V(\hat{\theta}) = I(\theta)^{-1} V_\theta(U(\theta)) I(\theta)^{-t}.$$

We often compute this expression only at the point estimate θ^* of θ, and we also approximate $I(\theta^*)$ with its observed value, the derivative $-U'(\theta^*)$. In order to compute $V(\hat{\theta})$ we also need an estimate of $V_\theta(U(\theta^*))$, which we obtain in the crudest of ways, by computing the average $n^{-1}\sum_i U(x_i, \theta^*)U(x_i, \theta^*)^t$. In all, this gives us an estimate of $V(\hat{\theta})$ which is referred to as the *robust variance* (another name often used is *sandwich variance*). It is robust in the sense that if we have an estimating equation derived from some probability model, this estimate is less sensitive to the validity of the modeling assumptions than the ordinary variance estimate is. When we use this approximation to compute confidence information, we refer to it as the quadratic approximation of the confidence function (around the estimate θ^*). The price paid for using the robust variance estimate is that in general it is not a very efficient estimator, and we may need rather large samples in order to obtain reasonable accuracy.

We may note that even though we assume that there is an underlying probability model which makes the stochastic variable $U(\theta)$ unbiased, this model is not really made explicit anywhere when we use the robust variance. It is there in the background to assure that when we estimate θ from the estimating equation, we are actually estimating a real parameter. We do not use the probability model in any more explicit way.

By construction the stochastic variable $U(\theta)$ is unbiased, but that does not necessarily mean that the estimate $\hat{\theta}$ is unbiased. It is not hugely biased, because

$$E_\theta(\hat{\theta}) \approx \theta + I(\theta)^{-1} E_\theta(U(\theta)) = \theta,$$

an approximation which is better the larger the sample is. However, higher-order approximations may reveal some deviance to this, with resulting bias in small samples. This is further discussed in Appendix 13.A.

For the rest of this chapter we will go through a few examples. All of these revisit discussions we have already had, but will provide some information about the relation between different approximations. We also use these examples to highlight some further important considerations, such as the importance of having the 'right' parameterization when we wish to use the quadratic approximation. In the next section we will discuss the application of this general theory to the special case of maximum likelihood estimation.

Example 13.1   In Example 12.4 we computed the two-sided confidence function for the log-rank test. The estimating function can be written as

$$U(\theta) = N_n(\infty) - \int_0^\infty \frac{Y_n(t)}{Y_n(t) + \theta Y_m(t)} dN_+(t),$$

and for its variance we have the estimate

$$\hat{V}(\theta) = \int_0^\infty \frac{Y_m(t)Y_n(t)\theta}{(Y_n(t) + \theta Y_m(t))^2} dN_+(t).$$

The confidence function for θ is then obtained by applying the CDF for a $\chi^2(1)$ distribution to $U(\theta)^2/\hat{V}(\theta)$. This ratio, as a function of θ, is shown in Figure 13.1 as the solid black curve; the corresponding 95% confidence interval for θ is obtained as the intersection of this curve with the 95th percentile for the $\chi^2(1)$ distribution (shown as the dashed horizontal line).

There are two more curves in Figure 13.1. The gray curve is obtained from the quadratic approximation based on θ. This requires that we compute the derivative of $U(\theta)$, which is related to the variance estimate above through $\hat{V}(\theta) = \theta U'(\theta)$. We therefore have that the robust variance estimate is $\hat{I}(\theta)^{-1}\hat{V}(\theta)\hat{I}(\theta)^{-1} = \theta\hat{V}(\theta)^{-1}$, with $\hat{I}(\theta) = -U'(\theta)$, and the ratio $(\theta - \theta^*)^2 \hat{V}(\theta^*)/\theta^*$ is what is shown as the gray curve in Figure 13.1. This is a quadratic function, symmetric around the point estimate θ^*, and we see that it is in poor agreement with the original function, defining confidence intervals that lie to the left of those derived from the original function. (The reason why we plot the percentiles and not confidence levels on the y-axis is precisely to highlight the quadratic nature of this curve.)

However, if we write $\theta = e^\beta$ and consider the expressions above as functions of β instead, we have that $\hat{V}(\beta) = U'(\beta)$. This does not change the black confidence curve in any way, because that approach is independent of how we parameterize. However, the quadratic approximation depends on what parameterization we use: the quadratic confidence curve for β is $(\beta - \beta^*)^2 \hat{V}(\beta^*)$. Plotted versus β, this would provide a quadratic curve; in Figure 13.1

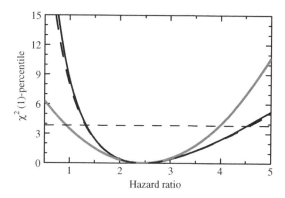

Figure 13.1 Confidence information about the proportional hazards parameter from the log-rank test, using three different approximations. For an explanation of the different curves, see Example 13.1.

it is plotted versus $\theta = e^\beta$ instead as the dashed curve. We see that although it differs somewhat from the original curve, it is a good approximation to it in any region of interest (which is defined by the 95th percentile). The lesson is clear: when using the quadratic approximation, it is important to choose the right parameterization. This, in turn, is related to the choice of parameter scale in which the distribution of the estimator is best described by a Gaussian distribution.

In the next example we revisit the problem of estimating confidence intervals for percentiles of a distribution (previously discussed in Section 6.5).

Example 13.2 Let $F_n(x)$ be the e-CDF for a continuous variable, based on a sample of size n. We wish to estimate the percentile $\theta = x_p$ of the corresponding population CDF $F(x)$. The estimating equation for this is

$$U(\theta) = F_n(\theta) - p = 0,$$

and we know that its variance is given by $V(U(\theta)) = V(F_n(\theta)) = F^c(\theta)^2 \tau(\theta)$, where $\tau(\theta) = F(\theta)/nF^c(\theta)$ if there is no censoring; in the presence of censored data we estimate it using the Greenwood estimator σ_n^2 of equation (11.7). The confidence functions in the complete and censored data cases are therefore $\chi_1(x)$ applied to

$$\frac{n(F_n(\theta) - p)^2}{p(1 - p)} \quad \text{and} \quad \frac{(F_n(\theta) - p)^2}{(1 - p)^2 \sigma_n^2},$$

respectively. This is the approach that was explored in Chapter 6.

To investigate the corresponding quadratic approximation we need to compute the derivative of $U(\theta)$. For this purpose we redefine the e-CDF $F_n(x)$ as its piecewise linear version instead of the staircase version. We then have that $U'(\theta) = F_n'(\theta)$, which is defined at all points θ except for the actual observations in the sample. This also implies that the information matrix is $I(\theta) = -F'(\theta)$, and it follows that $\hat\theta$ has the asymptotic variance $F'(\theta)^{-2}V(F_n(\theta))$. For complete data this means that the asymptotic distribution for $\hat{\theta}$ is given by

$$N\left(\theta, \frac{p(1-p)}{nF'(\theta)^2}\right),$$

with a minor modification for censored data.

We can also consider the problem of how to simultaneously estimate two percentiles $\theta = (x_p, x_q)$, where $p < q$. For this we define the two-dimensional estimating function

$$U(\theta) = (F_n(\theta_1) - p, F_n(\theta_2) - q),$$

which is such that $E(U'(\theta))$ is the diagonal matrix with entries $F'(\theta_1)$ and $F'(\theta_2)$. When $x < y$, the covariance between $F_n(x)$ and $F_n(y)$ is given by $F^c(x)F^c(y)\tau(x)$, from which we can deduce what the variance matrix for θ looks like, how we should estimate it, and how to obtain a two-dimensional confidence function for θ. In particular, we find that the asymptotic correlation between $\hat{x}_p$ and $\hat{x}_q$ is $\tau(x_p)/\tau(x_q)$, which for complete data reduces to $\sqrt{p(1-q)}/\sqrt{q(1-p)}$. This addresses the problem discussed in Box 7.8 in a more direct way.

In the next example we revisit the Mantel–Haenszel estimator of the odds ratio from an estimating equation perspective.

Example 13.3   Consider a single 2×2 table, for which the conditional distribution of the upper left element x_{11} given the margins is the (non-central) hypergeometric distribution with the odds ratio parameter θ. The estimating equation is

$$U(\theta) = x_{11}x_{22} - \theta x_{21}x_{12} = 0,$$

the solution of which is the empirical odds ratio. Its variance is given by

$$V_\theta(U(\theta)) = \frac{1}{2n^2}E_\theta((x_{11}x_{22} + \theta x_{21}x_{12})(x_{11} + x_{22} + \theta(x_{21} + x_{12}))).$$

Here we can either compute this expression explicitly as a function of θ, which is the true variance, or we can estimate it by inserting the observed values x_{ij} from the table cells and get an estimated variance (still a function of θ). Furthermore,

$$I(\theta) = -E_\theta(U'(\theta)) = -E_\theta(x_{21}x_{12}) = -\theta^{-1}E_\theta(x_{11}x_{22}),$$

and if we denote the empirical odds ratio by θ^*, we see that the robust covariance estimator for θ is given by $\theta^{*2}V_R(\theta^*)$, where

$$V_R(\theta) = \frac{1}{2n^2}(x_{11}x_{22})^{-2}(x_{11}x_{22} + \theta x_{21}x_{12})(x_{11} + x_{22} + \theta(x_{21} + x_{12})).$$

From this we also get the robust covariance estimate for the empirical log-odds ratio as $V_R(\theta^*)$. Another calculation shows that this is the variance for the logged odds ratio (see Section 4.5.1):

$$V_R(\theta^*) = \frac{1}{x_{11}} + \frac{1}{x_{12}} + \frac{1}{x_{21}} + \frac{1}{x_{22}}.$$

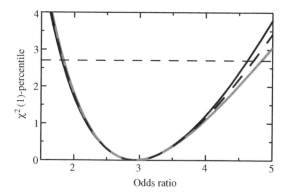

Figure 13.2 Deriving confidence intervals for the odds ratio in a 2×2 table. The three methods discussed in the text are illustrated.

We can graphically illustrate how this approximation works by comparing the two functions

$$Q_1(\theta) = U(\theta)^2 / V_\theta(U(\theta)) \quad \text{and} \quad Q_2(\theta) = (\ln(\theta) - \ln(\theta^*))^2 / V_R(\theta^*).$$

We actually have two versions of $Q_1(\theta)$: one in which we use the true variance and one in which we use the estimated variance. These three functions are all illustrated in Figure 13.2 for the first Hodgkin's lymphoma data set in Section 4.5.1. The solid black curve uses the function $Q_1(\theta)$ with the true variance as a function of θ, while the gray curve is this function using the estimate of this variance (still as a function of θ). The dashed curve, which is sandwiched between these, is $Q_2(\theta)$. The dashed horizontal line defines the intersection level for a 90% confidence interval, based on the $\chi^2(1)$ distribution. We see that these functions are much in agreement with each other at the lower confidence limit, but disagree somewhat at the upper limit. They agree with what we obtained in Figure 5.2 (dashed curves).

How to extend this to a general Mantel–Haenszel pooled odds ratio discussed in Section 5.5 is more or less immediate. The estimating function is the sum over strata,

$$U(\theta) = \sum_{i=1}^{n} \frac{1}{n_i}(x_{i11}x_{i22} - \theta x_{i12}x_{i21}),$$

where n_i is the total number of observations in table i. Applying the knowledge just obtained for a single cell, we see that the variance in equation (5.3) is derived from the robust variance for the estimating equation in the example, and is therefore what the quadratic approximation for the log-odds ratio is based on.

This step from a single 2×2 table to a series of tables, that is, the introduction of a stratified test, can be done in much larger generality. In fact, suppose we have a series of data sets, each defining an estimating equation $U_i(\theta) = 0$. We can then define a stratified test by defining the pooled estimating equation

$$U(\theta) = \sum_i w_i U_i(\theta) = 0$$

for some specified weights. In this way we can introduce both the stratified log-rank test discussed in Section 12.7 and a stratified Wilcoxon test. In the latter case one usually takes weights similar to those in the Mantel–Haenszel pooled odds ratio estimate.

Example 13.4   At this point we take a second look at the two quadratic forms discussed in Section 5.5, which we called the Mantel–Haenszel and Breslow–Day quadratic forms. If we have a series of N 2×2 tables with elements a_k, b_k, c_k, d_k, the estimating equation for the odds ratio for a single table is $U_k(\theta) = a_k - E_\theta(a_k) = 0$, where the expectation is computed assuming a hypergeometric distribution. The pooled estimator function is therefore $U(\theta) = \sum_k U_k(\theta)$, and a two-sided confidence function can be obtained by considering $U(\theta)^2/V(U(\theta))$, which is the Mantel–Haenszel quadratic form. We derive the confidence function for θ by using the fact that this quadratic form has approximately a $\chi^2(1)$ distribution.

The Breslow–Day quadratic form is the weighted least squares expression for θ. This is a case where the parameter to be estimated also appears in the variance, and we therefore may wisht to estimate the parameter using the GLS estimate. This produces precisely the same estimating equation as above, so the two quadratic forms $Q_{MH}(\theta)$ and $Q_{BD}(\theta)$ define the same test.

We could alternatively argue that $Q_{BD}(\theta) \in \chi^2(N)$ and derive a confidence function from this. However, similar to the discussion in Section 8.8, this contains two parts: the first addresses whether it is reasonable that there is a common θ, and the second what we know about such a θ. It takes more (individual strata cells need to be large; for $Q_{MH}(\theta)$ we only need to have either large individual strata cells or many strata, or a combination) to get accuracy when obtaining confidence intervals for θ, and it is therefore not recommended. Instead we use it for the other question: given the GLS estimate θ^* of θ, is this a reasonable summary of the different tables? For this we compare $Q_{BD}(\theta^*)$ to the $\chi^2(N-1)$ distribution.

Our final example in this section refers to the regression analysis in the Cox proportional hazards model.

Example 13.5   The Cox proportional hazards model solves the equation

$$U(\beta) = \frac{1}{n}\sum_{i=1}^{n}\int_0^\infty (z_i - \bar{z}(t, \beta))dN_i(t) = 0,$$

where

$$\bar{z}(t, \beta) = \partial_\beta \ln S^0(t, \beta), \quad S^0(t, \beta) = \frac{1}{n}\sum_{i=1}^{n} Y_i(t)e^{z_i\beta}.$$

If we introduce our standard martingales $\xi_i(t) = N_i(t) - \int_0^t Y_i(s)e^{z_i\beta}d\Lambda(s)$, the requirement that the estimating equation is an unbiased estimator of β means that we can write

$$U(\beta) = \frac{1}{n}\sum_{i=1}^{n}\int_0^\infty (z_i - \bar{z}(t, \beta))d\xi_i(t).$$

Its variance $V_\beta(U(\beta))$ is given by

$$E_\beta(U(\beta)^2) = E\left(\int_0^\infty (z - \bar{z}(t, \beta))^2 Y(t) e^{z\beta} d\Lambda(t)\right).$$

We also have that

$$U'(\beta) = -\sum_{i=1}^n \int_0^\infty \partial_{\beta\beta}^2 \ln S^0(t, \beta) dN_i(t),$$

and, under the assumption that the $\xi_i(t)$ are martingales, we see that the information matrix $I(\beta)$ is the same as $E_\beta(U(\beta)^2)$. An alternative estimate of $V_\beta(U(\beta))$ is

$$\frac{1}{n}\sum_{i=1}^n \left(\int_0^\infty (z_i(t) - \bar{z}(t, \beta)) d\hat{\xi}_i(t)\right)^2.$$

For the estimate $\hat{\xi}_i(t)$ we use the all-sample estimate of $\Lambda(t)$, which will be a function of β. If we compute this at the point estimate β^* and multiply by the inverse of $U'(\beta^*)$, we obtain the robust estimator for the Cox model estimating equation.

This last example provides us with a way to analyze time-to-event data using the Cox model also in situations where its model assumptions are not fulfilled, provided we use the robust variance estimate, and have a sufficiently large sample. In Section 13.4 we will apply this to the situation with recurrent events, where the assumption of independence of all events cannot be assumed to hold.

13.3   From maximum likelihood theory to generalized estimating equations

Our approach to parameter estimation has mostly been to define an equation that should hold true for the correct parameter value. Usually in statistics parameter estimation starts with likelihood theory, a methodology which we have also referred to occasionally. Here is the place to see how this theory relates to that of estimating equations.

When we design an experiment we should (in principle) also have a probability model for the outcome. The outcome is a sample, $x = (x_1, \ldots, x_n)$, and the probability model is a (multidimensional) CDF $F_\theta(x)$ for the data. The likelihood theory asks the following question: for which choice of θ is what we see most likely? This means that we introduce the likelihood function

$$L(\theta) = dF_\theta(x),$$

computed at the point corresponding to the sample we obtained. For a continuous model this is a density function, for a discrete model it is a probability function. Having already encountered a number of examples of this, we proceed directly to the general theory. To obtain the maximal value of the likelihood, which is also the maximum of $\ln L(\theta)$, we solve the *score equation*,

$$U(\theta) = (\ln L)'(\theta) = 0.$$

This implies that $\partial_\theta dF_\theta(x) = L'(\theta) = U(\theta)L(\theta) = U(\theta)dF_\theta(x)$ (where ∂_θ denotes differentiation with respect to θ), which in turn implies that $\partial_\theta E_\theta(X) = E_\theta(XU(\theta))$. The score equation is the estimating equation of maximum likelihood theory, to which we can apply the discussion we had in the previous section. However, it is an estimating equation with some special properties. First we confirm that it is unbiased, and therefore a valid estimating equation:

$$E_\theta(U(\theta)) = \int \frac{L'(\theta)}{L(\theta)} dF_\theta(x) = \partial_\theta \int dF_\theta(x) = 0.$$

If we differentiate this equation we obtain from the differentiation formula above that

$$E_\theta(U'(\theta)) + E_\theta(U(\theta)U(\theta)^t) = 0,$$

which means that in this case the information matrix equals the variance of the estimating function, $I(\theta) = V_\theta(U(\theta))$. This, in turn, implies that inference about θ is made from $U(\theta)^t I(\theta)^{-1} U(\theta)$, a stochastic variable whose distribution is often assumed (appealing to large-sample theory when our experiment is made up of a large number of independent sub-experiments) to be at least approximately $\chi^2(s)$, where s is the number of parameters.

The next problem is to obtain from this an expression for the variance of $\hat\theta$. We addressed this problem in the previous section, but we can do slightly more in this case, because $U(\theta)$ is a derivative. We use a second-order Taylor expansion of $\ln L(\theta)$ around the true value θ. Inserting the maximum likelihood estimate into this gives the approximation

$$\ln L(\hat\theta) \approx \ln L(\theta) + U(\theta)^t(\hat\theta - \theta) + \frac{1}{2}(\hat\theta - \theta)^t E_\theta(U'(\theta))(\hat\theta - \theta).$$

We also have that $0 = U(\hat\theta) \approx U(\theta) + E_\theta(U'(\theta))(\hat\theta - \theta)$, so

$$2(\ln L(\hat\theta) - \ln L(\theta)) \approx (\hat\theta - \theta)^t I(\theta)(\hat\theta - \theta). \tag{13.2}$$

If we can appeal to large-sample theory to deduce that $U(\theta)$ has an approximate Gaussian distribution, the distribution on the right-hand side is approximately $\chi^2(s)$, where s is the number of components of θ. Applied to the right-hand side, it follows that

$$\hat\theta \in AsN_s(\theta, I(\theta)^{-1}).$$

In summary, we have three different expressions,

$$U(\theta)^t I(\theta)^{-1} U(\theta), \quad (\hat\theta - \theta)^t I(\theta)(\hat\theta - \theta), \quad 2(\ln L(\hat\theta) - \ln L(\theta)),$$

all of which define (approximate) confidence functions if we apply the CDF $\chi_s(x)$ to the expression but replace the estimator $\hat\theta$ by the maximum likelihood estimate θ^*. Often we also replace the matrix function $I(\theta)$ by the observed matrix $-U'(\theta^*)$ in the first two expressions here. The last expression defines a test which is referred to as the *likelihood ratio test*.

So far we have assumed that we are analyzing the correct model, but if we combine the discussion here with that in the previous section, we can also use the score equation as an estimating tool when the model is not correct, as long as we mitigate the misspecification by using the robust variance estimate instead of the model-derived one. This makes the analysis less precise, compared to if the model had been correct, but there must be a price to pay for using the wrong model. As an example, suppose that we want to do a classical ANOVA on

---

Box 13.1 Profile likelihood and restricted maximum likelihood

---

If we estimate parameters with the maximum likelihood method in the presence of nuisance parameters, how do we eliminate these in a correct way from the analysis? The general theory assumes we carry them with us all the way. One alternative is profiling, which means that if $\theta = (\theta_1, \theta_2)$, where θ_2 represents nuisance parameters, we define the profile likelihood $P(\theta_1) = L(\theta_1, \hat{\theta}_2(\theta_1))$, where $\hat{\theta}_2(\theta_1)$ is the maximum likelihood estimate of θ_2 given that we know the value of θ_1. This is not a true likelihood, but sensible statistical analysis is often possible by imitating the likelihood ratio test and defining an approximate confidence region for θ_1 from

$$\{\theta_1; \ \chi_s(-2\ln(P(\theta_1)/P(\hat{\theta}_1))) \leq 1 - \alpha\}.$$

In the particular case where we have data that are $N(m, \sigma^2)$-distributed, and we want to do inference about m by profiling σ^2 out, a short computation shows that

$$\frac{P(m)}{P(\bar{x})} = \left(1 + \frac{t^2}{n-1}\right)^{-n/2}, \quad t = \sqrt{n}(\bar{x} - m)/s,$$

so the relative profiled likelihood is proportional to the density function for the $t(n-1)$ distribution.

If we instead regard σ as the parameter of interest and m as the nuisance parameter, the approximation is less satisfactory (though not bad). In that case there is an alternative method, called restricted maximum likelihood, which also works for more complicated Gaussian models, including the linear mixed effects model, when one wants to obtain unbiased inference about variance information from likelihood theory. The basic idea is that the residuals $r_i = x_i - \bar{x}$, follow a multivariate Gaussian distribution with mean zero and a variance matrix which is given by $\sigma^2(I - e^t e)$, where $e = (1, 1, \ldots, 1)$. This means that the maximum likelihood estimate $f^{-1} \sum_i r_i^2$ of σ^2 is unbiased, where $f = n - 1$ is the dimension of the space orthogonal to the line defined by e, and we end up with the standard sample variance s^2 as the estimate for σ^2. The general Gaussian case with a linear model for the mean and potentially a complicated covariance structure can be handled in the same way.

---

data that look far from Gaussian in distribution. We then use the estimating equation $U(\theta) = 0$ derived from the likelihood theory for the ANOVA model, but use the robust variance estimate when we compute confidence intervals and p-values.

Example 13.6 We wish to illustrate this by applying the standard ANOVA without the interaction term to the data discussed in Example 9.2. This means that we suggest a model

$$p = \theta_1 + \theta_2 x_2 + \theta_3 x_3$$

instead of the corresponding model for the log-odds. The outcome data, a dichotomous variable, are obviously not Gaussian in distribution. As said, we analyze this binary data with a standard ANOVA, but use the robust variance for confidence statements. (Details on how the robust variance estimator is computed in this case are given below.) Once this is done, we transform the estimate of p, with its associated confidence limits, into the log-odds by means of the transformation $p \to p/(1 - p)$. The following table compares the result of this analysis with that of the original logistic regression:

Parameter	Logistic regression		ANOVA	
	Estimate	95% CI	Estimate	95% CI
Patients, Study 1	0.532	(0.264, 0.800)	0.528	(0.267, 0.809)
Patients, Study 2	0.147	(−0.0761, 0.370)	0.149	(−0.0715, 0.373)
Control, Study 1	−0.268	(−0.529, −0.00638)	−0.265	(−0.527, −0.0114)
Control, Study 2	−0.653	(−0.805, −0.50)	−0.653	(−0.812, −0.502)

We see that the actual conclusion from the study is not sensitive to how we arrived at it. The methods differ, but they still provide very similar results.

The result in this example is very important, because it is the starting point for the method of *generalized estimating equations* (GEEs). This methodology is used in cases where we want to model clustered data, where clusters are independent but observations within clusters are not. A typical situation is when we have serial measurements of an outcome variable, so-called longitudinal data, in which the sets of data taken for individual patients are the clusters. There are different ways to approach the analysis of such data, one of which is based on specifying models for the data in the clusters, a kind of approach which we can think of as subject-specific and which we will discuss further in Section 13.5. With the GEE technique we instead take a population averaged approach in which we try to model the behavior of the population mean values as a function of time and other covariates. The simplest application of the GEE philosophy is to ignore all dependence between observations within individuals and estimate the model using a simple GLM, except that when we compute confidence statements, we use the robust variance estimate instead of the model-based one. Often the GEE technique is further refined by introducing some tentative, but simple, form of dependence between observations within clusters and modifying the estimating equation by weighting with the appropriate variance matrix, much as we did in Section 10.6. Also when we use this empirical model we still accept that this is not the correct model, and use the robust covariance estimate for variance estimation. Our hope is that if we can capture some of the variability this way, we may get a more efficient robust covariance estimator.

In Example 13.6 we used the robust variance estimator for a GLS estimate, as defined by equation (9.1). We conclude this section by providing some computational details that are applicable to models based on distributions from the exponential family.

Example 13.7   Consider the regression model of Y on X for which $E_\theta(Y|X = x) = f(\theta, x)$, and introduce the stochastic variable

$$U(\theta) = f'(\theta, X)^t (Y - f(\theta, X))/\sigma^2(\theta, X).$$

Since $E_\theta(U(\theta)|X = x) = 0$ for all x, we have that the variance of $U(\theta)$ is the expected value of the conditional variances $V_\theta(U(\theta)|X = x)$, given by

$$f'(\theta, x)^t V_\theta(Y|X = x) f'(\theta, x)/\sigma^4(\theta, x).$$

Since $\sigma^2(\theta, x) = V_\theta(Y|X = x)$, we have $V_\theta(U(\theta)|X = x) = f'(\theta, x)^t f'(\theta, x)/\sigma^2(\theta, x)$.

Next we consider the information matrix. The derivative of $U(\theta)$ when $X = x$ consists of $-f'(\theta, x)^t f'(\theta, x)/\sigma^2(\theta, x)$ plus a term involving $Y - f(\theta, x)$ for which the mean is zero. Its expected value (conditional on $X = x$) is therefore the same as above, and under the modeling assumption the information matrix is the same as the variance of the estimator. Finally, the model-free estimate of the variance of $U(\theta)$ is

$$\frac{1}{n}\sum_{i=1}^{n} U(x_i, \theta)U(x_i, \theta)^t = \frac{1}{n}\sum_{i=1}^{n} \sigma(\theta, x_i)^{-4} f'(\theta, x_i)^t f'(\theta, x_i)(y_i - f(\theta, x_i))^2.$$

Combining this with the information matrix above gives the robust variance for GLMs.

13.4   The analysis of recurrent events

To further illustrate that we can analyze a misspecified model for our data and still make an acceptable inference by using the robust variance estimator, we now study the problem with recurrent events that was introduced in Section 11.6. In that discussion we did not illustrate the analysis with any examples, because we wanted the robust variance introduced first. We now have it, so we will remedy that omission in this section. We noted in Section 11.6 that we cannot readily apply methods developed for single events in individuals to data consisting of recurrent events within individuals, because it requires a crucial assumption: all events must be independent, which would not be consistent with an expected patient heterogeneity in, for example, disease severity. As noted then, the simplest way to handle this problem is to reduce the data to the total number of events for each individual, ignoring when in time individual events occur. Instead we assume that there is, for each individual, a mean intensity over the period, defined by the expected number of events, divided by the observation time. If we assume that all individuals have the same constant hazard, with events occurring independently of each other, we could analyze these count data as a Poisson regression problem, adjusting for observation time. However, neither the assumption of homogeneity between patients, nor the within-patient independence of events, is likely to hold. This suggests the possibility of overdispersion, as illustrated in the next example.

Example 13.8   In a one-year study the investigators wanted to assess the effect on the occurrence of asthma exacerbations when a long-acting brochodilator drug (LABA) was added to an inhaled corticosteroid (ICS). The particular study had four treatment arms defined by two doses of ICS, separated by a factor of 4, each of which was studied with and without concomitant administration of the LABA. We will analyze this in two steps, as discussed in Chapter 9, and first analyze the number of exacerbations using a Poisson regression model approach.

In this step we estimate the group means of the exacerbation rates, adjusting for possible predictive factors for exacerbations. One set of predictors are related to the fact that patients with more severe asthma are expected to have more exacerbations. Both the FEV_1 as percentage of predicted normal[1] and the log of the ICS dose at enrollment were considered to be such variables. Other covariates included were body mass index and sex, together with the basic indicator variables for the four treatment groups. With these variables in a Poisson regression model, the output provides us with the adjusted exacerbations rates shown in Figure 13.3.

---

[1]This is a prediction of what FEV_1 should have been for a healthy individual of the same height, age and sex.

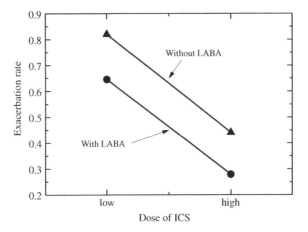

Figure 13.3 Estimated exacerbation rates from a Poisson regression model.

These rates are adjusted to a common value for each of the covariates, which means that they describe a situation where we have equal numbers of males and females in the groups, and the other covariates are set to their overall study mean.

A Poisson model implies that all events, within and between subjects, are independent. This is most likely a model misspecification, because heterogeneity is expected to increase the variance relative to what the standard Poisson model gives. There are two immediate remedies for this, based on the Poisson model:

1. We can introduce an overdispersion factor (see Section 9.7) in the Poisson model.

2. We can use the robust variance estimate.

The table below shows how these three models compare in terms of the standard error for the log-rates. (Note that the model analyzes log-rates, whereas we have plotted the rates in Figure 13.3). The ratio between the standard error for the model with and without overdispersion is constant, with the overdispersion parameter estimated to be 1.8. Going forward, we will use the method with the robust variance.

Treatment	log(rate)	Poisson	Standard error Overdispersion	Robust
ICS low	−0.198	0.084	0.149	0.121
ICS low+LABA	−0.437	0.093	0.164	0.151
ICS high	−0.820	0.109	0.192	0.139
ICS high+LABA	−1.278	0.134	0.237	0.169

This is the first step, and is about finding comparable mean group estimates. In the second step we want to compare these means in different ways, for which we may proceed in different ways, depending on what the question is. The standard way to express the results is in terms of hazard ratios: how much has the addition of a LABA, or an increase in the ICS dose, decreased the hazard? Such information is obtained directly from the Poisson regression by forming linear combinations of the model parameters, followed by an exponentiation. We will present the result in the next example.

An alternative second-step question may be this: how much should the ICS dose be increased, in order to achieve the same effect on the exacerbation rate as is obtained by adding the LABA? The two lines in Figure 13.3 represent two dose–response curves, one with and one without the LABA, and since they look parallel it is a natural question to ask what the relative (ICS) dose potency is. This means fitting two parallel lines to the vector of estimated rates λ^*. We therefore need to find parameters $\theta = (\theta_1, \theta_2, \theta_3)$ such that the function

$$(\theta_1, \theta_1 + \theta_2 \ln(\theta_3), \theta_1 + \theta_2 \ln(4), \theta_1 + \theta_2 \ln(4\theta_3)),$$

where θ_3 represents the relative dose potency, has an optimal fit to λ^*. This analysis should be performed weighted by the covariance matrix for λ^*, for which we do not have an estimate. The original Poisson analysis is done on the log-rates, so what we have is an estimate Σ^* of the covariance matrix for these. We therefore need to transform our analysis to the log scale. This means that we solve the GLS equation

$$f'(\theta)^t \Sigma^{*-1}(\ln(\lambda^*) - f(\theta)) = 0,$$

where $f(\theta)$ is the logarithm of the function above. The corresponding parameter estimates are $\theta_1 = 0.82$, $\theta_2 = -0.27$, and $\theta_3 = 1.85$, and we see that we estimate that the effect on the exacerbation rate of adding the LABA is equivalent to increasing the ICS dose by a factor of 1.85.

The analysis in this example is a population averaged approach; claims are related to overall group behavior and it uses the robust variance estimate (or an overdispersion parameter) to mitigate a misspecified statistical model. Overdispersion can arise in a number of ways which cannot be distinguished when we only have the aggregate count available. One likely cause is heterogeneity in disease severity, which we may model by the frailty model. This model assumes that rates are stable within patients during the whole study period, but may differ between them. If we assume a gamma distribution for the frailties, we replace the Poisson distribution with the negative binomial distribution, as was shown in Section 11.6.

Example 13.9   We now apply the negative binomial distribution to the data instead, using the same linear model for the mean as in the last example. The following table shows the group comparisons of primary interest, which are compared for the two models – the Poisson model with the robust variance, which is a population averaged approach, and the negative binomial model, which is a subject-specific approach.

Variable	Parameter	Estimate	95% CI	p-value
Poisson (robust variance)	ICS low±LABA	0.787	(0.578, 1.07)	0.20
	ICS high vs low	0.537	(0.397, 0.727)	0.00074
	ICS high±LABA	0.632	(0.444, 0.901)	0.033
Negative binomial	ICS low±LABA	0.761	(0.544, 1.07)	0.18
	ICS high vs low	0.481	(0.341, 0.678)	0.00045
	ICS high±LABA	0.592	(0.404, 0.868)	0.024

We see that the ratios are in reasonable agreement, and also that the precision in these estimates is similar for the two models.

So far we have not used all the available data, but only the reduced data set consisting of the total number of events experienced for each patient. One way to analyze the complete data is to use the estimating equation we would have used if all events were independent, and then use the robust variance estimate for inference. Alternatively, we can model the data with a frailty model, assuming that events within subject occur independently (given the degree of frailty).

Example 13.10   In order to analyze all exacerbation data (including precise timings), we next carry out a Cox regression, but instead of the standard model variance, we use the robust variance when we make inference about parameters. This is the upper half of the table below, which should be compared to the corresponding part of the table in Example 13.9. The main difference is that not only is the precision increased, but also there is a slight numerical increase in effects.

Variable	Parameter	Estimate	95% CI	p-value
Cox model	ICS low±LABA	0.825	(0.676, 1.01)	0.11
(robust	ICS high vs low	0.617	(0.495, 0.767)	0.00028
variance)	ICS high±LABA	0.658	(0.501, 0.865)	0.012
Frailty	ICS low±LABA	0.806	(0.660, 0.985)	0.077
model	ICS high vs low	0.615	(0.494, 0.766)	0.00026
	ICS high±LABA	0.654	(0.496, 0.863)	0.012

The lower half of this table shows a random effects analysis of all data, which is analogous to the previous use of the negative binomial distribution. It is an extension of the Cox regression model to accommodate gamma frailties, assuming independence between events within a subject, but allowing for subject-specific exacerbation rates. Such a model is called a shared frailty model, and will be briefly discussed in the next section. If we compare the result of this analysis with the corresponding analysis of total count only, we see numerically larger effects, together with increased precision. If we compare the two models in the table above, we see that they are in reasonable agreement, in terms of both estimates and precision.

We have so far not discussed how to describe the complete set of exacerbation data in order to find out if there is some pattern to when exacerbations occur. One way to do this is to describe the data on a mean level and compute the *mean cumulative function* (MCF), which is simply the average number of events that have occurred up to a given time point. For the Poisson model, its cumulative intensity $\Lambda(t)$ is also its mean, and therefore the MCF. In a heterogeneous world, where different individuals experience events, independent within individuals, according to cumulative intensities $\Lambda(t, \theta)$, with θs distributed as $P(\theta)$ in the population, the MCF becomes

$$m(t) = \int \Lambda(t, \theta) dP(\theta).$$

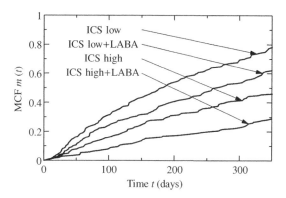

Figure 13.4 The estimated mean cumulative functions describing exacerbations for each treatment.

When the θs represent (proportional) frailties (i.e., when $\Lambda(t, \theta) = \theta\Lambda(t)$) this means that the MCF is actually proportional to $\Lambda(t)$, with the proportionality constant being the average value of the frailty distribution. In light of this, it is useful to plot the observed MCFs for different groups, in order to see if they appear to have such a simple relationship to each other.

The estimation of the MCF is done by a Nelson–Aalen type of estimator. We write $m(t)$ as $\int_0^t dm(s)$ and estimate $dm(t)$ by the number of observed events at time t, divided by the number of individuals at risk at that time. The estimated MCFs for the data above are shown in Figure 13.4, where we see that it is reasonable to consider these curves to be proportional to some underlying mean function. There is therefore some support for the notion that the frailty model is a reasonable explanation for much of the overdispersion.

We can also use the MCF and derive nonparametric tests similar to those in Chapter 12, using the MCF instead of the cumulative hazard function in the derivation. The (minor) problem here is to account for the dependence structure in the variance estimate. We will not pursue this; instead we will note a connection between this discussion and the one on competing risks. The MCF description corresponds to the Kaplan–Meier estimate in conventional survival analysis in that we estimate effects in an environment free of competing risks. This is probably relevant in our situation, since the censoring reasons mainly were planned ones (termination of study), or because patients were withdrawn by other (hopefully independent) reasons. However, in some other settings, for example if we look at recurrent tumors after a primary treatment course, it may be highly relevant not to ignore one particular competing risk, namely that of death. We want statements not in an environment free of death, but in its presence. This leads us into the same discussion as we had in Section 11.4, where we replaced the survival function with the cumulative incidence function. In this case we replace the MCF (which reduces to the survival function when there is only one event per patient) with something that reduces to the CIF when there is only one event per patient. If $F(t)$ is the CDF for the time-to-death variable and $\mu(t)$ the conditional mean number of events in the subset of individuals still alive at time t, the new MCF is

$$m(t) = \int_0^t F^c(s-)d\mu(s).$$

This function is estimated using the Kaplan–Meier estimate for survival and a Nelson–Aalen type of estimator for $d\mu(t)$. From there on we can obtain new tests. The whole approach is similar to the replacement of the conventional log-rank test with Gray's test in a competitive environment.

13.5   Defining and estimating mixed effects models

The recurrence data of the previous section is an example of clustered data, which means that the observations come in clusters, or groups, so that all data are not independent. The cluster may be the data from a patient on whom we repeatedly measure an outcome variable, such as on different doses of a particular drug. Alternatively, it may be measurements taken at different time points, as is the case in a longitudinal study. It could also be an account of all the exacerbations, with timings, a patient experiences during the period he is observed. The general setup is that we have n clusters (say, patients), and for each of these we have a varying number n_i of observations y_{ij} of the outcome variable. Let $y_i = (y_{i1}, \ldots, y_{in_i})$ denote the whole vector of observations for patient i.

We have encountered data of this kind in two examples so far. The first was the dose–response discussion in Chapter 10, where we have, for each of 20 individuals, five different observations at different dose levels. For each individual we had a dose–response function of a specific form, but with a subject-specific parameter, the ED_{50}. The other example is the toxicological experiment in rats in Example 12.1. So far, when we have analyzed these data, we have done so under the assumption that the survival times for the 150 rats are all independent. However, the data come from 50 litters with three pups in each. Rats from the same litter are more closely related genetically than rats from different litters, which means that there could be some correlation between lifetimes for rats within litter. In other words, there might be a litter-specific frailty, ignorance of which might affect our result. In situations like these, the likelihood is built up as a product of likelihoods for individual clusters. (We assume that clusters are independent.) We will now discuss what the likelihood for such clusters looks like in a few situations. Once we have done that, we can apply the general maximum likelihood theory to the estimation of whatever model parameters are part of the problem. The model that is applicable to the dose–response case was defined in Example 10.6.1, and we start with it.

Example 13.11   In the notation above, the model that was specified in Example 10.6.1 is that for each individual i, the $n_i = 5$ observations $y_{ij} \in N(f(\theta, D_j), \sigma^2)$ are independent, with the function $f(\theta, D) = D/(e^\theta + D)$ describing the dose response for an individual. We also assumed that $\theta \in N(\mu, \eta^2)$. Given that we know θ and σ^2, the density of the distribution for the vector y_i is given by

$$p_i(y_i|\theta, \sigma^2) = (2\pi\sigma^2)^{-n_i/2} e^{-Q(y_i, \theta)/2\sigma^2}, \quad \text{where } Q(y_i, \theta) = \sum_{j=1}^{n_i}(y_{ij} - f(\theta, D_j))^2.$$

The likelihood for a randomly sampled individual is obtained by taking the average of θ, weighted according to the $N(\mu, \eta^2)$ distribution:

$$L_i(\mu, \eta, \sigma) = \int p_i(y_i|\theta, \sigma^2)d\Phi\left(\frac{\theta - \mu}{\eta}\right).$$

A change of variable in this integral shows us that the total likelihood can be written as

$$L(\mu, \sigma, \eta) = e^{-n \ln(\sigma)} \prod_i \int e^{-Q(y_i, \mu + \sqrt{2}\eta\xi)/2\sigma^2 - \xi^2} d\xi,$$

up to a multiplicative constant. Here each integral is one-dimensional, so it is relatively simple to compute it numerically to a reasonable precision using an appropriate integration method. In order to estimate parameters, and derive confidence information about these, we use maximum likelihood theory. When we do so, we find that the estimate of the mean of ED_{50} is 0.79 with 95% confidence interval (0.41, 1.50), which is compatible with the known value of 1 (recall that these were simulated data). Moreover, we estimate the between-subject variability (η) to be 1.42, and the within-subject variability (σ) to be 0.11, both of which are in good agreement with the true numbers ($\sqrt{2}$ and 0.1, respectively). This is the method we referred to in Section 10.6 as being the preferred method for the analysis of these data.

Example 13.12 We now construct the rat model in a similar way. For litter i we have three intensity functions $\Lambda_{ij}(t)$, $j = 1, 2, 3$, for the three pups. The assumption we make is that $\Lambda_{ij}(t) = \eta z_j \Lambda_0(t)$, where $\Lambda_0(t)$ is a reference intensity common to all, z_j is one if the rat is a control and it equals θ if the rat is treated with the drug (θ is the hazard ratio, describing the treatment effect), and η is a litter-specific frailty. Within litter we assume that lifetimes for pups are independent so, if we knew the frailty of a particular litter to be η, the survival function for that litter would be

$$\prod_j e^{-\eta z_j \Lambda_0(t_j)} = e^{-\eta \sum_j z_j \Lambda_0(t_j)}.$$

Since we do not know η, we average over it to get

$$\int_0^\infty e^{-\eta \sum_j z_j \Lambda_0(t_j)} dP(\eta) = \mathcal{L}\left(\sum_j z_j \Lambda_0(t_j)\right),$$

where $\mathcal{L}(s)$ is the Laplace transform of the frailty distribution $P(\eta)$. The corresponding probability density is obtained by differentiation, and repeating some of the calculations we did earlier gives us the log-likelihood

$$L(\theta, \alpha) = \sum_{i=1}^n \left(\delta_j \ln \left(\sum_{j=1}^{n_i} z_j d\Lambda_0(t_j) \right) + \ln \left((-1)^{\delta_j} \mathcal{L}^{(\delta_j)} \left(\sum_{j=1}^{n_i} z_j \Lambda_0(t_j) \right) \right) \right).$$

The parameter θ, which describes the treatment effect, is hidden in the z_j, and α defines the distribution $P(\eta)$. A model of this kind is called a shared frailty model; in this case the frailty is shared between the pups in a litter. If we have a parametric representation of $\Lambda_0(t)$, parameter estimation is a classical maximum likelihood problem, but we can also carry out the analysis in the more general case with an unspecified baseline hazard.

If we recall the close connection between the Poisson distribution and the Cox regression model, it should come as no surprise that a popular choice is to take $P(\eta)$ as a gamma distribution with mean one. There are different numerical methods available for the analysis of such a model which, when applied to our rat data, gives almost the same estimate for θ as

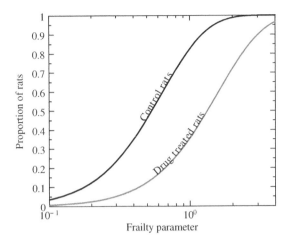

Figure 13.5   The distributions of rat frailty in the two groups in the toxicological experiment.

was obtained in the analysis when clusters were ignored, namely 2.5 with 95% confidence interval (1.3, 4.7). The variance of the heterogeneity distribution is estimated to be 0.47.

To illustrate what this model implies, look at Figure 13.5 which shows the frailty distributions for each group. We see how the control animals have a frailty distribution that lies to the left of that of the drug-treated ones. The drug effect is, according to the model, a multiplicative constant, so on the log scale we should obtain two parallel (horizontally shifted) CDFs.

As already mentioned, another example of a shared frailty model is the recurrent exacerbation data discussed in the previous section, in which frailty for exacerbations is shared within patients.

Both these examples illustrate mixed effects models of clustered data. Such models can be constructed for a variety of data, where we start the modeling with a description of what happens on the individual level by defining a function of some parameters. Some of these parameters may take different values in different individuals, and are then called random parameters, whereas other parameters are the same for all individuals and referred to as fixed parameters. As before, the random parameters are described by a distribution, which contains some unknown model parameters that we need to estimate. In addition to this, the outcome variable will also have some associated variability that requires a distribution describing it, again specified up to some unknown model parameters. When there is more than one random parameter we usually need to account for the dependence between them, which leads us to assume that the distribution for the random parameters (or some simple transformation thereof, such as taking the logarithm) is a multidimensional Gaussian distribution, because this is more or less the only model that can easily describe the correlation between parameters. But there is a problem. The likelihoods of the clusters so defined are multidimensional integrals which in most cases cannot be explicitly computed. We therefore need to resort to numerical methods, and numerical methods for higher-dimensional integrals are computer-intensive. This has led to the introduction of various approximations, approximations that do not necessarily provide us with what we are led to believe from our initial model description. These approximations will be the subject of the rest of this section.

Let a model be defined in such a way that the likelihood $L_i(\theta_i, \alpha)$ for a cluster is a function of two parameters (usually vectors): θ_i is to be thought of as a regression parameter, whereas α is an auxiliary parameter, mainly expressing within-cluster variance and correlations. We do not specify what this model looks like; it can be derived from a binomial or Poisson model, or some other family from the exponential family of distributions, or it can be some model for time-to-event data. The most important case is when we assume that the data within cluster are described by a Gaussian likelihood, and we discuss this case in some more detail. The important difference between θ_i and α is that the former may differ between clusters, whereas α is fixed. In simpler cases we assume that we can write $\theta_i = \theta + \xi$, where θ is a fixed constant and ξ has a distribution described by a CDF $P_\Sigma(\xi)$ in the population of clusters. This distribution is assumed to have a known mean and otherwise defined by a parameter vector Σ. (The reason why we choose this symbol here is that in the most important cases Σ defines a variance matrix. We will then use the same symbol for both the matrix and the parameterization that defines it.) Based on the individual likelihoods, we can now define an overall likelihood, applicable to a cluster which is randomly sampled from the population:

$$L_i(\theta, \alpha, \Sigma) = \int L_i(\theta + \xi, \alpha) dP_\Sigma(\xi).$$

This looks simple, but is computationally challenging.

To be more specific, we consider in some more detail the important case where $P_\Sigma(\xi)$ is a $N(0, \Sigma)$ distribution. In such a case we can make a transformation of variables in the integral and write the likelihood as

$$\prod_{i=1}^{n} \int L_i(\theta + \sqrt{2\Sigma}\xi, \alpha) e^{-|\xi|^2} d\xi$$

(the meaning of $\sqrt{\Sigma}$ is not important for our discussion, but refers to a Cholesky decomposition of Σ). This is an integral which is numerically tractable in one or two dimensions, but is challenging in general; if we need n points to compute a one-dimensional integral, we will need n^d points to compute a d-dimensional one, a number that grows fast with dimension (this is referred to as the curse of dimensionality). So when we try to estimate the parameters of such a model, it is not the theory of how we subsequently obtain knowledge about the true parameters that is the problem. The general maximum likelihood theory will do that for us. The problem is to find a feasible numerical method that will provide us with useful estimates of the parameters.

In order to illustrate some approaches to this problem, we make further restrictions. We assume that for cluster (think subject) i we have an outcome vector variable y_i which has a multidimensional Gaussian distribution, such that:

• its mean is given by a regression function $f_i(\theta_i)$ which is fully computable for each subject, provided we know θ_i (the definition of the function may contain measured covariates);

• its variance matrix is of the form $\sigma^2 \Lambda_i$, where we assume Λ_i is known (the identity matrix if we assume that different observations within cluster are independent). The assumption that we know Λ_i is one of convenience only.

With these assumptions we can now write the likelihood for cluster i as

$$L_i(\theta_i, \sigma^2) = e^{-H_i(\theta_i, \sigma^2)/2},$$

where

$$H_i(\theta_i, \sigma^2) = \sigma^{-2}(y_i - f_i(\theta_i))^t \Lambda_i^{-1}(y_i - f_i(\theta_i)) + \ln(2\pi \det(\sigma^2 \Lambda_i)).$$

As above, we assume that the distribution of the θ_i in the population of clusters has a Gaussian distribution with mean θ and variance matrix Σ. Collecting all these assumptions together, we find that the full likelihood is

$$L(\theta, \sigma^2, \Sigma) = \prod_i \int e^{-Q_i(\theta, \sigma^2, \Sigma, \xi)/2} d\xi,$$

where

$$Q_i(\theta, \sigma^2, \Sigma, \xi) = H_i(\theta + \xi, \sigma^2) + \xi^t \Sigma^{-1} \xi + \ln(\det 2\pi \Sigma).$$

Even though we have restricted our assumptions considerably, this is still a complicated multidimensional integral for which we in most cases do not have a closed expression that allows its direct integration. However, in the special case of quasi-linearoty, in which

$$f_i(\theta + \xi) = g_i(\theta) + C_i(\theta)\xi,$$

the log-likelihood is proportional to

$$\sum_{i=1}^n ((y_i - g_i(\theta))^t V_i(\theta, \omega)^{-1}(y_i - g_i(\theta)) + \ln(\det 2\pi V_i(\theta, \omega))),$$

where we have introduced the notation $\omega = (\Sigma, \sigma^2)$ for the coefficients that describe variability from different sources, and $V_i(\theta, \omega) = \sigma^2 \Lambda_i + C_i(\theta)\Sigma C_i(\theta)^t$. Note how $V_i(\theta, \omega)$ is made up of the within-subject variability and the between-subject variability. On this likelihood we can apply the general theory of maximum likelihood estimation and obtain estimates and confidence descriptions on all the parameters in the model. For fixed ω it is almost a weighted least squares problem in θ – almost, because the variance also contains θ, which makes the problem less straightforward than it appears at first sight, something we have encountered before.

There is one further restriction we can make to get a problem that is almost devoid of (serious) numerical problems: to assume that $C_i(\theta) = C_i$ does not depend on θ. In such a case we have no θ in V_i, and for given ω the estimation of θ is a straightforward (weighted) least squares problem. In fact, we can profile θ out using this observation, which gives a (profiled) likelihood in ω only. If we further assume that $g_i(\theta) = x_i\theta$, where x_i is a covariate (row) vector for subject i, we have the *linear mixed effects model*. For this case we can find an explicit solution for θ given ω, so the computation of the profiled likelihood does not involve an internal numerical optimization, as it does when $g_i(\theta)$ is not a linear function in θ. Linear mixed effects models are important models in statistics, but this is not the place to discuss them; as for the Cox regression model, there are many books available on this subject. Our objective with this discussion is different, so we return to the general case with a nonlinear regression function.

Nonlinear mixed effects model are common in biostatistics, in particular for longitudinal data, where we have series of measurements of an outcome variable at different time points. A well-advertised and important application of these models are the so-called population pharmacokinetic models in which serial measurements of plasma concentration data are modeled. The regression equation is in such cases derived from basic pharmacokinetic considerations, usually a simple compartment model, and the main purpose of the analysis is usually to find explanatory variables for the variability of some of the important pharmcokinetic parameters, including clearance and volume. The success of the population pharmacokinetic approach is to a large extent due to special software, NONMEM, which provided an early solution to the challenge of estimating the parameters in a nonlinear mixed effects model. However, its estimation methods do not use the full maximum likelihood, including multidimensional integrals, instead it employs various approximations, which we want to take a closer look at.

In order to indicate how these approximation methods work, we write down the score equation for the quasi-linear case above. It consists of one set of equations derived from the differentiation of the log-likelihood with respect to θ, and another set of equations derived from differentiating it with respect to ω. The structure of the system is

$$\sum_{i=1}^{n} g_i'(\theta)^t V_i(\theta, \omega)^{-1}(y_i - g_i(\theta)) + (\partial_\theta v_i)^t F_i(\theta, \omega) = 0, \tag{13.3}$$

$$\sum_{i=1}^{n} (\partial_\omega v_i)^t F_i(\theta, \omega) = 0. \tag{13.4}$$

The exact expressions for v_i and F_i are not important for our discussion, it is the general structure of these equations that matters. (If you feel you need to know these functions, and understand the notation, we have that $v_i(\theta, \omega) = \text{vec}(V_i(\theta, \omega))$ and

$$F_i(\theta, \omega) = \frac{1}{2}(V_i(\theta, \omega)^{-1} \otimes V_i(\theta, \omega)^{-1})(s_i(\theta) - v_i(\theta, \omega)),$$

where $s_i(\theta) = \text{vec}((y_i - g_i(\theta))(y_i - g_i(\theta))^t)$.) By solving this system we can find the maximum likelihood estimate of both θ and ω. If $C_i(\theta)$ does not depend on θ, the function $v_i(\theta, \omega)$ does not depend on θ, and the second term in equation (13.3) vanishes. This is why we can profile θ out in this case. Also when this is not the case, the estimating equations for the quasi-linear case are explicit enough (do not involve numerical calculation of integrals) for this to be a feasible numerical problem.

This is not the case for the general nonlinear mixed effects model, which involves multi-dimensional integrals. One way to attack such problems is to try to approximate the integrand of the likelihood integral by a (multidimensional) Gaussian density function, which is equivalent to a method for the numerical approximation of integrals which is due to Laplace. A conceptually different approach is to approximate the regression function with a quasi-linear version. The latter is what the software package NONMEM does, and it can be done to different degrees of complexity, including the following:

First order. This is the simplest method in which we expand $f_i(\theta + \xi)$ around $\xi = 0$ and keep the linear part only. This defines a quasi-linear method with $g_i(\theta) = f_i(\theta)$ and $C_i(\theta) = f_i'(\theta)$.

Conditional first order. This is done in the same way as the first-order method, except that we now linearize around a ξ other than zero. It is an iterative process, in which we start with the first-order case. Once the parameters are estimated for that case, we compute post-hoc estimates of individual ξs as empirical Bayes estimates in the same way as was discussed in Example 10.6.1. Then we linearize the individual regression function around this estimate instead (so approximations are subject-specific), and repeat the process (i.e., apply the quasi-linear model to this linearized system). This procedure is then repeated a number of times until the θs do not change from iteration to iteration. (This method is closely related to the Laplace method mentioned above, but we leave it to the references to explain how.)

Example 13.13   Consider the data discussed in Section 10.6, but assume that we now want to fit the more general dose–response function

$$\frac{D^{\theta_2}}{e^{(\theta_1+\xi)\theta_2} + D^{\theta_2}}.$$

This means that we introduce into the model the Hill parameter, here denoted by θ_2. It is only θ_1 that is random; the slope is fixed, but unknown (in the previous analyses it was held fixed to the known value one).

We can do a reasonably exact analysis of these data since the integral is one-dimensional, which gives us the estimates $\theta_1 = -0.235$, $\theta_2 = 1.002$. This is rather close to the known truth ($\theta_1 = 0, \theta_2 = 1$). If we use the first-order NONMEM method, however, the estimate becomes $\theta_1 = -0.129, \theta_2 = 0.732$, which provides us with a different slope of the function. This estimated slope we recognize from our discussion in Section 10.6 as (essentially) the parameter value that describes the population mean value curve. The first-order method therefore estimates the population means and is not really a description of the individual curves, as we might have expected. If we use the conditional first-order approach for estimation we get parameter estimates similar to those of the maximum likelihood method with numerical integration.

13.6   Comments and further reading

Traditionally statistical theory has been based on likelihood methods or least squares. Putting the estimating equation at center appears to be a more recent approach, sometimes referred to as a quasi-likelihood approach. A systematic treatise with a number of non-elementary examples can be found in Heyde (1997). Likelihood theory, on the other hand, was developed by R. A. Fisher in the early 1920s, and can be found in all major textbooks on statistical theory. From a historic perspective, it was the development of likelihood theory that defined statistics as a subject in itself. For a comprehensive discussion of GEE models, in particular for longitudinal data, see Diggle et al. (2002). This book describes the variance models that are most often used, and how to estimate parameters in such models.

Lindsey (1998) gives a discussion on what constitutes overdispersion in models such as Poisson models. Extensions of the Cox proportional hazards model are discussed with examples in Therneau and Grambsch (2000) and Hougaard (2000). The particular case of frailty models is also the subject of these books, and between them they outline two different numerical methods for the estimation of the gamma shared frailty model. Recurrent events are the subject of Nelson (2003), as well as the review article by Cook and Lawless (2002). For the analysis of recurrent events in the presence of a terminating event such as death, see Ghosh and Lin (2000). For mathematical details on the use of the Cox equation as a GEE for situations with model misspecification, see Lin et al. (2000). Finally, the original publication of the data we used to illustrate recurrent events with is Pauwels et al. (1997).

There are nowadays quite a few books on nonlinear mixed effects models. An early one which both discusses theory and contains many illustrative examples is Davidian and Giltinan (1995). Another one is Pinheiro and Bates (2000), which discusses the linear theory in much more detail than the first book. As mentioned in the text, the first software to handle nonlinear mixed effects models was NONMEM (Beal and Sheiner, 2004), but today such models can be analyzed in many important statistical software packages. A review of how the different models used in NONMEM are derived from the likelihood is also given in Wang (2007). More sophisticated estimation methods also exist, using other approximations to the integrals involved (Pinheiro and Bates, 1995).

References

Beal, S.L. and Sheiner, L.B. (2004) NONMEM®users guide. *University of California, NONMEM Project Group.*

Cook, R.J. and Lawless, J.F. (2002) Analysis of repeated events. *Statistical Methods in Medical Research,* **11**(2), 141–166.

Davidian, M. and Giltinan, D.M. (1995) *Nonlinear Models for Repeated Measurement Data,* vol. 62 of *Monographs on Statistics and Applied Probability.* New York: Chapman & Hall.

Diggle, P.J., Heagerty, P., Liang, K.Y. and Zeger, S.L. (2002) *Analysis of Longitudinal Data,* vol. 25 of *Oxford Statistical Science Series* second edn. Oxford: Oxford University Press.

Firth, D. (1993) Bias reduction of maximum likelihood estimates. *Biometrika,* **80**(1), 27–38.

Ghosh, D. and Lin, D.Y. (2000) Nonparametric analysis of recurrent events and death. *Biometrics,* **56**(2), 554–562.

Heyde, C.C. (1997) *Quasi-likelihood and Its Applications: A General Approach to Optimal Parameter Estimation,* Springer Series in Statistics. New York: Springer.

Hougaard, P. (2000) *Analysis of Multivariate Survival Data,* Statistics for Biology and Health. New York: Springer.

Lin, D.Y., Wei, L.J., Yang, I. and Ying, Z. (2000) Semiparametric regression for the mean and rate functions of recurrent events. *Journal of the Royal Statistical Society, Series B,* **62**(4), 711–730.

Lindsey, J.K. (1998) Counts and times to events. *Statistics in Medicine,* **17**(15/16), 1745–1751.

McCullagh, P. and Nelder, J.A. (1989) *Generalized Linear Models,* Monographs on Statistics & Applied Probability second edn. London: Chapman & Hall.

Nelson, W.B. (2003) *Recurrent Events Data Analysis for Product Repairs, Disease Recurrences, and Other Applications.* Philadelphia: Society for Industrial and Applied Mathematics.

Pauwels, R.A., Löfdahl, C.G., Postma, D.S., Tattersfield, A.E., O'Byrne, P., Barnes, P.J. and Ullman, A. (1997) Effect of inhaled formoterol and budesonide on exacerbations of asthma. *New England Journal of Medicine,* **337**, 1405–1411.

Pinheiro, J.C. and Bates, D.M. (1995) Approximations to the log-likelihood function in nonlinear mixed-effects models. *Journal of Computational and Graphical Statistics,* **4**(1), 12–35.

Pinheiro, J.C. and Bates, D.M. (2000) *Mixed-Effects Models in S and S-PLUS.* New York: Springer.

Therneau, T.M. and Grambsch, P.M. (2000) *Modeling Survival Data: Extending the Cox Model,* Statistics for Biology and Health. New York: Springer.

Wang, Y. (2007) Derivation of various NONMEM estimation methods. *Journal of Pharmacokinetics and Pharmacodynamics,* **34**(5), 575–593.

13.A Appendix: Formulas for first-order bias

Why is it that for almost all models, except for the all-important special case of ANOVAs, the estimator has a built-in bias? Let $U(\theta)$ be an estimating function, which by definition means that $E_\theta(U(\theta)) = 0$, and let $\hat{\theta}$ denote the solution to the estimating equation $U(\theta) = 0$. To simplify notation, assume that we have a single parameter (the argument is the same in general, but more complicated to write down). If we write

$$E_\theta(\hat{\theta}) = \theta + b(\theta),$$

when is the bias term $b(\theta)$ zero? A second-order Taylor expansion leads to

$$U(\theta) + U'(\theta)(\hat{\theta} - \theta) + \frac{1}{2}U''(\theta)(\hat{\theta} - \theta)^2 \approx U(\hat{\theta}) = 0,$$

and we start the discussion by making a few unrealistic assumptions: (1) that the mean of the expression on the left is zero; (2) that $U''(\theta)$ is a deterministic function. In such a case we have that

$$2E_\theta(U'(\theta)(\hat{\theta} - \theta)) + U''(\theta)V_\theta(\hat{\theta}) = 0.$$

In the first term, we write $U'(\theta) = -I(\theta) + (U'(\theta) + I(\theta))$ and then recall that to a first-order approximation we have that $\hat{\theta} = \theta + I(\theta)^{-1}U(\theta)$ (equality here is a third assumption). Using all this, we find that

$$E_\theta(U'(\theta)(\hat{\theta} - \theta)) = -I(\theta)b(\theta) + E_\theta(U'(\theta)I(\theta)^{-1}U(\theta)) + E_\theta(I(\theta)I(\theta)^{-1}U(\theta))$$
$$= -I(\theta)b(\theta) + I(\theta)^{-1}E_\theta(U'(\theta)U(\theta)),$$

from which we deduce that

$$b(\theta) = \frac{1}{2}I(\theta)^{-1}(2I(\theta)^{-1}E_\theta(U'(\theta)U(\theta)) + U''(\theta)V_\theta(\hat{\theta})).$$

This formula becomes an approximation if we drop the assumptions made, but we need to replace $U''(\theta)$ by its expected value $E_\theta(U''(\theta))$. We see that the bias is derived from the fact that $U(\theta)$ has a non-zero curvature. If there is no curvature, there is no bias, which is the case for the conventional ANOVAs, based on Gaussian data.

In the special case of maximum likelihood estimators we can differentiate the formula $E_\theta(U'(\theta)) = -E_\theta(U(\theta)^2)$ to obtain

$$E_\theta(U''(\theta)) + E_\theta(U'(\theta)U(\theta)) + 2E_\theta(U(\theta)U'(\theta)) + E_\theta(U(\theta)^3) = 0.$$

In this case we also have that $V(\hat{\theta}) = I(\theta)^{-1}$, so the bias formula for the maximum likelihood estimate becomes

$$b(\theta) = -\frac{1}{2}I(\theta)^{-2}[E_\theta(U'(\theta)U(\theta) + E_\theta(U(\theta)^3)].$$

Example 13.A   To see what this means for the estimation of a (single) natural parameter in a distribution from the exponential family without dispersion parameter, we recall that the score function is $U(\theta) = x - \kappa'(\theta)$. This means that

$$E_\theta(U'(\theta)U(\theta)) = -\kappa''(\theta)E_\theta(U(\theta)) = 0, \quad E_\theta(U(\theta)^3) = \kappa'''(\theta),$$

where $\kappa'''(\theta)$ is the skewness of the distribution, and we can therefore write the bias as $b(\theta) = -\frac{1}{2}\kappa'''(\theta)/\kappa''(\theta)^2$. With a sample of size n, we multiply $\kappa(\theta)$ by n and obtain the first-order bias

$$b(\theta) = -\frac{\kappa'''(\theta)}{2n\kappa''(\theta)^2}.$$

As a example, consider the odds $\theta = \ln(p/(1-p))$ for a binomial distribution with parameter p. Since $\kappa(\theta) = -\ln(1 + e^\theta)$, the formula above becomes

$$b(\theta) = -\frac{e^\theta - e^{-\theta}}{2n} = -\frac{\sinh\theta}{n}.$$

We see that the magnitude of the bias is largest when θ is small or large. This implies that when we apply the logistic model to rare event data, there is more (first-order) bias in the estimates than in situations when the events are more common.

In order to investigate this for GLMs in general we assume that $\theta = g(\beta)$ is a function of another parameter β. The estimating equation now is $U^*(\beta) = g'(\beta)U(g(\beta))$ and some calculations show that the bias formula for the maximum likelihood estimator becomes

$$-\frac{g''(\beta)E(U(\theta)^2) + g'(\beta)^2(E(U'(\theta)U(\theta)) + E(U(\theta)^3))}{2g'(\beta)^3 I(\theta)^2} = -\frac{g''(\beta)}{2g'(\beta)^3 I(\theta)} - \frac{b(\theta)}{g'(\beta)},$$

with $\theta = g(\beta)$. Without investigating this formula in any detail, it shows that how large the first-order bias in the maximum likelihoof estimate is, depends on the precise choice of parameterization. It explains why one sometimes improves precision in estimation by analyzing the model in the logarithm of the parameters, instead of the parameters themselves. The formulas above have assumed a single parameter, but are easily extended to the case where there are more parameters, using matrix algebra. This leads us to general expressions for the bias in the maximum likelihood estimate in GLM models (McCullagh and Nelder, 1989).

An alternative approach to correction of bias is to adjust the estimating function, so that the estimate we obtain becomes less biased (Firth, 1993). The key observation in order to achieve this is that $U(\theta + b(\theta)) \approx U(\theta) + U'(\theta)b(\theta)$. Together with the observation that $E_\theta(U'(\theta)) = -I(\theta)$, we see that what we want to do is to find an estimating function which is close to $U(\theta) - I(\theta)b(\theta)$. Inserting the first-order bias above, we get the required estimator. It can be shown that this method is equivalent to maximizing a penalized likelihood for GLMs.

Index

Understanding Biostatistics, First Edition. Anders Källén.
© 2011 John Wiley & Sons, Ltd. Published 2011 by John Wiley & Sons, Ltd.

STATISTICS IN PRACTICE

Human and Biological Sciences

Berger – Selection Bias and Covariate Imbalances in Randomized Clinical Trials
Berger and Wong – An Introduction to Optimal Designs for Social and Biomedical Research
Brown and Prescott – Applied Mixed Models in Medicine, Second Edition
Carstensen – Comparing Clinical Measurement Methods
Chevret (Ed) – Statistical Methods for Dose-Finding Experiments
Ellenberg, Fleming and DeMets – Data Monitoring Committees in Clinical Trials: A Practical Perspective
Hauschke, Steinijans & Pigeot – Bioequivalence Studies in Drug Development: Methods and Applications
Källén – Understanding Biostatistics
Lawson, Browne and Vidal Rodeiro – Disease Mapping with WinBUGS and MLwiN
Lesaffre, Feine, Leroux & Declerck – Statistical and Methodological Aspects of Oral Health Research
Lui – Statistical Estimation of Epidemiological Risk
Marubini and Valsecchi – Analysing Survival Data from Clinical Trials and Observation Studies
Molenberghs and Kenward – Missing Data in Clinical Studies
O'Hagan, Buck, Daneshkhah, Eiser, Garthwaite, Jenkinson, Oakley & Rakow – Uncertain Judgements: Eliciting Expert's Probabilities
Parmigiani – Modeling in Medical Decision Making: A Bayesian Approach
Pintilie – Competing Risks: A Practical Perspective
Senn – Cross-over Trials in Clinical Research, Second Edition
Senn – Statistical Issues in Drug Development, Second Edition
Spiegelhalter, Abrams and Myles – Bayesian Approaches to Clinical Trials and Health-Care Evaluation
Walters – Quality of Life Outcomes in Clinical Trials and Health-Care Evaluation
Whitehead – Design and Analysis of Sequential Clinical Trials, Revised Second Edition
Whitehead – Meta-Analysis of Controlled Clinical Trials
Willan and Briggs – Statistical Analysis of Cost Effectiveness Data
Winkel and Zhang – Statistical Development of Quality in Medicine

Earth and Environmental Sciences

Buck, Cavanagh and Litton – Bayesian Approach to Interpreting Archaeological Data
Chandler and Scott – Statistical Methods for Trend Detection and Analysis in the Environmental Sciences
Glasbey and Horgan – Image Analysis in the Biological Sciences
Haas – Improving Natural Resource Management: Ecological and Political Models
Helsel – Nondetects and Data Analysis: Statistics for Censored Environmental Data
Illian, Penttinen, Stoyan, H and Stoyan D–Statistical Analysis and Modelling of Spatial Point Patterns
McBride – Using Statistical Methods for Water Quality Management
Webster and Oliver – Geostatistics for Environmental Scientists, Second Edition
Wymer (Ed) – Statistical Framework for Recreational Water Quality Criteria and Monitoring

Printed and bound by CPI Group (UK) Ltd, Croydon, CR0 4YY

27/10/2024

14580150-0001